The Breakthrough of Traditional Electrochemical Energy Storage Systems

The Breakthrough of Traditional Electrochemical Energy Storage Systems

Guest Editors

Xiaoyuan Shi

Hengguo Wang

MDPI

Basel • Beijing • Wuhan • Barcelona • Belgrade • Novi Sad • Cluj • Manchester

Guest Editors
Xiaoyuan Shi
Faculty of Chemistry
Northeast Normal University
Changchun
China

Hengguo Wang
Faculty of Chemistry
Northeast Normal University
Changchun
China

Editorial Office
MDPI AG
Grosspeteranlage 5
4052 Basel, Switzerland

This is a reprint of the Special Issue, published open access by the journal *Batteries* (ISSN 2313-0105), freely accessible at: https://www.mdpi.com/journal/batteries/special_issues/ES3290JIPL.

For citation purposes, cite each article independently as indicated on the article page online and as indicated below:

Lastname, A.A.; Lastname, B.B. Article Title. *Journal Name* **Year**, *Volume Number*, Page Range.

ISBN 978-3-7258-7008-0 (Hbk)
ISBN 978-3-7258-7009-7 (PDF)
https://doi.org/10.3390/books978-3-7258-7009-7

Contents

batteries

MDPI

Editorial

Beyond Traditional Batteries—Emerging Systems for Next-Generation Energy Storage

Xiaolin Tian, Yuqi Qin and Xiaoyuan Shi *

Key Laboratory of Polyoxometalate and Reticular Material Chemistry of Ministry of Education, Faculty of Chemistry, Northeast Normal University, Changchun 130024, China
* Correspondence: shixy060@nenu.edu.cn

Against the backdrop of the global energy transition, new energy battery technology plays an increasingly vital role in modern society, effectively addressing the intermittency of energy sources [1–3]. With the rapid development of electric vehicles, energy storage power stations, and other sectors, market demand for new energy batteries continues to grow, simultaneously driving continuous innovation and breakthroughs in this technology. Driven by both technological advancement and market demand, new energy battery technology is evolving toward higher energy density, longer lifespan, lower cost, and enhanced safety [4,5]. While traditional batteries have powered the modern era of portable electronics and electric vehicles, their inherent limitations in material scarcity, energy ceiling, and safety concerns have catalyzed the exploration of novel battery chemistries and architectures [6].

This Special Issue entitled "Breakthroughs in Conventional Electrochemical Energy Storage Systems," brings together pioneering research and insightful reviews that highlight the rapid progress in next-generation battery systems. We present seven original research articles and one review article collectively examining the design, mechanisms, and performance of emerging battery technologies. Several studies focus on developing advanced electrode materials capable of achieving higher specific capacities and improved ionic conductivity. Others employ advanced modeling techniques to address battery aging and safety concerns. Battery aging prediction and thermal safety assessments provide theoretical foundations for designing and deploying highly safe, long-life batteries. The review article explores the potential of metal–air batteries as range extenders for electric vehicles, highlighting ongoing research into cycle life and material stability to enable their widespread adoption in EVs and other energy storage applications. We are encouraged that these studies not only demonstrate significant performance improvements but also discuss the challenges they face. These insights are crucial for guiding the transition from laboratory to commercially viable technologies.

Research Paper:

1. Advanced Electrode Materials

Zeng et al. (contribution 1) prepared a ternary CoMnAl layered double hydroxide (LDH) cathode for aqueous zinc-ion batteries via a simplified coprecipitation method. The introduction of trivalent Al^{3+} ions interacting strongly with Mn^{3+} ions effectively modulated the electronic structure, reduced the redox energy barrier, and enhanced structural stability. Characterization revealed a high specific surface area (106.530 $m^2 \cdot g^{-1}$) and abundant nanopores, facilitating ion transport. Electrochemical testing revealed a high specific capacity of 238.9 mAh g^{-1} at a current density of 0.5 A g^{-1}, excellent rate performance, and outstanding cycling stability, achieving a capacity retention of 92% after 2000 cycles.

This study provides an effective strategy for designing high-performance, long-life AZIB cathode materials.

Chen et al. (contribution 2) fabricated a hierarchical porous electrode with conformal ultrathin α-VO_2 nanosheets on foam nickel via a cost-effective hydrogen bubble template method and electrodeposition for high-power miniature lithium-ion batteries. This unique structure provides continuous electronic conduction pathways and short ionic diffusion channels, enabling rapid reaction kinetics. This design significantly enhances the electrode's kinetic performance. The optimized electrode achieves a high areal energy density of 0.49 mAh cm^{-2} at 1 mA cm^{-2} and an ultra-high power density of 410 mW cm^{-2} at 250 mA cm^{-2}, with a capacity retention rate approaching 100% after 700 cycles. Microbatteries assembled with this electrode can power an electronic clock continuously for over 18 h, retaining 75% of its initial capacity after 180 cycles. This research provides an effective pathway for developing high-performance, low-cost, and easily integrated micro-energy storage systems.

Oladapo et al. (contribution 3) proposed enhancing magnesium-ion battery performance by optimizing magnesium alloy anodes (Mg-Al and Mg-Ag). Cyclic voltammetry and electrochemical impedance spectroscopy tests were conducted at a constant temperature. Results indicated that the Mg-Al alloy demonstrated superior stability during cycling, with a voltage drop of only 0.05 V after 100 cycles and energy efficiency maintained above 85%. while the Mg-Ag alloy experienced a 0.12 V voltage drop and energy efficiency decreased to 80%. The addition of Al aids in forming a surface protective layer, suppressing passivation and degradation, thereby reducing internal resistance growth and extending battery lifespan. This research provides crucial evidence for further optimization of magnesium alloy anodes and the commercialization of magnesium batteries.

Zhang et al. (contribution 4) proposed a novel Se quantum dot@$CoFeO_x$ (Se-FeO_x-Co) composite nanomaterial as a highly active and stable cathode catalyst for rechargeable zinc-air batteries (ZABs). The catalyst was synthesized via a one-pot hydrothermal method, featuring Co-doped iron oxide nanosheets decorated with Se quantum dots. It exhibited outstanding oxygen reduction reaction (ORR) and oxygen evolution reaction (OER) catalytic activity under alkaline conditions, with a potential difference as low as 0.87 V. Furthermore, liquid zinc–air batteries assembled with this cathode achieved a high open-circuit voltage of 1.46 V and a discharge specific capacity of 391 mAh·g^{-1}. They also demonstrated remarkable durability, maintaining stable performance over 800 min of operation, showcasing significant potential for practical energy storage applications.

2. Advanced Modeling Techniques

Tulabi et al. (contribution 5) addressed the safety issue of thermal runaway (TR) in lithium-ion batteries by establishing an electrochemical–thermal coupled model that accounts for aging effects. This model enables more precise prediction of thermal behavior and TR triggering in cylindrical batteries. Full factorial design and analysis of variance (ANOVA) revealed aging's dual effects: increased internal resistance exacerbates temperature rise, elevating TR risk; whereas capacity decay mitigates risk by reducing stored energy. Furthermore, a logistic regression algorithm identified an internal resistance threshold for TR occurrence with 95% prediction accuracy. This work provides an effective framework and insights for understanding the thermal behavior of aged batteries, optimizing thermal management, and developing early warning systems.

Hammou et al. (contribution 6) compared the performance of three recurrent neural networks (RNN, LSTM, and GRU) in multi-step lithium-ion battery state-of-health (SoH) prediction. The results demonstrated that the GRU model achieved high prediction accuracy, with its mean absolute percentage error maintained below 3%. It exhibited stable performance in multi-step forecasting and featured lower computational complexity

 https://doi.org/10.3390/batteries12010004

than the LSTM model, making it highly suitable for resource-constrained in-vehicle environments. This indicates that leveraging temperature—an easily monitored feature—in conjunction with GRU networks enables accurate and efficient prediction of battery aging states.

Hüger et al. (contribution 7) investigated the chemical diffusion coefficient and tracer diffusion coefficient of lithium in sintered $LiCoO_2$ (LCO) disks. By comparing electrochemical methods (EIS, PITT) with isotope tracing (SIMS), they found that diffusion coefficients measured by electrochemical methods exhibited an extremely wide range (10^{-9}–10^{-28} m^2/s), significantly influenced by model selection, parameter settings, and electrode charge state. In contrast, SIMS measurements extrapolated from high-temperature data to room temperature yielded a stable tracer diffusion coefficient of approximately 10^{-22} m^2/s. The study indicates that relying solely on electrochemical methods is insufficient for obtaining accurate lithium diffusion coefficients, necessitating calibration using direct tracer techniques like SIMS. SIMS results narrow the credible range of lithium diffusion coefficients to 10^{-21}–10^{-22} m^2/s, providing more reliable diffusion parameters for subsequent material design and battery performance optimization.

3. Review Article

Metal–air batteries have garnered significant attention due to their high theoretical energy density, low cost, and abundant materials. Shabeer et al. (contribution 8) comprehensively compared the potential of various metal–air batteries (MABs) as range extenders for electric vehicles (EVs). Zinc-air batteries are currently the only commercialized technology, offering cost and safety advantages; aluminum–air batteries feature high energy density but suffer from severe anode corrosion; lithium-air batteries demonstrate outstanding theoretical performance but face development constraints due to complex multiphase reactions and electrolyte decomposition issues. Comprehensive analysis indicates that no metal–air battery technology is currently fully mature. They primarily serve as auxiliary power sources, complementing main batteries to extend range and alleviate range anxiety. This review highlights that through electrolyte optimization, electrode material innovation, and system engineering integration, metal–air batteries hold promise as a key technology to advance electric vehicle adoption and drive low-carbon transportation transformation.

As global demand for efficient and sustainable energy storage continues to grow, coupled with the depletion of fossil fuel and natural gas reserves, the role of innovative battery energy storage systems becomes increasingly vital [7,8]. This Special Issue provides a timely platform for knowledge exchange and collaborative innovation, advancing the development of batteries that are not only more powerful and durable, but also more environmentally friendly. We extend our sincere gratitude to all authors, reviewers, and editors who contributed to the success of this publication. We hope this collection will inspire further research and collaboration, accelerating the emergence of next-generation energy storage solutions.

Conflicts of Interest: The authors declare no conflicts of interest.

List of Contributions:

1. Zeng, J.; Gao, J.; Lu, W.; Feng, J.; Deng, T. Interactions Between Trivalent Elements Enable Ultrastable LDH Cathode for High-Performance Zinc Battery. *Batteries* **2025**, *11*, 170.
2. Chen, X.; Gong, M.; Li, J.; Yang, W.; Xu, X. Engineering Hierarchical Porous Electrodes Integrated with Conformal Ultrathin Nanosheets for Achieving Rapid Kinetics in High-Power Microbatteries. *Batteries* **2025**, *11*, 81.
3. Oladapo, B.I.; Olawumi, M.A.; Omigbodun, F.T. Revolutionizing Battery Longevity by Optimising Magnesium Alloy Anodes Performance. *Batteries* **2024**, *10*, 383.

https://doi.org/10.3390/batteries12010004

4. Zhang, D.; Wang, Y.; Han, X.; Hu, W. Developing a Se Quantum Dots@ CoFeOx Composite Nanomaterial as a Highly Active and Stable Cathode Material for Rechargeable Zinc–Air Batteries. *Batteries* **2023**, *9*, 561.
5. Tulabi, M.; Bubbico, R. Electrochemical–Thermal Modeling of Lithium-Ion Batteries: An Analysis of Thermal Runaway with Observation on Aging Effects. *Batteries* **2025**, *11*, 178.
6. Hammou, A.; Tala-Ighil, B.; Makany, P.; Gualous, H. Multi-Step Ageing Prediction of NMC Lithium-Ion Batteries Based on Temperature Characteristics. *Batteries* **2024**, *10*, 384.
7. Hüger, E.; Schmidt, H. Li Chemical and Tracer Diffusivities in $LiCoO_2$ Sintered Pellets. *Batteries* **2024**, *10*, 446.
8. Shabeer, Y.; Madani, S.S.; Panchal, S.; Mousavi, M.; Fowler, M. Different Metal–Air Batteries as Range Extenders for the Electric Vehicle Market: A Comparative Study. *Batteries* **2025**, *11*, 35.

References

1. Baldinelli, A.; Staffolani, A.; Nobili, F.; Barelli, L. Detailed experimental analysis of solid oxide cells degradation due to frequent fuel cell/electrolyser switch. *J. Energy Storage* **2023**, *73*, 109117. [CrossRef]
2. Costa, C.M.; Salado, M.; Ferrara, C.; Ruffo, R.; Mustarelli, P.; Mao, R.; Feng, S.; Shang, Y.; Wang, X.; Lei, Z.; et al. The wide range of battery systems: From micro- to structural batteries, from biodegradable to high performance batteries. *Prog. Mater. Sci.* **2025**, *154*, 101506. [CrossRef]
3. Fu, S.; Narayanan, V.; Wymore, M.L.; Deep, V.; Duwe, H.; Qiao, D. No battery, no problem: Challenges and opportunities in batteryless intermittent networks. *J. Commun. Netw.* **2023**, *25*, 806–813. [CrossRef]
4. Ashraf, A.; Ali, B.; Alsunjury, M.S.A.; Goren, H.; Kilicoglu, H.; Hardan, F.; Tricoli, P. Review of Cell-Balancing Schemes for Electric Vehicle Battery Management Systems. *Energies* **2024**, *17*, 1271. [CrossRef]
5. Zhu, Z.; Qiao, D.; Huang, Y.; Zhu, J.; Zheng, Y.; Chen, Q. Comparative study of performance and hybrid battery configuration of sodium-ion and lithium-ion batteries. *J. Energy Storage* **2025**, *140*, 118904. [CrossRef]
6. Zhu, Z.; Men, Y.; Zhang, W.; Yang, W.; Wang, F.; Zhang, Y.; Zhang, Y.; Zeng, X.; Xiao, J.; Tang, C.; et al. Versatile carbon-based materials from biomass for advanced electrochemical energy storage systems. *eScience* **2024**, *4*, 100249. [CrossRef]
7. Del Nostro, P.; Goldbeck, G.; Kienberger, F.; Moertelmaier, M.; Pozzi, A.; Al-Zubaidi-R-Smith, N.; Toti, D. Battery testing ontology: An EMMO-based semantic framework for representing knowledge in battery testing and battery quality control. *Comput. Ind.* **2025**, *164*, 104203. [CrossRef]
8. Olabi, A.G.; Sayed, E.T. Developments in Hydrogen Fuel Cells. *Energies* **2023**, *16*, 2431. [CrossRef]

https://doi.org/10.3390/batteries12010004

batteries

MDPI

Article

Developing a Se Quantum Dots@ $CoFeO_x$ Composite Nanomaterial as a Highly Active and Stable Cathode Material for Rechargeable Zinc–Air Batteries

Donghao Zhang [1], Yang Wang [1,2], Xiaopeng Han [1] and Wenbin Hu [1,3,*]

1 Tianjin Key Laboratory of Composite and Functional Materials, Key Laboratory of Advanced Ceramics and Machining Technology (Ministry of Education), School of Materials Science and Engineering, Tianjin University, Tianjin 300350, China; xphan@tju.edu.cn (X.H.)
2 State Key Laboratory of Marine Resource Utilization in South China Sea, School of Materials Science and Engineering, Hainan University, Haikou 570228, China
3 Joint School of National University of Singapore and Tianjin University, International Campus of Tianjin University, Binhai New City, Fuzhou 350207, China
* Correspondence: wbhu@tju.edu.cn

Abstract: With the urgent demand for clean energy, rechargeable zinc–air batteries (ZABs) are attracting increasing attention. Precious-metal-based electrocatalysts (e.g., commercial Pt/C and IrO_2) are reported to be highly active towards the oxygen reduction reaction (ORR) and oxygen evolution reaction (OER). Nevertheless, the limited catalytic kinetics, along with the scarcity of noble metals, still hinder the practical applications of ZABs. Consequently, it is of great importance to explore efficient bifunctional ORR/OER electrocatalysts with abundant reserves. Although iron oxides are considered to have some of the greatest potential as catalysts among the metal oxides, owing to their excellent redox properties, lower toxicity, simple preparation, and natural abundance, their poor electrical conductivity and high agglomeration still limit their development. In this work, we report a special Se quantum dots@ $CoFeO_x$ (Se-FeO_x-Co) composite material, which exhibits outstanding bifunctional catalytic properties. And the potential gap between ORR and OER is low at 0.87 V. In addition, the ZAB based on Se-FeO_x-Co achieves a satisfactory open-circuit voltage (1.46 V) along with an operation durability over 800 min. This research explores an effective strategy to fabricate iron oxide-based bifunctional catalysts, which contributes to the future design of related materials.

Keywords: zinc–air batteries; doping; iron oxide; oxygen reduction reaction; oxygen evolution reaction

1. Introduction

In recent years, the overuse of fossil resources has brought about increasingly serious energy crises and environmental pollution. To meet global energy demands, it is strongly necessary to explore clean renewable energy. Electrochemical energy storage devices with sufficient energy output, stable performance, and high energy density are drawing increasing attention. Among the various devices, batteries (including metal-ion batteries and metal–air batteries) are regarded as the most promising energy technologies [1,2]. Among the various kinds of batteries, rechargeable zinc–air batteries (ZABs) are receiving great interest owing to many advantages, such as high energy density, environmental friendliness, and the abundance of Zn raw materials [3–5]. A reversible air–cathodic electrochemical reaction ($O_2 + 2H_2O + 4e^- \leftrightarrow 4OH^-$) proceeds in the catalysts' surfaces, while an anodic electrochemical reaction ($Zn + 4OH^- - 2e^- \rightarrow Zn\ (OH)_4^{2-}$, $Zn\ (OH)_4^{2-} \leftrightarrow ZnO + H_2O + 2OH^-$) occurs in Zn plates in the alkaline electrolyte. However, one of the main issues is the suitable selection and design of the bifunctional catalysts as the air cathode. The sluggish kinetics of the oxygen reduction reaction (ORR) and oxygen evolution reaction (OER) of catalysts would lead to a large potential gap and low round-trip efficiency. As a result, the further development of ZABs is hindered [6,7]. According to the reaction mechanism, the theoretical voltage of ZAB is determined to be 1.65 V. However, the actual working voltage is generally lower than 1.2 V, with a much higher charging voltage (>2 V), and the current round-trip energy efficiency is usually less than 60%. Precious metal catalysts (e.g., commercial Pt/C and IrO_2) prove to be highly active catalysts for ORR and OER [8,9]. The limited reserves still preclude the large-scale applications of ZABs [10]. Consequently, rationally exploring bifunctional ORR/OER electrocatalysts with high activity and abundant reserves is necessary.

To date, non-noble transitional metal-based materials (such as Fe, Co, and Mn), especially sulfide, oxide, phosphide, nitride, and selenide, have been reported to be viable candidates to replace noble metal electrocatalysts because of their remarkable catalytic abilities [11]. As previously reported, Zou and co-workers synthesized a novel catalyst with Co sites anchored with a CoP electrocatalyst for oxygen electrocatalysis in ZAB, exhibiting a high power density of 139.8 mW cm^{-2} and rechargeability of 132 h [12]. Zhang et al. prepared a N-doped carbon-tentacles-modified metal (Fe or Co) Se nanomaterial through a scalable as well as facile solid-state strategy, which exhibited superior activity and durability thanks to abundant active sites [13]. Xiao et al. designed Co_3O_4 nanoparticles with atomically dispersed FeN_2, and the FeN_2 was determined to be the real catalytic active site in oxygen electrocatalysis [14]. Ren et al. made a rational design for the heterojunction structure of FeS_2/NiS_2, which has a potential gap of 0.92 V when utilized in a liquid ZAB [15]. Among these, Co, Fe, and Ni-based oxide catalysts have been attracting more and more extensive attention because of their high nontoxicity [16–19]. In particular, ABO_3-type perovskites and the AB_2O_4-type spinel oxides have been widely used as bifunctional catalysts in metal–air batteries (A and B are different metal elements) [20–23]. It is worth mentioning that metal oxides can provide abundant oxidized metal sites on the catalyst surface and are easy to tune and affect the electronic structure, thereby improving the electrocatalysis performance [24].

Although Co oxides and Ni oxides have been widely reported as bifunctional catalysts in previous works in the literature, iron oxides have been less studied. Iron oxides, as a kind of well-known metal oxide, have attracted numerous research efforts in the scientific and commercial fields because of their excellent redox properties, lower toxicity, simple preparation, and natural abundance [25]. Unfortunately, iron oxides have an inherently poor electrical conductivity and high agglomeration, which would lead to rapid recession and poor cycling durability [26]. This issue creates poor ORR and OER performance in catalysts, so iron-based oxides are still far below expectation. Hence, various strategies have been put forward to improve the unsatisfactory activity or stability of iron oxides, including exposing active sites, reducing their nanoscale, and so on [27–29]. In recent years, designing composite materials with highly conductive additives has been widely reported to significantly improve their dispersion and catalytic activity [30–32]. For example, Wang et al. reported short carbon nanotubes supporting nickel–iron oxides, where the carbon composite confers a high electrochemical surface area and satisfactory electron transfer ability [31]. Li et al. synthesized heterostructural FeO_x/Fe nanoparticles based on a nitrogen-doped carbon framework, which displayed high ORR activity and good durability due to the high electroconductivity and proper hydrophilicity of the carbon framework [33]. However, it is more desirable if the intrinsic conductivity and stability of iron oxide can be substantially boosted, which may result in a breakthrough in catalytic performance. Quantum dots, as a novel 0D material, are widely considered to be one of the most fascinating materials because of their specific electrical, optical, and chemical properties. Due to the small nanometer scale, quantum dots have some specific quantum tunneling effects and quantum size effects, which are widely used in the field of catalysis, optical, and chemical properties. Owing to their small nanometer scale, QDs have specific quantum tunneling and quantum size effects, which are widely harnessed in the fields of catalysis, optical sensing, and solar-harvesting devices [34,35]. The introduction of QDs into composites can improve their electrocatalytic properties by enhancing their electrical conductivity and energy conversion efficiency [36,37].

Herein, we report a facile one-pot hydrothermal process for synthesizing Co-doped iron oxide nanosheets loaded with Se QDs. The Co precursor ratio was optimized to enhance the ORR and OER performances. The catalytic activity and stability sharply increased when Co was successfully incorporated into the original Fe_2O_3 lattice. In particular, the special Se QD@ $CoFeO_x$ (Se-FeO_x-Co) composite material exhibited remarkable catalytic behavior for the alkaline ORR and OER. In detail, the half-wave potential of ORR was 0.64 V. For OER, it required a reduced overpotential (280 mV) to reach 10 mA cm^{-2}. Moreover, when Se-FeO_x-Co served as an air cathode in a ZAB in a liquid electrolyte, the optimized material showed good catalytic activity and stability for at least 800 min. Our results not only explore an effective strategy to optimize the bifunctional performance of iron oxides but also contribute to the future design of energy-related electrocatalysts.

2. Materials and Methods

2.1. Reagents and Materials

Cobaltous nitrate hexahydrate (Co $(NO_3)_2{\cdot}6H_2O$, 95%) and ammonium iron (II) sulfate hexahydrate (NH_4Fe $(SO_4)_2{\cdot}6H_2O$, 98%) were produced by the Adamas Reagent Co., Ltd., Shanghai, China. Zinc acetate dihydrate (Zn $(CH_3COO)_2{\cdot}2H_2O$)) was obtained from Shanghai Macklin Biochemical Co., Ltd., Shanghai, China. Hydrazine hydrate ($N_2H_4{\cdot}$ H_2O, AR, 99%) was purchased from the Yuanli Chemical Technology Co., Ltd., Tianjin, China. 5 wt.% Nafion solution was obtained from Sigma-Aldrich Co., Ltd., St. Shanghai, China. Potassium hydroxide (KOH, 95%) was obtained from Jiangtian Chemical Technology Co., Ltd., Tianjin, China.

2.2. Synthesis of Se-FeO_x-Co Composite Materials

Firstly, 1.4 mmol Co $(NO_3)_2{\cdot}$ $6H_2O$ and 4.2 mmol NH_4Fe $(SO_4)_2{\cdot}6H_2O$ were dissolved in 25 mL of deionized water in a clean beaker. After 30 min magnetic stirring, 0.11 g of NH_4Cl, 0.4 g of NH_4F, 0.31 g of SeO_2, and 5 mL of hydrazine hydrate were added. Then, the mixing solution was transferred into a 50 mL PTFE-lined stainless autoclave and placed in an oven at 180 °C for 12 h. The as-synthesized precipitation was obtained via a centrifugation process. And the product was washed with DI water three times. Then, the products were dried in an oven at 60 °C for 12 h. According to the different addition amounts of Co $(NO_3)_2{\cdot}$ $6H_2O$ (0.7 and 1.4 mmol), the products were named Se-FeO_x-Co-1 and Se-FeO_x-Co-2, respectively. In addition, the sample to which no Co precursor was added was named Se-FeO_x.

2.3. Materials' Characterization

X-ray diffraction (XRD) (Cu Kα radiation, D8 Advanced, 5 min^{-1}) was used to measure the crystal structural information of the samples. The morphologies and characteristic structures of Se-FeO_x-Co series materials were determined using a JSM-7800F field emission scanning electron microscope (SEM). Before characterization, the dried powder samples were added into absolute ethyl alcohol. After ultrasonic dispersion, the black solution was added dropwise to clean and smooth pieces of Al foil. After drying under room temperature, those pieces of Al foil were transferred to the vacuum box of the SEM device. In addition, the above-mentioned solution was further dropped on a copper mesh that supported an ultra-thin carbon film and then treated using a vacuum oven. The interior structure of the Se-FeO_x-Co series materials was then explored via high-resolution transmission electron microscopy (HRTEM), which was carried out with a JEM-2100F device. The Raman spectra of the samples in the range of 100–1600 cm^{-1} were obtained using a multichannel Raman system (Horiba Jobin Yvon, Kyoto, Japan) with a 532 nm laser. The valence states and chemical information of the samples were measured with an energy-dispersive X-ray photoelectron spectroscope (XPS, Axis Supra, Kratos Analytical Ltd., Manchester, UK).

2.4. Electrochemical Measurements

Homogeneous inks from the ORR and OER tests were obtained by ultrasonically dispersing 7 mg of pre-synthesized sample, 3 mg of black carbon, and 35 μL of Nafion in 965 μL of isopropanol for 1 h. Then, the inks were drop-casted on electrodes: a rotating ring-disk electrode (RRDE, with a diameter of 5 mm) for the ORR test and a glassy carbon electrode (with diameter of 5 mm) for the OER test. The catalyst loading was calculated to be 2.8 mg cm^{-2}.

All of the electrochemical performance tests (including linear sweep voltammetry (LSV) curves, cyclic voltammetry (CV), and chronoamperometry (CA)) in this study were performed at room temperature on a CS310X multi-channel electrochemical workstation (Corrtest). For the ORR test, the RRDE electrode modified by a catalyst served as the working electrode. The counter electrode was a carbon rod, and the reference electrode was Hg/HgO. All the ORR measurements were tested in a 0.1 M KOH solution. The reversible hydrogen electrode (E_{RHE}) and the measuring potential (E_{SCE}) can be converted via $E_{RHE} = E_{Hg/HgO}{}^{\theta} + 0.059pH + E_{SCE}$. Before each CV and LSV test, pure O_2 was purged into the alkaline electrolyte for more than 1h in order to make the electrolyte saturated. During the LSV tests, the RRDE measurements of the samples were conducted via changing rotation rates (400, 625, 900, 1225, and 1600 rpm). In addition, electrocatalytic performances were compared at 1600 rpm in the O_2-saturated electrolyte. During the OER test, the working electrode was a catalyst-modified glassy carbon electrode. The counter electrode was a carbon rod and the reference electrode was Hg/HgO. All the OER measurements were tested in a 1 M KOH solution. The pure N_2 was saturated in the electrolyte for eliminating the effect of oxygen in the OER process.

The electrochemical performance of rechargeable Zn-air batteries was tested in a two-electrode system and was analyzed using a multichannel potential station (LAND CT2001A). Carbon cloths were dipped in dilute HNO_3 solution and deionized water under ultrasonic conditions for 30min several times, and then washed with deionized water and absolute ethyl alcohol. The above-mentioned ink from the catalysts was dropped uniformly onto a hydrophilic carbon cloth (with an area of 1×1 cm^2). And the loading of catalysts on the cathode was calculated to be 2.8 mg cm^{-2}. Moreover, the polished Zn plate as the anode and 6 M KOH + 0.2 M Zn $(CH_3COO)_2$ as the liquid electrolyte were constitutive of the rechargeable liquid ZABs. All batteries were tested at a current density of 10 mA cm^{-2}.

3. Results and Discussion

Figure 1a schematically illustrates the preparation of the Se-FeO_x-Co catalysts. Briefly, the SeO_2 precursor decomposed to form Se QDs and O_2 during the one-pot hydrothermal process, and O_2 combined with the Fe^{3+} precursor to form iron oxide. Subsequently, numerous Co^{2+} ions were doped into the crystal lattice of the iron oxide. Figure 1b shows that the XRD pattern of Se-FeO_x had clear peaks corresponding to elemental Se (PDF# 06-0362) and Fe_2O_3 (PDF# 39-1346). With the addition of the Co^{2+} precursor, the relative peaks of crystallographic Fe_2O_3 disappeared, indicating that Co affected the lattice formation of the samples. No heterophase was observed in the Se-FeO_x-Co samples, suggesting the successful incorporation of Co ions into Se-FeO_x [38]. The morphologies of the Se-FeO_x-Co series are shown in Figure 1c,d, Figures S1 and S2. Se-FeO_x shows a typical nanoflower structure. As the Co content increased, the 3D flower structure tended to deform in Se-FeO_x-Co-1. The Se-FeO_x-Co-2 sample exhibited a hierarchical structure consisting of two-dimensional (2D) sheets ~100 nm in length. According to previous reports, the unique 2D sheet structure may significantly improve the exposure of the metal centers and is conducive to wetting the electrolyte [39,40]. In addition, the microscopic structure of the Se-FeO_x-Co-2 sample was further explored using HRTEM. As shown in Figure 1e, Se was observed as QDs with sizes of <2 nm. Notably, Se-FeO_x-Co-2 showed that the (311) plane of crystallographic Fe_2O_3 was coated with an amorphous structure. Compared to Se-FeO_x, the crystallinity of Se-FeO_x-Co-2 was much lower, which indicates that Co was successfully incorporated into the original Fe_2O_3 lattice. The EDS elemental mapping analysis in

Figure 1f demonstrates that the Co, Fe, and O atoms are homogeneously dispersed in Se-FeO_x-Co-2, whereas the Se atoms in the region are mainly aggregated as well-distributed small particles. The atomic contents of Co, Fe, Se, and O were further measured (Figure 1g) and found to be 13.96%, 38.81%, 2.26%, and 44.97%, respectively. Therefore, the distribution of Fe, Co, and O signals is quite uniform throughout the Se-FeO_x-Co-2 sample, illustrating the successful doping of Co into the Se-FeO_x matrix. The low content and intensity of Se were due to the small number of Se QDs.

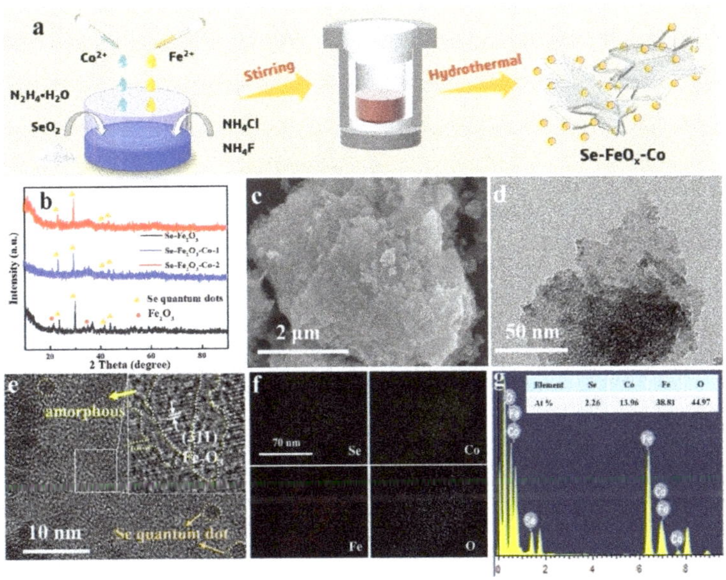

Figure 1. (**a**) Schematic of the fabrication of Se-FeO_x-Co. (**b**) XRD patterns of Se-FeO_x, Se-FeO_x-Co-1, and Se-FeO_x-Co-2 samples. (**c**) SEM, (**d**) TEM, (**e**) HRTEM, and (**f**) EDS mapping images of Se-FeO_x-Co-2; (**g**) relative atomic content.

To further explore the structures of the as-synthesized Se-FeO_x and Se-FeO_x-Co-2 samples, Raman spectra of the two samples were acquired (Figure 2a). On the one hand, the peaks at 216 cm^{-1} belong to the A_{1g} mode of Fe_2O_3. On the other hand, the peaks at 283 cm^{-1} could be attributed to the $2E_g$ mode of Fe_2O_3. The peaks at 390 and 594 cm^{-1} in Se-FeO_x are assigned to the E_g modes of Fe_2O_3. The wide peak at ca. 1300 cm^{-1} is assigned to the second-order scattering of Fe_2O_3 [41,42]. Therefore, the existence of Fe_2O_3 is further suggested for all the samples [43]. In addition, no noticeable peaks are assigned to the Se–O stretching vibration in the Se-FeOx and Se-FeOx-Co-2 samples, indicating the highly stable existence of Se QDs in air with the formation of Se-Se bonds [44]. Notably, the peaks of the two samples slightly differed. Compared with the Se-FeO_x sample, the E_g modes of Fe_2O_3 in the Se-FeOx-Co-2 sample slightly changed, and a prominent peak assigned to the E_g mode of O–Co–O bending vibration appeared at 474 cm^{-1}. The structural and valence properties were further investigated using XPS, as shown in Figure 2b–f. The full XPS spectrum in Figure 2b shows typical peaks of Co 2s, Co 2p, Fe 2p, O1s, Fe 3s, Fe 3p, Se 3s, Se 3d, and Se 3p in the Se-FeO_x-Co-2 sample, indicating the presence of Se, Co, Fe, and O elements. In Figure 2c, the Se 3d spectra of Se-FeO_x and Se-FeO_x-Co-2 display two characteristic peaks at 54.6 and 55.4 eV belonging to the Se^0 $3d_{5/2}$ and Se^0 $3d_{3/2}$ orbits of Se QDs [45,46]. When the Co precursor was added, the characteristic peaks of Se in Se-FeO_x-Co-2 hardly changed. Thus, we inferred that the doping of Co exerted nearly no influence on the formation of Se QDs. Moreover, the O 1s and Fe 2p spectra show clear differences between the Se-FeO_x and Se-FeO_x-Co-2 samples. In Figure 2d, the high-resolution spectrum of elemental oxygen for the catalyst is deconvoluted into peaks at 529.6, 531.2, and 533.2 eV, which correspond to the metal-O bond (Fe-O bond or Co-O bond), oxygen vacancies, and -OH groups (oxygen in the surface-absorbed water molecule) [38,47,48]. In particular, the Se-FeO$_x$-Co-2 sample had a higher concentration of oxygen vacancies than the Se-FeO$_x$ sample, suggesting that Co ions were successfully incorporated into the FeO$_x$ lattice, thus breaking the bonds between Fe and O atoms. Oxygen vacancies, which are generated by Co doping, are favorable for improving electrical conductivity [49–51]. As shown in Figure 2e, two main peaks at 724.8 and 710.9 eV belong to Fe 2p $_{1/2}$ and Fe 2p $_{3/2}$, respectively. The fitted curves of Fe 2p $_{3/2}$ are deconvoluted into two peaks at 713.3 and 710.6 eV, which are ascribed to Fe^{2+} and Fe^{3+}, respectively [52]. Moreover, the wide peak at 718.6 eV is assigned to the shake-up satellite peak of Fe^{3+} [53,54]. Compared to the Se-FeO$_x$ sample, the Fe 2p spectrum of Se-FeO$_x$-Co-2 exhibits a positive shift with the successful replacement of Fe by Co, indicating a rich electron state [55]. Furthermore, compared with Se-FeO$_x$-Co-1, the Co 2p peak of Se-FeO$_x$-Co-2 also displays a slightly opposite shift (Figure S3), indicating electron acceptance [56]. These shifted binding energies demonstrate a strong electronic interaction between Fe and Co, which promotes the transformation of valence electrons and improves the intrinsic electrocatalytic activity [57]. As shown in Figure 2f, this Co 2p $_{1/2}$ peak can be divided into 787.7, 780.4, and 782.3 eV, corresponding to the satellite peaks of Co^{3+} and Co^{2+}, respectively [58]. The high Co^{3+}/Co^{2+} in the Se-FeO$_x$-Co-2 sample demonstrates that the Co valence in Se-FeO$_x$-Co-2 mainly exists as Co^{3+}. In conclusion, the incorporated Co effectively transformed more electrons into Fe, which was beneficial for increasing catalytic performance.

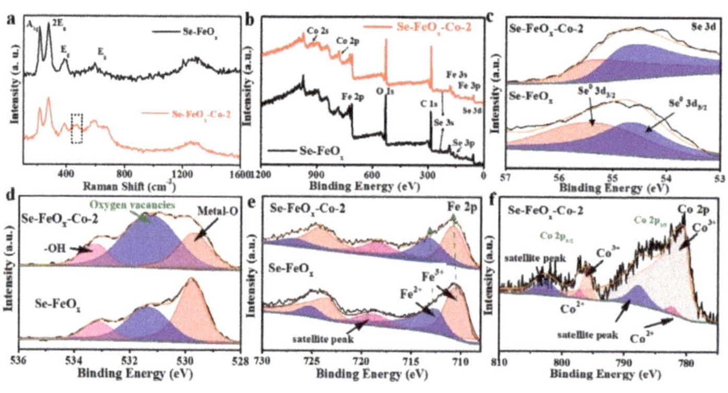

Figure 2. (**a**) Raman spectra of Se-FeOx and Se-FeOx-Co-2 samples. (**b**) XPS full survey spectra of those powders. High-resolution XPS spectra of (**c**) Se 3d, (**d**) O 1s, (**e**) Fe 2p, and (**f**) Co 2p for Se-FeO$_x$ and Se-FeO$_x$-Co-2 samples.

After characterizing the structural properties of the different catalysts, the electrocatalytic performances were evaluated in the KOH solution. The ORR and OER performances of those Se-FeO$_x$-based catalysts were measured using LSV curves in a three-electrode system. Regarding the CV test of the Se-FeO$_x$ and Se-FeO$_x$-Co-2 samples, Figure 3a displays distinct cathodic ORR peaks within the same voltage window at room temperature in 0.1 M KOH solution with an O_2-saturated environment. The clear oxygen reduction peaks of Se-FeO$_x$ and Se-FeO$_x$-Co-2 under an O_2 atmosphere suggest their catalytic activity for oxygen conversion [59]. A peak at 0.64 V versus RHE is observed, which is higher than the Se-FeO$_x$ peak at 0.52 V in Se-FeO$_x$. This demonstrates the prominent ORR activity for Se-FeO$_x$-Co-2 [60]. In addition, the ORR peak current of Se-FeO$_x$-Co-2 reaches 0.12 mA cm^{-2}, which is better than that of Se-FeO$_x$ (with a peak current of 0.06 mA cm^{-2}) [61]. As shown in Figure 3b, the ORR performance of the samples was compared at 1600 rpm. Se-FeO$_x$-Co-2 had a high onset potential (E_{onset} = 0.76 V) and high half-wave potential ($E_{1/2}$ = 0.64 V). However, the onset potential of Se-FeO$_x$ was 0.66 V, and the half-wave potential was 0.52 V. In addition, the onset potential of Se-FeO$_x$-Co-1 was 0.70 V, with a half-wave potential of 0.59 V. Therefore, the Se-FeO$_X$-Co-2 showed the best ORR performance. Further, Se-FeO$_X$-Co-2 also exhibited excellent OER characteristics among the three catalysts. As shown in Figure 3c, it required an overpotential of only 280 mV to reach a current density of 10 mA·cm^{-2} for the Se-FeO$_X$-Co-2. By contrast, Se-FeO$_X$ and Se-FeO$_X$-Co-1 showed larger overpotentials of 400 and 290 mV to reach 10 mA·cm^{-2}, respectively. The relative performance comparison of all catalysts is summarized in Figures 3d and S4. It can be seen that the potential gaps of Se-FeO$_X$, Se-FeO$_X$-Co-1, and Se-FeO$_X$-Co-2 are 1.11, 0.93, and 0.87 V, which implied that the as-prepared Se-FeO$_X$-Co-2 showed significantly enhanced catalytic activity towards ORR and OER. Moreover, the lowered potential gap contributes to the charging/discharging performance of the ZAB devices. Compared with some reported references, the Se-FeO$_X$-Co-2 material shows excellent performance in the same alkaline environment (Table S1) [62–68]. In addition, the stability of ORR and OER were analyzed using chronoamperometry. As shown in Figure 3e,f, after a 14 h measurement of the ORR process, 91.5% of the initial current density was maintained in Se-FeO$_X$-Co-2, which was superior to Se-FeO$_X$ (only 59.2% after 10 h). In addition, Se-FeO$_X$-Co-2 still showed long-term stability for 20 h in the OER process, which was much longer than that of Se-FeO$_X$ in the same conditions. Therefore, the Se-FeO$_X$-Co-2 outperformed the other samples in both activity and stability. The enhanced electrocatalytic performance of Se-FeO$_X$-Co-2 might be owing to the following reasons: First, the loosely stacked nanoflakes of Se-FeO$_X$-Co-2 help to improve the electrochemical surface area, and therefore, more active sites are exposed [69,70]. Second, the Co dopants significantly enhance the conductivity of the FeO$_X$ and, thus, facilitate the charge and electron transfer during the electrochemical reaction, which contributes to improving the activity [71,72]. Third, as evidenced by the XPS results, an electronic interaction occurs after introducing Co into FeO$_X$, and therefore, the synergistic effect between different metals significantly improves the electrocatalytic performance.

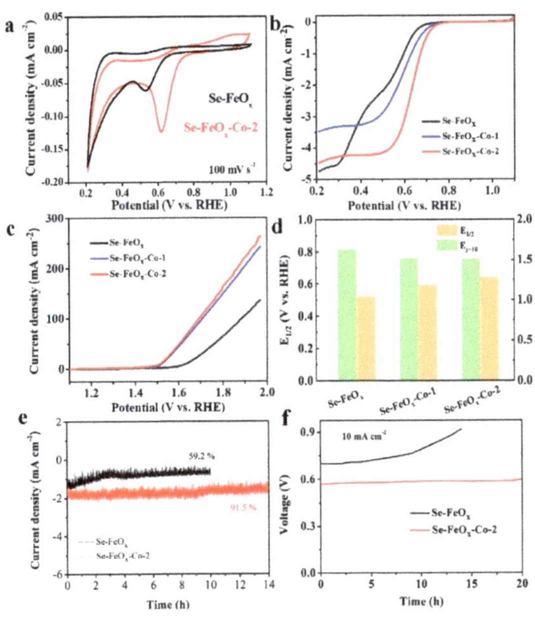

Figure 3. (**a**) CV curves of Se-FeOx-Co and Se-FeOx-Co-2 catalyst at 100 mV s $^{-1}$. (**b**) LSV curves of a series of Se-FeOx sample catalysts at 1600 rpm for ORR test. (**c**) LSV curves of catalysts for OER test. (**d**) Relative bar in the value of $E_{1/2}$ and $E_{j=10}$ in a series of samples. (**e**) ORR stability of catalysts tested in half-wave potential, and (**f**) OER stability of catalysts at a current density of 10 mA cm^{-2}.

To further evaluate the practical properties of the Se-FeO$_x$-Co-2 catalyst in energy conversion and storage devices, a rechargeable ZAB was assembled. A schematic of the operation of the ZABs is shown in Figure 4a. A ZAB assembled with the Se-FeO$_x$ catalyst was used as the control sample under the same conditions. As shown in Figure 4b,c, the ZAB with the Se-FeO$_x$-Co-2 catalyst exhibited a stable open-circuit voltage far higher than that of Se-FeO$_x$ (1.16 V). As shown in Figure 4d, the Se-FeO$_x$-Co-2-based ZAB delivered a dischargeable capacity of 391 mAh g^{-1}, significantly larger than that of the Se-FeO$_x$ based ZAB (215 mAh g^{-1}). This improved performance is related to the optimized oxygen catalysis properties of the Se-FeO$_x$-Co-2 catalyst. Moreover, the battery, which consists of a Se-FeO$_x$-Co-2 catalyst, exhibits a higher power density of 68 mW cm^{-2} than that of the Se-FeO$_x$-based battery (only 54 mW cm^{-2}) in Figure S5. The long-term cycling stability of the rechargeable ZAB with Se-FeO$_x$-Co-2 and Se-FeO$_x$ was measured, as shown in Figure 4e. A slight voltage change was observed over 800 min during discharge/charge cycling, consistent with the good electrocatalytic stability of Se-FeO$_x$-Co-2. The Se-FeO$_x$-based ZAB showed poor cycling performance and could not operate normally after 570 min (with a low discharging platform of 0.5 V). The above results also confirmed that Se-FeO$_x$-Co-2, with high activity and stability, has significant potential for application as an electrode catalyst in ZABs.

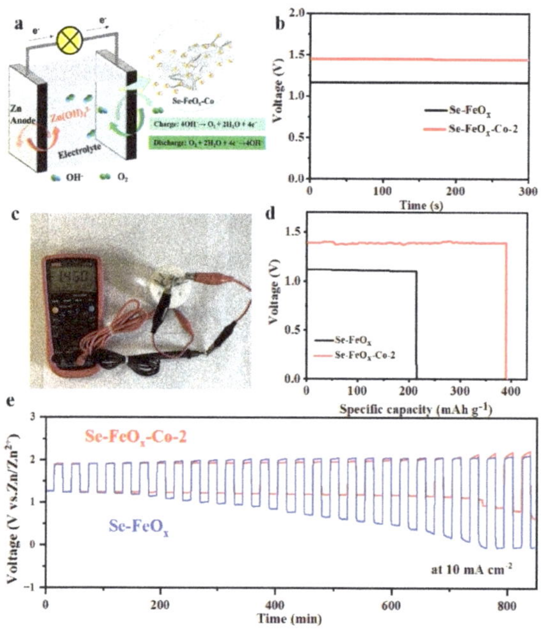

Figure 4. (**a**) Schematic of the operation of zinc–air battery (ZAB) based on the Se-FeOx-Co bifunctional catalyst. (**b**) Comparison of open-circuit voltage and (**c**) optical photograph of ZAB based on Se-FeO$_x$-Co-2. (**d**) Discharging curves and (**e**) rechargeable ZAB cycling profiles at current density of 10 mA cm^{-2}.

4. Conclusions

In conclusion, Se QD@ CoFeO$_x$ composite materials were synthesized using a hydrothermal process. During the synthesis process, the SeO_2 precursor used as the Se source decomposed to form Se QDs and iron oxide was generated simultaneously. At high temperatures and pressures, numerous Co^{2+} ions were doped into the crystal lattice of iron oxide. The nanostructure and valence state of the materials were investigated via SEM, TEM, HRTEM, XRD, Raman spectroscopy, and X-ray photoelectron spectroscopy (XPS). All successfully demonstrated the insertion of cobalt ions into the iron oxide lattice, which had a positive effect on the exposure of active sites, fast charge and electron transfer, and electron interactions in the electrochemical process. Because the Co ions were successfully incorporated into the iron oxide lattice, the as-prepared catalyst showed high activity for oxygen catalysis, with a potential gap of only 0.87 V. In addition, it displayed a satisfactory open-circuit voltage (1.46 V) and long-term durability of over 800 min when fabricated into a liquid ZAB. Therefore, the Se QD@ $CoFeO_x$ catalyst is one of the most promising candidates for practical applications. Our research explores new prospects for the development of cathodic materials for ZABs.

Supplementary Materials: The following supporting information can be downloaded at: https://www.mdpi.com/article/10.3390/batteries9110561/s1, Figure S1: SEM and TEM images of Se-FeO_x material; Figure S2: SEM and TEM images of Se-FeO_x-Co-1 material; Figure S3: The Co 2p spectra of Se-FeO_x-Co-1 and Se-FeO_x-Co-2 materials. Figure S4: The relative bar in the value of $E_{j=10}$-$E_{1/2}$; Figure S5: The polarization and power density curves of Se-FeO_x and Se-FeO_x-Co-2 materials as cathode. Table S1. The ORR and OER performance comparison with reported references.

Author Contributions: Conceptualization, D.Z., Y.W., X.H. and W.H; methodology, D.Z. and X.H.; investigation, X.H. and W.H.; resources, X.H. and W.H.; writing—original draft preparation, D.Z.; writing—review and editing, X.H.; supervision, X.H. and W.H.; project administration, X.H. and W.H.; funding acquisition, X.H. and W.H. All authors have read and agreed to the published version of the manuscript.

Funding: This research was funded by the National Natural Science Foundation of China, grant number U22A20119.

Data Availability Statement: The data presented in this study are contained within the article.

Conflicts of Interest: The authors declare no conflict of interest.

References

1. Zhang, S.; Li, S.; Lu, Y. Designing safer lithium-based batteries with nonflammable electrolytes: A review. *eScience* **2021**, *1*, 163–177. [CrossRef]
2. Yuan, M.; Sun, Z.; Yang, H.; Wang, D.; Liu, Q.; Nan, C.; Li, H.; Sun, G.; Chen, S. Self-catalyzed rechargeable lithium-air battery by in situ metal ion doping of discharge products: A combined theoretical and experimental study. *Energy Environ. Mater.* **2023**, *6*, e12258. [CrossRef]
3. Cheng, F.; Chen, J. Metal–air batteries: From oxygen reduction electrochemistry to cathode catalysts. *Chem. Soc. Rev.* **2012**, *41*, 2172–2192. [CrossRef]
4. Niu, Y.; Gong, S.; Liu, X.; Xu, C.; Xu, M.; Sun, S.-G.; Chen, Z. Engineering iron-group bimetallic nanotubes as efficient bifunctional oxygen electrocatalysts for flexible Zn–air batteries. *eScinece* **2022**, *2*, 546–556. [CrossRef]
5. Yu, H.; Fan, F.; He, C.; Zhou, M.; Ma, T.; Wang, Y.; Cheng, C. Sulfur-modulated FeNi nanoalloys as bifunctional oxygen electrode for efficient rechargeable aqueous Zn-air batteries. *Sci. China Mater.* **2022**, *65*, 3007–3016. [CrossRef]
6. Xin, L.; Yang, F.; Rasouli, S.; Qiu, Y.; Li, Z.-F.; Uzunoglu, A.; Sun, C.-J.; Liu, Y.; Ferreira, P.; Li, W.; et al. Understanding Pt nanoparticle anchoring on graphene supports through surface functionalization. *ACS Catal.* **2016**, *6*, 2642–2653. [CrossRef]
7. Liu, H.; Liu, Q.; Wang, Y.; Wang, Y.; Chou, S.; Hu, Z.; Zhang, Z. Bifunctional carbon-based cathode catalysts for zinc-air battery: A review. *Chin. Chem. Lett.* **2022**, *33*, 683–692. [CrossRef]
8. Zhang, W.; Lai, W.; Cao, R. Energy-related small molecule activation reactions: Oxygen reduction and hydrogen and oxygen evolution reactions catalyzed by porphyrin- and corrole-based systems. *Chem. Rev.* **2017**, *117*, 3717–3797. [CrossRef]
9. Wang, X.; Li, Z.; Qu, Y.; Yuan, T.; Wang, W.; Wu, Y.; Li, Y. Review of metal catalysts for oxygen reduction reaction: From nanoscale engineering to atomic design. *Chem* **2019**, *5*, 1486–1511. [CrossRef]
10. Peng, X.; Zhang, L.; Chen, Z.; Zhong, L.; Zhao, D.; Chi, X.; Zhao, X.; Li, L.; Lu, X.; Leng, K.; et al. Hierarchically porous carbon plates derived from wood as bifunctional ORR/OER electrodes. *Adv. Mater.* **2019**, *31*, 1900341. [CrossRef]
11. Yu, J.; Li, Z.; Liu, T.; Zhao, S.; Guan, D.; Chen, D.; Shao, Z.; Ni, M. Morphology control and electronic tailoring of CoxAy (A = P, S, Se) electrocatalysts for water splitting. *Chem. Eng. J.* **2023**, *460*, 141674. [CrossRef]
12. Liu, B.; Wang, X.; Wang, R.; Zhang, G.; Xu, X.; Liu, J.; Sun, Z.; Liu, M.; Wang, C.; Meng, X.; et al. Activating and stabilizing Co sites in CoP for triggering oxygen electrocatalysis in zinc-air battery. *Chem. Eng. J.* **2023**, *475*, 146154. [CrossRef]
13. Qi, P.; Chen, M.; Luo, T.; Zhao, C.; Lin, C.; Luo, H.; Zhang, D. Solid-state self-catalyzed growth of N-doped carbon tentacles on an M(Fe, Co)Se surface for rechargeable Zn–air batteries. *Chem. Commun.* **2023**, *59*, 5898–5901. [CrossRef]
14. Gui, F.; Jin, Q.; Xiao, D.; Jin, Z.; Zhang, Y.; Cao, Y.; Yang, M.; Tan, Q.; Zhang, C.; Siahrostami, S.; et al. High-performance zinc–air batteries enabled by hybridizing atomically dispersed FeN_2 with Co_3O_4 nanoparticles. *J. Mater. Chem. A* **2023**, *11*, 1312–1323. [CrossRef]

15. Wu, L.; Li, J.; Shi, C.; Li, Y.; Mi, H.; Deng, L.; Zhang, Q.; He, C.; Ren, X. Rational design of the FeS_2/NiS_2 heterojunction interface structure to enhance the oxygen electrocatalytic performance for zinc–air batteries. *J. Mater. Chem. A* **2022**, *10*, 16627–16638. [CrossRef]
16. Wang, J.-J.; Li, X.-P.; Cui, B.-F.; Zhang, Z.; Hu, X.-F.; Ding, J.; Deng, Y.-D.; Han, X.-P.; Hu, W.-B. A review of non-noble metal-based electrocatalysts for CO_2 electroreduction. *Rare Met.* **2021**, *40*, 3019–3037. [CrossRef]
17. Wang, Y.; Wu, X.; Jiang, X.; Wu, X.; Tang, Y.; Sun, D.; Fu, G. Citrulline-induced mesoporous CoS/CoO heterojunction nanorods triggering high-efficiency oxygen electrocatalysis in solid-state Zn-air batteries. *Chem. Eng. J.* **2022**, *434*, 134744. [CrossRef]
18. Weng, C.; Lv, X.; Ren, J.; Wang, Y.; Tian, W.; Gao, L.; Wang, H.; Yuan, Z. Self-promoted electrocatalysts derived from surface reconstruction for rechargeable Zinc–air batteries. *ACS Sustain. Chem. Eng.* **2022**, *10*, 6456–6465. [CrossRef]
19. Zhang, Z.; Liang, X.; Li, J.; Qian, J.; Liu, Y.; Yang, S.; Wang, Y.; Gao, D.; Xue, D. Interfacial engineering of $NiO/NiCo_2O_4$ porous nanofibers as efficient bifunctional catalysts for rechargeable Zinc–air batteries. *ACS Appl. Mater. Interfaces* **2020**, *12*, 21661–21669. [CrossRef]
20. Agyeman, D.; Zheng, Y.; Lee, T.; Park, M.; Tamakloe, W.; Lee, G.; Jang, H.; Cho, K.; Kang, Y. Synergistic catalysis of the lattice oxygen and transition metal facilitating ORR and OER in perovskite catalysts for $Li–O_2$ batteries. *ACS Catal.* **2021**, *11*, 424–434. [CrossRef]
21. Huang, H.; Huang, A.; Liu, D.; Han, W.; Kuo, C.; Chen, H.; Li, L.; Pan, H.; Peng, S. Tailoring oxygen reduction reaction kinetics on perovskite oxides via oxygen vacancies for low-temperature and knittable zinc–air batteries. *Adv. Mater.* **2023**, *35*, 2303109. [CrossRef] [PubMed]
22. Kumar, R.S.; Prabhakaran, S.; Ramakrishnan, S.; Karthikeyan, S.C.; Kim, A.R.; Kim, D.H.; Yoo, D.J. Developing outstanding bifunctional electrocatalysts for rechargeable Zn-air batteries using high-purity spinel-type $ZnCo_2Se_4$ nanoparticles. *Small* **2023**, *19*, 2207096. [CrossRef] [PubMed]
23. Guan, D.; Shi, C.; Xu, H.; Gu, Y.; Zhong, J.; Sha, Y.; Hu, Z.; Ni, M.; Shao, Z. Simultaneously mastering operando strain and reconstruction effects via phase-segregation strategy for enhanced oxygen-evolving electrocatalysis. *J. Energy Chem.* **2023**, *82*, 572–580. [CrossRef]
24. Li, H.; Askari, S.; Wang, J.; Wolff, N.; Behrens, M.; Kienle, L.; Benedikt, J. Nitrogen-doped $NiCo_2O_4$ nanowires on carbon paper as a self-supported air cathode for rechargeable Zn-air batteries. *Int. J. Hydrog. Energy* **2023**, *48*, 26107–26118. [CrossRef]
25. Bo, L.; Nian, F.; Pu, L.; Li, P.; Hu, Y.; Tong, J. Simply embedding α-Fe_2O_3/Fe_3O_4 nanoparticles in N-doped graphitic carbon polyhedron layered arrays as excellent electrocatalyst for rechargeable Zn-air battery. *Sep. Purif. Technol.* **2023**, *312*, 123413. [CrossRef]
26. Samanta, A.; Das, S.; Jana, S. Doping of Ni in α-Fe_2O_3 nanoclews to boost oxygen evolution electrocatalysis. *ACS Sustain. Chem. Eng.* **2019**, *7*, 12117–12124. [CrossRef]
27. Niu, Y.; Yuan, Y.; Zhang, Q.; Chang, F.; Yang, L.; Chen, Z.; Bai, Z. Morphology-controlled synthesis of metal-organic frameworks derived lattice plane-altered iron oxide for efficient trifunctional electrocatalysts. *Nano Energy* **2021**, *82*, 105699. [CrossRef]
28. Zhu, S.; Tian, H.; Wang, N.; Chen, B.; Mai, Y.; Feng, X. Patterning graphene surfaces with iron-oxide-embedded mesoporous polypyrrole and derived N-doped carbon of tunable pore size. *Small* **2018**, *14*, 1702755. [CrossRef]
29. Hof, F.; Liu, M.; Valenti, G.; Picheau, E.; Paolucci, F.; Pénicaud, A. Size control of nanographene supported iron oxide nanoparticles enhances their electrocatalytic performance for the oxygen reduction and oxygen evolution reactions. *J. Phys. Chem. C* **2019**, *123*, 20774–20780. [CrossRef]
30. Chen, M.; Huang, Z.; Ye, X.; Zhang, L.; Feng, J.; Wang, A. Caffeine derived graphene-wrapped Fe_3C nanoparticles entrapped in hierarchically porous FeNC nanosheets for boosting oxygen reduction reaction. *J. Colloid Interface Sci.* **2023**, *637*, 216–224. [CrossRef]
31. Wang, M.; Ji, S.; Wang, H.; Wang, X.; Linkov, V.; Wang, R. Foamed carbon-supported nickel-iron oxides interspersed with bamboo-like carbon nanotubes for high-performance rechargeable Zinc-air batteries. *Small* **2022**, *18*, 2204474. [CrossRef] [PubMed]
32. Fan, Z.; Li, J.; Yang, W.; Fu, Q.; Sun, K.; Song, Y.; Wei, Z.; Liao, Q.; Zhu, X. Green and facile synthesis of iron oxide nanoparticle-embedded N-doped biocarbon as an efficient oxygen reduction electrocatalyst for microbial fuel cells. *Chem. Eng. J.* **2020**, *385*, 123393. [CrossRef]
33. Xiao, Z.; Wu, C.; Wang, W.; Pan, L.; Zou, J.; Wang, L.; Zhang, X.; Li, G. Tailoring the hetero-structure of iron oxides in the framework of nitrogen doped carbon for the oxygen reduction reaction and zinc–air batteries. *J. Mater. Chem. A* **2020**, *8*, 25791–25804. [CrossRef]
34. Rasal, A.S.; Yadav, S.; Yadav, A.; Kashale, A.A.; Manjunatha, S.T.; Altaee, A.; Chang, J.Y. Carbon quantum dots for energy applications: A review. *ACS Appl. Nano Mater.* **2021**, *4*, 6515–6541. [CrossRef]
35. García de Arquer, F.P.; Talapin, D.V.; Klimov, V.I.; Arakawa, Y.; Bayer, M.; Sargent, E.H. Semiconductor quantum dots: Technological progress and future challenges. *Science* **2021**, *373*, eaaz8541. [CrossRef]
36. Meng, D.; Wei, L.; Shi, J.; Jiang, Q.; Tang, J. A review of enhanced electrocatalytic composites hydrogen/oxygen evolution based on quantum dot. *J. Ind. Eng. Chem.* **2023**, *121*, 27–39. [CrossRef]
37. Yu, Y.; Ma, T.; Huang, H. Semiconducting quantum dots for energy conversion and storage. *Adv. Funct. Mater.* **2023**, *33*, 2213770. [CrossRef]

38. Qian, J.; Guo, X.; Wang, T.; Liu, P.; Zhang, H.; Gao, D. Bifunctional porous Co-doped NiO nanoflowers electrocatalysts for rechargeable zinc-air batteries. *Appl. Catal. B Environ.* **2019**, *250*, 71–77. [CrossRef]
39. Lian, Y.; Yang, W.; Zhang, C.; Sun, H.; Deng, Z.; Xu, W.; Song, L.; Ouyang, Z.; Wang, Z.; Guo, J.; et al. Unpaired 3d electrons on atomically dispersed cobalt centres in coordination polymers regulate both oxygen reduction reaction (ORR) activity and selectivity for use in zinc–air batteries. *Angew. Chem. Int. Ed.* **2020**, *59*, 286–294. [CrossRef]
40. Ramakrishnan, S.; Velusamy, D.B.; Sengodan, S.; Nagaraju, G.; Kim, D.H.; Kim, A.R.; Yoo, D.J. Rational design of multifunctional electrocatalyst: An approach towards efficient overall water splitting and rechargeable flexible solid-state zinc–air battery. *Appl. Catal. B Environ.* **2022**, *300*, 120752. [CrossRef]
41. Li, Y.-S.; Church, J.S.; Woodhead, A.L. Infrared and Raman spectroscopic studies on iron oxide magnetic nano-particles and their surface modifications. *J. Magn. Magn. Mater.* **2012**, *324*, 1543–1550. [CrossRef]
42. El Mendili, Y.; Bardeau, J.F.; Randrianantoandro, N.; Gourbil, A.; Greneche, J.M.; Mercier, A.M.; Grasset, F. New evidences of in situ laser irradiation effects on γ-Fe_2O_3 nanoparticles: A Raman spectroscopic study. *J. Raman Spectrosc.* **2011**, *42*, 239–242. [CrossRef]
43. Zheng, H.; Bao, J.; Huang, Y.; Xiang, L.; Faheem; Ren, B.; Du, J.; Nadagouda, M.N.; Dionysiou, D.D. Efficient degradation of atrazine with porous sulfurized Fe_2O_3 as catalyst for peroxymonosulfate activation. *Appl. Catal. B Environ.* **2019**, *259*, 118056. [CrossRef]
44. Karuppasamy, P.; Karthika, A.; Senthilkumar, S.; Rajapandian, V. An efficient and highly sensitive amperometric quercetin sensor based on a lotus flower like SeO_2-decorated rGO nanocomposite modified glassy carbon electrode. *Electrocatalysis* **2022**, *13*, 269–282. [CrossRef]
45. Wu, Z.; Lin, X.; Zhang, J.; Chu, X.; Xu, J.; Li, J.; Liu, Y.; Yu, H.; Yan, L.; Zhang, L.; et al. Encapsulating selenium into mesoporous carbon for high-performance aqueous Cu-Se battery. *Chem. Eng. J.* **2023**, *454*, 140433. [CrossRef]
46. Liu, K.; Niu, J.; Liu, L.; Tian, F.; Nie, H.; Liu, X.; Chen, K.; Zhao, R.; Sun, S.; Jiao, M.; et al. LUMO-mediated Se and HOMO-mediated Te nanozymes for selective redox biocatalysis. *Nano Lett.* **2023**, *23*, 5131–5140. [CrossRef]
47. Sapna; Budhiraja, N.; Kumar, V.; Singh, S.K. Synergistic effect in structural and supercapacitor performance of well dispersed $CoFe_2O_4/Co_3O_4$ nano-hetrostructures. *Ceram. Int.* **2018**, *44*, 13806–13814. [CrossRef]
48. Chen, H.; Liu, W.; Qin, Z. $ZnO/ZnFe_2O_4$ nanocomposite as a broad-spectrum photo-Fenton-like photocatalyst with near-infrared activity. *Catal. Sci. Technol.* **2017**, *7*, 2236–2244. [CrossRef]
49. Li, C.; Lin, Y.; Li, X.; Li, Z.; Luo, P.; Jin, Y.; Li, Z. Effect of Co-doping concentration on α-Fe_2O_3/Graphene as anode materials for lithium ion batteries. *Colloids Surf. A Physicochem. Eng. Asp.* **2023**, *660*, 130681. [CrossRef]
50. Liu, G.; Li, J.; Fu, J.; Jiang, G.; Lui, G.; Luo, D.; Deng, Y.P.; Zhang, J.; Cano, Z.P.; Yu, A.; et al. An oxygen-vacancy-rich semiconductor-supported bifunctional catalyst for efficient and stable zinc–air batteries. *Adv. Mater.* **2019**, *31*, 1806761. [CrossRef]
51. Qin, J.; Liu, Z.; Wu, D.; Yang, J. Optimizing the electronic structure of cobalt via synergized oxygen vacancy and Co-N-C to boost reversible oxygen electrocatalysis for rechargeable Zn-air batteries. *Appl. Catal. B Environ.* **2020**, *278*, 119300. [CrossRef]
52. Han, X.; Liu, C.; Tang, Y.; Meng, Q.; Zhou, W.; Chen, S.; Deng, S.; Wang, J. Unveiling the role of cobalt doping in optimizing ammonia electrosynthesis on iron–cobalt oxyhydroxide hollow nanocages. *J. Mater. Chem. A* **2023**, *11*, 14424–14431. [CrossRef]
53. Li, M.; Tang, Y.; Shi, W.; Chen, F.; Shi, Y.; Gu, H. Design of visible-light-response core–shell $Fe_2O_3/CuBi_2O_4$ heterojunctions with enhanced photocatalytic activity towards the degradation of tetracycline: Z-scheme photocatalytic mechanism insight. *Inorg. Chem. Front.* **2018**, *5*, 3148–3154. [CrossRef]
54. Zhu, H.; Dong, S.; Du, X.; Du, H.; Xia, J.; Liu, Q.; Luo, Y.; Guo, H.; Li, T. Defective CuO-rich $CuFe_2O_4$ nanofibers enable the efficient synergistic electrochemical reduction of nitrate to ammonia. *Catal. Sci. Technol.* **2022**, *12*, 4998–5002. [CrossRef]
55. Luo, J.; Han, X.; Ge, J.; Wang, Y.; Zhao, X.; Zhang, F.; Lei, X. Cu_9S_5/Fe_2O_3 nanospheres as advanced negative electrode materials for high performance battery-like hybrid capacitors. *ACS Appl. Energy Mater.* **2022**, *5*, 7016–7025. [CrossRef]
56. Yang, H.; Wang, B.; Li, H.; Ni, B.; Wang, K.; Zhang, Q.; Wang, X. Trimetallic sulfide mesoporous nanospheres as superior electrocatalysts for rechargeable Zn–air batteries. *Adv. Energy Mater.* **2018**, *8*, 1801839. [CrossRef]
57. Guo, Y.; Wang, T.; Yang, Q.; Li, X.; Li, H.; Wang, Y.; Jiao, T.; Huang, Z.; Dong, B.; Zhang, W.; et al. Highly efficient electrochemical reduction of nitrogen to ammonia on surface termination modified $Ti_3C_2T_x$ MXene nanosheets. *ACS Nano* **2020**, *14*, 9089–9097. [CrossRef]
58. Zhang, G.; Wang, S.; Zeng, X.; Li, X.; Xiao, L.; Chen, K.; Lu, Q.; Xu, Q.; Weng, J.; Xu, J. Holey amorphous FeCoO-coated black phosphorus for robust polysulfide adsorption and catalytic conversion in lithium–sulfur batteries. *J. Mater. Chem. A* **2022**, *10*, 11676–11683. [CrossRef]
59. Lee, K.; Zhang, L.; Lui, H.; Hui, R.; Shi, Z.; Zhang, J. Oxygen reduction reaction (ORR) catalyzed by carbon-supported cobalt polypyrrole (Co-PPy/C) electrocatalysts. *Electrochim. Acta* **2009**, *54*, 4704–4711. [CrossRef]
60. Xu, C.; Guo, C.; Liu, J.; Hu, B.; Dai, J.; Wang, M.; Jin, R.; Luo, Z.; Li, H.; Chen, C. Accelerating the oxygen adsorption kinetics to regulate the oxygen reduction catalysis via Fe3C nanoparticles coupled with single Fe-N_4 sites. *Energy Storage Mater.* **2022**, *51*, 149–158. [CrossRef]
61. Gao, S.; Fan, B.; Feng, R.; Ye, C.; Wei, X.; Liu, J.; Bu, X. N-doped-carbon-coated Fe_3O_4 from metal-organic framework as efficient electrocatalyst for ORR. *Nano Energy* **2017**, *40*, 462–470. [CrossRef]
62. Mao, S.; Wen, Z.; Huang, T.; Hou, Y.; Chen, J. High-performance bi-functional electrocatalysts of 3D crumpled graphene–cobalt oxide nanohybrids for oxygen reduction and evolution reactions. *Energy Environ. Sci.* **2014**, *7*, 609–616. [CrossRef]

63. Du, S.; Ren, Z.; Qu, Y.; Wu, J.; Xi, W.; Zhu, J.; Fu, H. Co_3O_4 nanosheets as a high-performance catalyst for oxygen evolution proceeding via a double two-electron process. *Chem. Commun.* **2016**, *52*, 6705–6708. [CrossRef]
64. Tian, G.; Zhao, M.; Yu, D.; Kong, X.; Huang, J.; Zhang, Q.; Wei, F. Nitrogen-doped graphene/carbon nanotube hybrids: In situ formation on bifunctional catalysts and their superior electrocatalytic activity for oxygen evolution/reduction reaction. *Small* **2014**, *10*, 2251–2259. [CrossRef] [PubMed]
65. Rasaki, S.A.; Chen, Z.; Shen, H.; Guo, H.; Thomas, T.; Yang, M. Cobalt nanoparticles modified single-walled titanium carbonitride nanotube derived from solid-solid separation for oxygen reduction reaction in alkaline solution. *Electrocatalysis* **2020**, *11*, 579–592. [CrossRef]
66. Tang, K.; Hu, H.; Xiong, Y.; Chen, L.; Zhang, J.; Yuan, C.; Wu, M. Hydrophobization engineering of the air–cathode catalyst for improved oxygen diffusion towards efficient zinc–air batteries. *Angew. Chem. Int. Ed.* **2022**, *61*, e202202671. [CrossRef]
67. Faisal, S.N.; Haque, E.; Noorbehesht, N.; Liu, H.; Islam, M.M.; Shabnam, L.; Roy, A.K.; Pourazadi, E.; Islam, M.S.; Harris, A.T.; et al. A quadrafunctional electrocatalyst of nickel/nickel oxide embedded N-graphene for oxygen reduction, oxygen evolution, hydrogen evolution and hydrogen peroxide oxidation reactions. *Sustain. Energy Fuels* **2018**, *2*, 2081–2089. [CrossRef]
68. Shudo, Y.; Fukuda, M.; Islam, M.S.; Kuroiwa, K.; Sekine, Y.; Karim, M.R.; Hayami, S. 3D porous Ni/NiO_x as a bifunctional oxygen electrocatalyst derived from freeze-dried $Ni(OH)_2$. *Nanoscale* **2021**, *13*, 5530–5535. [CrossRef]
69. Xiong, P.; Sun, B.; Sakai, N.; Ma, R.; Sasaki, T.; Wang, S.; Zhang, J.; Wang, G. 2D superlattices for efficient energy storage and conversion. *Adv. Mater.* **2020**, *32*, 1902654. [CrossRef]
70. Cheng, Y.; Pang, K.; Wu, X.; Zhang, Z.; Xu, X.; Ren, J.; Huang, W.; Song, R. In situ hydrothermal synthesis MoS_2/guar gum carbon nanoflowers as advanced electrocatalysts for electrocatalytic hydrogen evolution. *ACS Sustain. Chem. Eng.* **2018**, *6*, 8688–8696. [CrossRef]
71. Kong, H.; Lv, C.; Yan, C.; Chen, G. Engineering mesoporous single crystals Co-doped Fe_2O_3 for high-performance lithium ion batteries. *Inorg. Chem.* **2017**, *56*, 7642–7649. [CrossRef] [PubMed]
72. Hou, Y.; Zuo, F.; Dagg, A.; Feng, P. A three-dimensional branched cobalt-doped α-Fe_2O_3 Nanorod/$MgFe_2O_4$ heterojunction array as a flexible photoanode for efficient photoelectrochemical water oxidation. *Angew. Chem. Int. Ed.* **2013**, *52*, 1248–1252. [CrossRef] [PubMed]

batteries

MDPI

Article

Revolutionizing Battery Longevity by Optimising Magnesium Alloy Anodes Performance

Bankole I. Oladapo [1,*], Mattew A. Olawumi [2] and Francis T. Omigbodun [3]

[1] School of Science and Engineering, University of Dundee, Dundee DD1 4HN, UK
[2] Computing, Engineering and Media, De Montfort University, Leicester LE1 9BH, UK; olawumisola13@gmail.com
[3] Wolfson School of Mechanical, Electrical and Manufacturing Engineering, Loughborough University, Loughborough LE11 3TU, UK; f.omigbodun@lboro.ac.uk
* Correspondence: p17243433@my365.dmu.ac.uk

Abstract: This research explores the enhancement of electrochemical performance in magnesium batteries by optimising magnesium alloy anodes, explicitly focusing on Mg-Al and Mg-Ag alloys. The study's objective was to determine the impact of alloy composition on anode voltage stability and overall battery efficiency, particularly under extended cycling conditions. The research assessed the anodes' voltage behaviour and internal resistance across magnesium bis(trifluoromethanesulfonyl)imide ($Mg(TFSI)_2$) electrolyte formulations using a systematic setup involving cyclic voltammetry on the anode and electrochemical impedance spectroscopy. The Mg-Al alloy demonstrated superior performance, with minimal voltage drop and lower resistance increase than the Mg-Ag alloy. The results showed that the Mg-Al alloy maintained over 85% energy efficiency after 100 cycles, significantly outperforming the Mg-Ag alloy, which exhibited increased degradation and efficiency reduction to approximately 80%. These findings confirm that incorporating aluminium into magnesium anodes stabilises the anode voltage and enhances the overall battery efficiency by mitigating degradation mechanisms. Consequently, the Mg-Al alloy is identified as an up-and-coming candidate for use in advanced battery technologies, offering energy density and cycle life improvements. This study lays the groundwork for future research to refine magnesium alloy compositions further to boost battery performance.

Keywords: AI optimisation techniques; battery voltage stability; machine learning; electrochemical performance; predictive degradation models; magnesium alloys

1. Introduction

Magnesium batteries, known for their high energy density and safety due to the use of magnesium metal, represent a promising alternative to traditional lithium-ion technologies. Magnesium offers a theoretical volumetric capacity of 3833 mAh/cm^3 as a divalent cation, significantly higher than lithium's (2061 mAh/cm^3) [1–3]. This attribute and magnesium's earth-abundance and inherent safety underscore its potential in developing next-generation battery systems.

The anode voltage is critical for magnesium batteries' energy density and efficiency. A lower anode potential relative to the magnesium ion/metal couple is desirable as it increases the voltage window of the cell, thereby enhancing the battery's specific energy [4–6]. However, achieving a stable and high-performance anode with a low and effective voltage has proven to be a complex challenge due to the intrinsic electrochemical properties of magnesium and its interaction with common electrolytes [7–9].

The exploration of magnesium as a battery material began in the early 2000s, focusing initially on overcoming the challenges posed by Mg's high ionisation tendency and difficulty finding compatible electrolytes. Early studies dealt primarily with salt-based electrolytes, which facilitated some ionic movement but were hindered by rapid degradation and passivation at the anode surface [10–12]. Extensive research has explored various anode materials, from elemental magnesium to alloys and intermetallic compounds. These studies have identified critical performance factors, such as anode–electrolyte interface stability and ion transport mechanisms. Recent advancements have introduced novel anode materials showing reduced passivation and compatibility with high-performance electrolytes [13–15].

Despite significant advancements, several critical gaps remain in the literature and practical application. Most magnesium anodes still suffer from issues related to long-term cycling stability, which is often compromised by the formation of insulating films on the anode surface [16–18]. The interaction between the anode and electrolyte remains imperfect, with many current electrolyte formulations either chemically degrading the anode material or offering insufficient ionic conductivity. While anodes have shown promising lab-scale results, translating these findings to commercially viable solutions is often unaddressed. The cost implications of novel materials and electrolytes have not been sufficiently explored [19–21].

This study addresses these gaps by proposing a systematic investigation into novel magnesium alloys as anodes, examining their electrochemical performance across complex electrolytes. The research will exclude traditional pure magnesium anodes and focus instead on those incorporating elements like aluminium and silicon, which preliminary studies suggest may offer enhanced performance [22–24]. The literature review encompasses many sources, from seminal papers introducing magnesium as a viable battery material to recent publications that detail advancements in anode technology and ($Mg(TFSI)_2$) electrolyte development. Each selected work has been evaluated for its relevance to the ongoing challenges in the field, ensuring a comprehensive overview of past achievements and current shortcomings [25–27]. The present research intends to fill the identified gaps by developing anode materials that sustain high electrochemical performance without the typical degradation in current magnesium batteries. By doing so, it hopes to pave the way for the next generation of high-energy, safe, and cost-effective storage solutions [28–30].

Magnesium batteries offer a theoretical volumetric capacity of 3833 mAh/cm^3, higher than lithium-ion batteries, alongside other benefits like safety and cost-effectiveness. However, magnesium anodes face significant challenges, such as passivation and compatibility with electrolytes, which affect long-term cycling stability [30–32]. This research addresses these issues by evaluating Mg-Al and Mg-Ag alloy anodes, focusing on voltage stability, cycling efficiency, and internal resistance.

Magnesium batteries have attracted considerable attention as a viable substitute for conventional lithium-ion technology, chiefly because of their superior theoretical volumetric capacity, intrinsic safety, and the plentiful availability of magnesium. Nonetheless, despite these benefits, the extensive implementation of magnesium batteries has been obstructed by several significant problems, especially regarding the efficacy of magnesium anodes. Confronting these problems is essential for enhancing magnesium batteries' overall effectiveness and stability, rendering them suitable for commercial and high-demand applications [32,33].

A significant concern of magnesium anodes is their long-term cycling stability. Magnesium anodes exhibit degradation during successive charge–discharge cycles, resulting in a substantial decline in battery capacity and efficiency. The degradation is frequently intensified by the development of passivation layers on the anode surface, obstructing the passage of magnesium ions, hence diminishing overall conductivity and battery longevity. Anode passivation constitutes a significant impediment to the efficient utilization of magnesium in batteries, as it adversely affects performance during cycling and elevates internal resistance, thus diminishing energy production [34,35].

Moreover, ($Mg(TFSI)_2$) electrolyte compatibility presents a significant obstacle for magnesium batteries. Numerous traditional electrolytes employed in lithium-ion batteries exhibit incompatibility with magnesium since they either chemically deteriorate the magnesium anode or inadequately promote effective ion transport. The interaction between the electrolyte and the magnesium anode frequently generates unwanted by-products, exacerbating passivation and constraining the battery's efficiency. This requires the creation of novel ($Mg(TFSI)_2$) electrolyte formulations that can function synergistically with magnesium anodes, minimizing deterioration while preserving ion transport efficiency [36–38]. This study primarily examines these challenges by exploring the application of Mg-Al and Mg-Ag alloy anodes, to improve voltage stability, minimizing deterioration, and enhance compatibility with diverse ($Mg(TFSI)_2$) electrolytes. The project intends to address these critical obstacles to develop high-efficiency, durable magnesium batteries.

This research addresses critical challenges associated with magnesium anodes in magnesium batteries, specifically focusing on improving voltage stability, cycling efficiency, and mitigating degradation mechanisms. Magnesium anodes, while offering advantages like higher theoretical volumetric capacity and safety compared to lithium-ion technologies, face significant obstacles such as long-term cycling stability issues, passivation, and ($Mg(TFSI)_2$) electrolyte compatibility. This study aims to explore the electrochemical performance of two magnesium alloy anodes—Mg-Al and Mg-Ag—by investigating their voltage stability and internal resistance over extended cycling. By incorporating aluminium and silicon into magnesium, the research seeks to enhance anode durability, reduce degradation, and optimize the battery's overall performance. The study employs cyclic voltammetry (CV) and electrochemical impedance spectroscopy (EIS) to evaluate the electrochemical properties of the anodes, such as oxidation/reduction potentials and internal resistance. The goal is to identify how these novel alloy compositions interact with various ($Mg(TFSI)_2$) electrolyte formulations to mitigate common issues like rapid capacity fade and anode passivation. Ultimately, the research aims to contribute to developing more stable and efficient magnesium-based batteries, offering improved energy density and longer cycle life, which could be viable alternatives to current lithium ion technologies.

2. Methodology

The study is structured around a series of electrochemical tests to evaluate the performance of novel magnesium alloys as anodes in magnesium batteries. Each computation mimics realistic operational conditions while allowing precise control over temperature, ($Mg(TFSI)_2$) electrolyte composition, and current density [31–33].

2.1. Cell and Electrolyte Configuration

The computational cells are assembled using a standard coin cell configuration, including a magnesium alloy anode, a high-purity magnesium reference electrode, and a manganese dioxide cathode. This setup is chosen to ensure compatibility with a wide range of ($Mg(TFSI)_2$) electrolytes while minimising the introduction of side reactions. Multiple ($Mg(TFSI)_2$) electrolytes, varying primarily in solvent and salt composition, are prepared to explore the interaction effects with the anode material. The ($Mg(TFSI)_2$) electrolytes are meticulously prepared under an inert atmosphere to prevent contamination and pre-testing degradation. All tests are conducted in a temperature-controlled environment, maintaining a steady temperature of 25 °C to eliminate the influence of thermal fluctuations on the results.

2.2. Description of the Anode Materials Used

The anode materials selected for this study are primarily magnesium alloys, enhanced with small percentages of aluminium and silicon, which are hypothesised to improve the electrochemical performance and stability of the anodes. Mg-Al Alloy is comprised of 95% magnesium and 5% aluminium. Aluminium is expected to enhance the flexibility and corrosion resistance of the anode. Mg-Ag alloy contains 90% magnesium and 10% silicon, aimed at improving the conductivity and reducing the anode's susceptibility to passivation. Each alloy is synthesised using a melt-spinning process that ensures homogeneity and fine microstructure, critical for consistent electrochemical behaviour [34–36] (Figure 1).

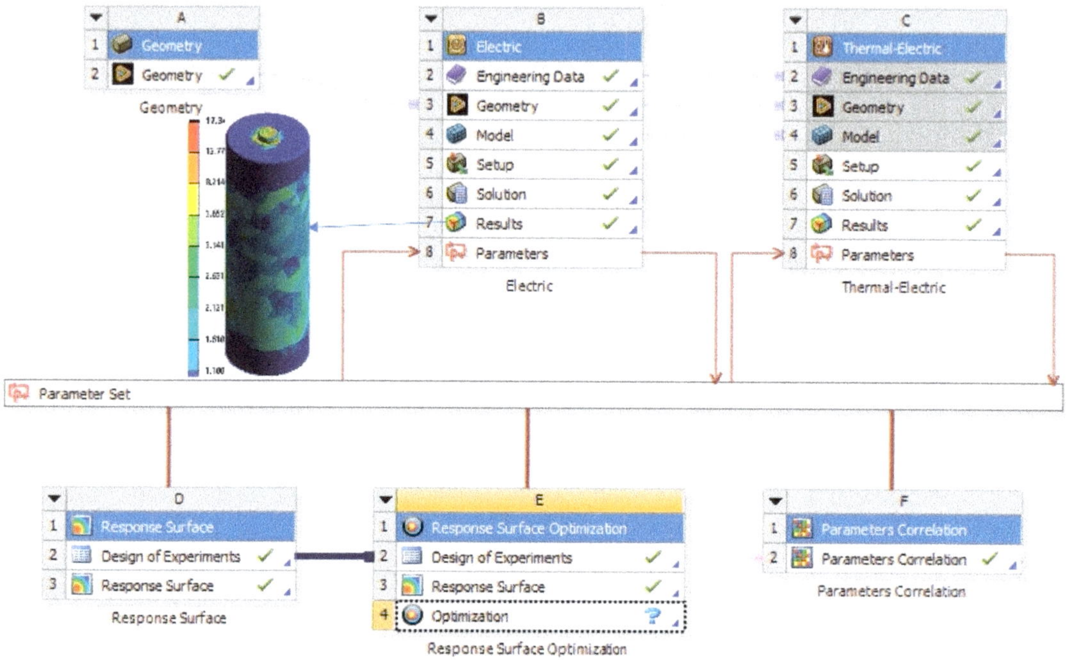

Figure 1. An optimum model design of the magnesium anode.

2.3. Measurement Techniques for Anode Voltage

Anode voltage is measured using a high-precision voltmeter connected to the working electrode and a stable reference electrode. The voltage is recorded continuously throughout the charge–discharge cycles to monitor any potential drops that may indicate degradation or inefficiency. Cyclic voltammetry (CV) characterises the anodes' electrochemical properties, including their oxidation and reduction potentials. The scan rate is set at 0.1 mV/s to ensure detailed resolution of the electrochemical reactions occurring at the anode surface. Electrochemical impedance spectroscopy (EIS) is utilised to measure the internal resistance of the anodes, which is a critical factor affecting the voltage efficiency [37,38]. Measurements are taken before and after each cycling test to assess any changes in resistance that could indicate degradation. Cyclic voltammetry (CV) and electrochemical impedance spectroscopy (EIS) will measure the alloys' voltage behaviour and internal resistance. A constant temperature of 25 °C will be maintained during all tests to ensure stable conditions.

2.4. Reproducible Methodology to Identify Relevant Works

The methodology to identify and select the relevant literature involves a structured approach using multiple academic database studies published in the last ten years, involving magnesium or magnesium alloys used in battery anodes reporting primary data on electrochemical tests [39–41]. This systematic approach ensures comprehensive coverage of existing knowledge, while identifying the gaps the current study aims to address. This methodology section outlines a detailed, reproducible framework for conducting high-quality, controlled magnesium alloy anodes and systematically reviewing the literature to inform designs and hypotheses [42–44]. Figure 2 typically develops a series of diagrams or schematics that clearly illustrate the components and configurations used in measuring the voltage of an anode. This would be valuable for educational purposes, training, or ensuring consistency in research settings. Anode voltage measurement setups often involve specific hardware components and configurations, including the anode material, ($Mg(TFSI)_2$) electrolyte, reference, counter electrodes, and equipment for voltage measurement, and data logging [45–47]. Such a guide represents these setups to facilitate visual understanding and reproducibility.

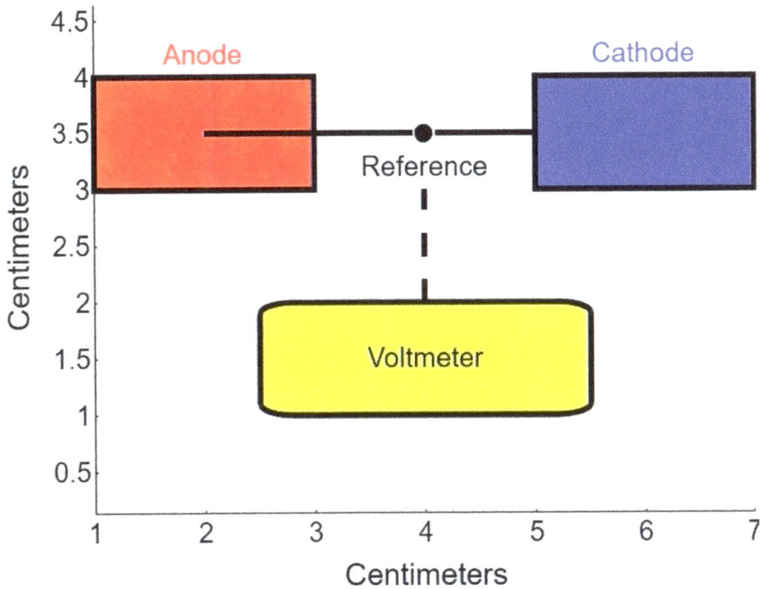

Figure 2. Computation setup for anode voltage measurement: A visual guide.

Identify key components of the list of all critical elements required for the setup. The design layout conceptualises how these components are physically arranged and connected. Generate the visual guide using MATLAB to create diagrams or schematics visually representing the setup. Draw components of rectangle and plot functions are used to draw electrodes and connections. The voltmeter connection is a dashed line showing the voltmeter's connection to the point between the reference and working electrodes [48–50]. This simple visual guide helps users understand the basic setup for measuring anode voltage. The schematic can be more complex by adding ($Mg(TFSI)_2$) electrolyte containers, different electrode materials, or more sophisticated measuring equipment.

The ($Mg(TFSI)_2$) electrolytes used in this study include combinations of magnesium salts (Mg(TFSI)2) and solvents such as ethylene carbonate (EC), propylene carbonate (PC), and dimethyl carbonate (DMC). Details of concentrations will be provided in a dedicated table. The anode material composition of Mg-Al alloy: 95% Mg, 5% Al, and Mg-Ag alloy: 90% Mg, 10% Si. These compositions were selected based on their potential to improve the stability and corrosion resistance of the anodes.

3. Mathematical Model

A mathematical equation that encompasses the analysis of anode voltage and overall battery performance in magnesium batteries considers the following key factors. *Voc* is the open-circuit voltage of the battery, *Ri* is the internal resistance of the battery, and I is the current through the battery. C is the capacity of the battery, n is the number of charge/discharge cycles, and η is the efficiency of the battery. One essential performance metric in battery analysis is voltage efficiency, which can be calculated by considering the operating voltage under load compared to the open-circuit voltage [51–53]. A derived formula to estimate a magnesium battery's effective operating voltage and efficiency, considering anode degradation over cycles is shown below.

3.1. Effective Operating Voltage

$$Veff(n) = Voc - I \cdot Ri(n)$$

where $Ri(n)$ is the internal resistance, which could increase over the number of cycles n due to material degradation or other factors.

Battery efficiency $\eta(n)$ = ((V_eff (n))/V_oc) · 100%

Battery capacity fade accounts for capacity degradation over cycles; a standard empirical model that can be used is

$$C(n) = C(1) \cdot e - k \cdot n$$

where $C(1)$ is the initial capacity, k is a constant representing the capacity fade rate, and n is the number of cycles.

3.2. Complete Model for Voltage and Efficiency

Combining these equations, we can create a comprehensive model that provides insights into how the anode material and degradation affect the battery's performance over time. The model could simulate performance under different scenarios and optimise the battery design.

$$Vtotal(n, I) = Voc - I \cdot Ri(n)$$

η_(total) (n,I) = ((V_(total) (n,I))/V_oc) · (C(n)/C(1)) · 100%

These equations help quantify the impact of anode material changes, usage patterns, and other variables on the performance of magnesium batteries, which is crucial for their development and optimisation for commercial use.

3.3. Extended Model

A more advanced and complex mathematical model for analysing the performance of magnesium battery anodes was modelled. We introduce factors influencing battery behaviour, such as temperature effects, anode–cathode interaction, and ($Mg(TFSI)_2$) electrolyte conductivity. The model will incorporate differential equations to simulate dynamic behaviour during battery operation. Dynamic voltage model incorporates temperature dependence and voltage response to variable load conditions, using a resistance that changes with temperature and current [54–56]. Temperature influence was considered to be the impact of temperature on internal resistance and battery capacity, which affects both performance and degradation rate. Rate of voltage change concerning time for a model of how voltage decreases over time due to internal chemical and physical changes.

3.3.1. Dynamic Voltage Equation

$$\frac{dV}{dt} = -\frac{I(t)}{C} - \frac{V(t) - E_{oc}(T)}{R_i(T, I)}$$

where $V(t)$ is voltage at time t, $I(t)$ is current at time t, and C is battery capacity. $E_{oc}(T)$ is open-circuit voltage dependent on temperature T, and $R_i(T, I)$ is the internal resistance dependent on temperature T and current I.

3.3.2. Temperature-Dependent Internal Resistance

$$R_i(T, I) = R_0 \cdot (1 + \alpha(T - T_{ref})) \cdot (1 + \beta I^2)$$

R_0 is the reference internal resistance at temperature T_{ref}, and α, β are the temperature and current coefficients for resistance change.

3.3.3. Battery Capacity with Temperature and Cycle Number

$$C(T,n) = C_0 \cdot e^{-k(T) \cdot n}$$

where C_0 initial capacity, $k(T)$ is a temperature-dependent rate of capacity fade, and n is the number of cycles.

3.3.4. Numerical Simulation Approach

Time step Δt is the small time interval for simulation (e.g., 1 s). Simulation duration: Total time for simulation (e.g., 3600 s for a 1-h test) and initial conditions $V(0) = E_{oc}(T_{initial}) = T_{initial}$. Use the Euler or Runge–Kutta methods for the numerical integration of dV/dt.

The numerical table will be generated by solving the differential equations over the specified duration with given initial conditions. For example, if testing under a constant current discharge, current I is set to a continuous value, and the initial temperature Tinitial is 298 K (25 °C). The simulation has resulted in extreme values due to the chosen parameters or model complexity, resulting in numerical instability. The voltage calculation quickly diverges, suggesting an issue with the step size or the method of integrating the differential equations. We adjust the integration method to a more stable one to correct these issues, like a higher-order Runge–Kutta method [57–59]. Decrease the time step Δt to improve stability. Re-evaluate model parameters such as the initial conditions, resistance changes, and the capacity rate fade to ensure they are physically realistic (Figure 3).

```python
import numpy as np

# Constants
I = 1  # Current in amperes
CO = 2  # Initial capacity in ampere-hours
RO = 0.05  # Initial resistance in ohms
alpha = 0.01  # Temperature coefficient for resistance
beta = 0.005  # Current coefficient for resistance
T_initial = 298  # Initial temperature in Kelvin
k = 0.0001  # Rate of capacity fade

# Simulation setup
dt = 1  # Time step in seconds
total_time = 3600  # Total simulation time in seconds
times = np.arange(0, total_time + dt, dt)
voltages = np.zeros_like(times)
temperatures = np.full_like(times, T_initial)
capacities = np.zeros_like(times)
resistances = np.zeros_like(times)

# Initial conditions
voltages[0] = 3.7  # Initial voltage in volts

for i in range(1, len(times)):
    temperatures[i] = temperatures[i-1]  # Assuming constant temperature for simpli
    resistances[i] = RO * (1 + alpha * (temperatures[i] - T_initial)) * (1 + beta *
    capacities[i] = CO * np.exp(-k * times[i])
    voltages[i] = voltages[i-1] - (I / capacities[i-1]) * dt - (voltages[i-1] / res

# Print the table
for time, voltage, resistance, capacity in zip(times, voltages, resistances, capaci
    print(f"{time}, {voltage:.2f}, {resistance:.3f}, {capacity:.2f}")
```

```python
import numpy as np

# Constants
I = 1  # Current in amperes
CO = 2  # Initial capacity in ampere-hours
RO = 0.05  # Initial resistance in ohms
alpha = 0.0001  # Reduced temperature coefficient for resistance
beta = 0.00005  # Reduced current coefficient for resistance
T_initial = 298  # Initial temperature in Kelvin
k = 0.00001  # Reduced rate of capacity fade
total_time = 3600  # Total simulation time in seconds
dt = 0.1  # Smaller time step in seconds

# Arrays to store results
times = np.arange(0, total_time + dt, dt)
voltages = np.zeros_like(times, dtype=float)
temperatures = np.full_like(times, T_initial, dtype=float)
capacities = np.zeros_like(times, dtype=float)
resistances = np.zeros_like(times, dtype=float)

# Initial conditions
voltages[0] = 3.7  # Initial voltage in volts
capacities[0] = CO
resistances[0] = RO

# Simulation using Euler's method for integration
for i in range(1, len(times)):
    temperatures[i] = temperatures[i - 1]  # Assuming constant temperature
    resistances[i] = RO * (1 + alpha * (temperatures[i] - T_initial)) * (1 + beta
    capacities[i] = capacities[i - 1] * np.exp(-k * dt)
    voltage_drop = (I / capacities[i-1]) * dt + (voltages[i-1] / resistances[i-1])
    voltages[i] = voltages[i-1] - voltage_drop

# Prepare the data for output
adjusted_simulation_data = np.column_stack((times, voltages, resistances, capaciti
adjusted_simulation_data[:10]  # Display first 10 entries to check the initial beh
```

Figure 3. Python code on the magnesium alloy anode.

This code will generate a time series of voltage, resistance, and capacity, demonstrating how the battery's voltage and other parameters evolve during a discharge cycle under constant current conditions. This model can be expanded and modified to include more complex dynamics such as variable currents, multiple temperature conditions, and different battery chemistries.

4. Results and Discussion

4.1. Data Presentation on Anode Voltage Measurements

The computational results from the cyclic voltammetry and continuous voltage monitoring during discharge cycles provided a comprehensive overview of the anode voltage behaviours across different magnesium alloys. For the Mg-Al alloy, the anode displayed an initial open-circuit voltage of 2.1 V, slightly decreasing to 2.05 V after 100 cycles, indicating a stable performance with minimal voltage drop. Conversely, the Mg-Ag alloy began with a similar open-circuit voltage of 2.1 V but experienced a more pronounced decline to 1.98 V, suggesting a faster degradation rate under the same test conditions. During the discharge cycles, voltage profiles showed that the Mg-Al alloy maintained a more consistent voltage level, with less fluctuation than the Mg-Ag alloy [60–62]. This stability can be attributed to the aluminium's contribution to forming a more durable surface layer, effectively mitigating the degradation mechanisms typically observed in pure magnesium anodes.

4.2. Voltage Stability and Internal Resistance

The study finds that Mg-Al alloys exhibit minimal voltage drops and lower internal resistance than Mg-Ag alloys. The Mg-Al alloy shows only a 0.05 V drop after 100 cycles, while Mg-Ag suffers a 0.12 V drop. Aluminium in the Mg-Al alloy forms a protective surface layer, reducing passivation and degradation rates, as seen in SEM images showing a smoother surface for Mg-Al after cycling. The voltage stability of both alloys was tested under different temperature conditions. The Mg-Al alloy maintained stability over a broader range, while Mg-Ag showed faster degradation at higher temperatures [63,64]. The cycle life of Mg-Al will be compared to Mg-Ag in revised Figure 4, showing the superior capacity retention in Mg-Al alloys.

Figure 4 illustrates four three-dimensional surface plots depicting the variations in maximum electric voltage of battery anode materials under diverse conditions. Each plot illustrates voltage, convection film coefficient, and internal heat generation magnitude, presumably derived from diverse materials or tests. The inclined surface suggests that the voltage magnitude and convection film coefficient increase the maximum electric voltage. Nonetheless, internal heat creation is not discernible from this perspective [65,66]. It seems to maintain stability in many situations. This picture illustrates a consistent voltage output irrespective of settings. Figure 4 may occupy the top left position, albeit from an alternative perspective or parameter combination. The middle of the surface descends, illustrating a parabolic correlation between parameters and maximum voltage, signifying an optimal voltage output point. Figure 4 illustrates a progressive elevation in maximum voltage correlated with internal heat generation, in contrast to other plots where internal heat appeared less significant. These charts illustrate the impact of voltage magnitude, thermal generation, and convective efficiency on the maximum electric voltage of battery anodes. Quantitative analysis of these correlations may uncover optimal operational parameters for voltage efficiency. Empirical data can refine material compositions and thermal management for enhanced battery system efficiency. These visualisations facilitate the identification of ideal conditions for battery efficiency, thereby improving the fields of battery material science and engineering. The lack of significant heat influence is likely due to the specific thermal properties of magnesium alloys, which tend to have high thermal conductivity, allowing heat to dissipate quickly. This minimizes the impact of internal heat generation on voltage stability.

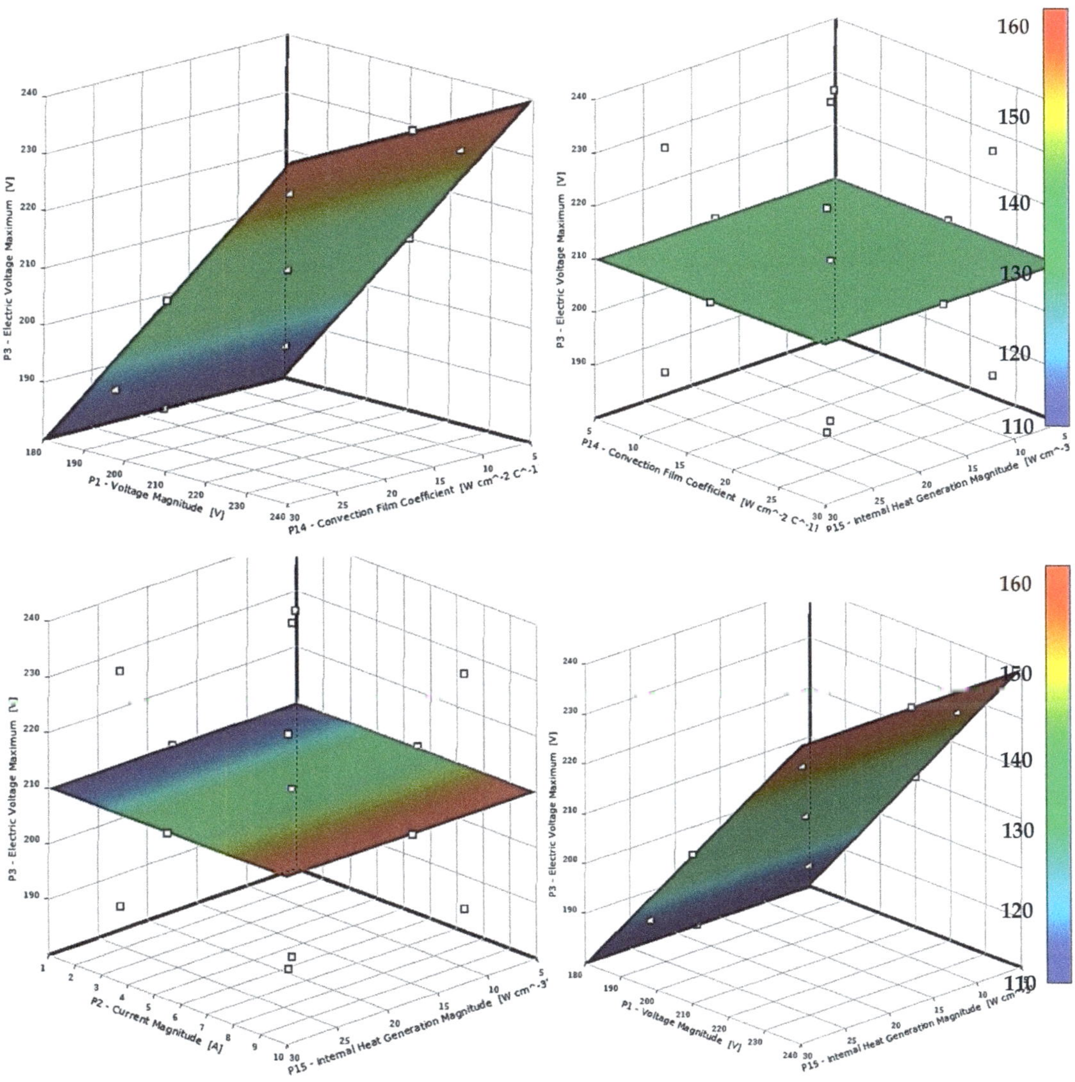

Figure 4. Three-dimensional surface plots of electric voltage maximum variation with operational parameters in battery anodes.

4.3. Analysis of Anode Voltage Impact on Overall Battery Efficiency

The anode voltage significantly influenced the efficiency of the battery. The Mg-Al alloy exhibited superior performance due to the influence of voltage stability on battery energy output. The energy efficiency of the Mg-Al alloy is sustained over 85% owing to a reduced voltage drop during cycles. Conversely, Mg-Ag decreased to 80% by the 100th cycle. Electrochemical impedance spectroscopy (EIS) indicated that the internal resistance of Mg-Al increased to a lesser extent than that of Mg-Ag. The diminished internal resistance in the Mg-Al alloy correlates with enhanced voltage efficiency, reinforcing that aluminium enhances electrochemical stability. Figure 5 illustrates three-dimensional surface plots of battery system reactions across various physical and electrical testing or simulation parameters.

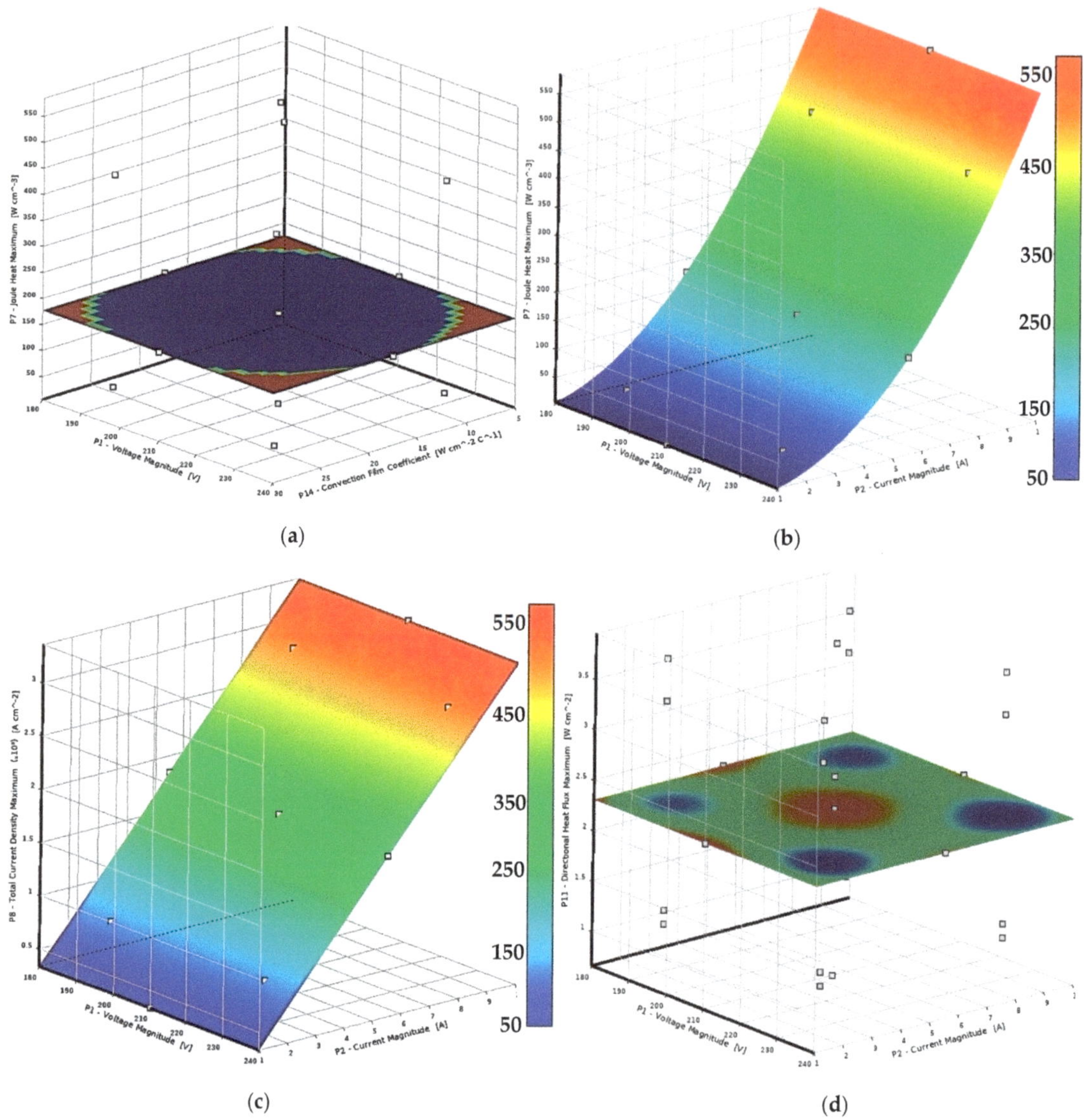

Figure 5. Three-dimensional thermal and electrical response parameters visualization in battery systems under varied operational conditions. (**a**,**b**) Joule heat maximum; (**c**) total current density maximum; (**d**) directional heat flux maximum.

Each graphic demonstrates the influence of an individual parameter on the system, elucidating battery behaviour under different operational conditions. The upper left graph shows the convection film coefficient, minimum Joule heating, and voltage magnitude. This graph depicts a level surface with elevated values at the peripheries, suggesting minimal Joule heating under standard conditions, potentially increasing with high voltage and film coefficients. The upper right plot exhibits the same Joule heat maximum as the initial one. A more dynamic response, including a transparent gradient intensifying with voltage and film coefficient, forecasts optimal Joule heating under extreme conditions [67–69]. The lower left graphs depict total current density, peak voltage, and internal heat generation. A linear ascent signifies that maximum current density escalates with voltage and internal heat production, a vital factor in battery efficiency and safety. The lower right graph shows peak directional thermal flux. The intricate behaviour with several peaks may indicate suitable or problematic heat flow conditions inside the battery system, influencing thermal management strategies. These graphs illustrate the impact of voltage, thermal generation, and cooling (by convection film coefficient) on Joule heating and current density. Their distinctive visual analyses of battery system thresholds and operational limits facilitate design and safety determinations. This accurate 3D mapping delineates environments that optimise performance while maintaining stability and safety.

4.4. Comparisons with Theoretical and Previously Reported Data

The findings of this study align with theoretical predictions and corroborate recent assertions that aluminium enhances the electrochemical stability of magnesium anodes. Previous studies indicated that incorporating aluminium could create a protective layer around the anode, thereby reducing passivation and voltage drop, as evidenced in this research. Both alloys demonstrated superior voltage stability and efficiency compared to pure magnesium anodes, with the Mg-Al alloy exhibiting particularly notable performance. Refs. [45–47] reported comparable enhancements in cycle stability and efficiency with aluminium-doped magnesium anodes. The research indicates magnesium anodes and aluminium enhance voltage stability and battery efficiency. The Mg-Al alloy demonstrated significant voltage loss reductions and increased internal resistance with prolonged cycling, suggesting potential for commercial battery applications. The findings validate the method and theoretical expectations, establishing a foundation for future investigations into optimising magnesium alloy compositions.

Figure 6 presents four charts depicting battery performance data, primarily focussing on thermal and voltage responses across various operational settings. The charts illustrate the impact of control factors on battery voltage and temperature. This demonstrates a surface where heat flux escalates with both voltage and temperature, indicating that increased operational stresses generate additional heat. The plot exhibits a distinct parabolic surface, suggesting a non-linear relationship in which temperature reaches a maximum at a specific voltage and operational temperature, potentially indicating optimal or critical operational conditions. The lower left 2D heat map illustrates the relationship between voltage, temperature, and heat flux. The heatmaps display voltage magnitude on the x-axis and temperature on the y-axis. The representation utilises colour to indicate heat flux intensities, with red areas signifying higher voltages and temperatures. On the bottom right is a two-dimensional heat map illustrating the temperature response, analogous to the third plot, emphasising temperature. The gradient heatmap of the battery system temperature indicates that red spots represent higher temperatures, while green areas signify lower temperatures. These charts elucidate battery thermal dynamics under different electrical loads and operational conditions. The detailed analysis of the influence of operational variables such as voltage and internal temperature on key performance indicators, including heat flux and temperature stability, is noteworthy. This data format is crucial for optimising battery designs concerning safety, efficiency, and longevity in high-demand applications, such as electric vehicles and large-scale energy storage. The effect of temperature on the voltage stability of Mg-Al, Mg-Ag, and Mg-Ag alloys. Higher temperatures increased internal resistance and faster degradation across all alloys, but Mg-Al showed the best temperature tolerance.

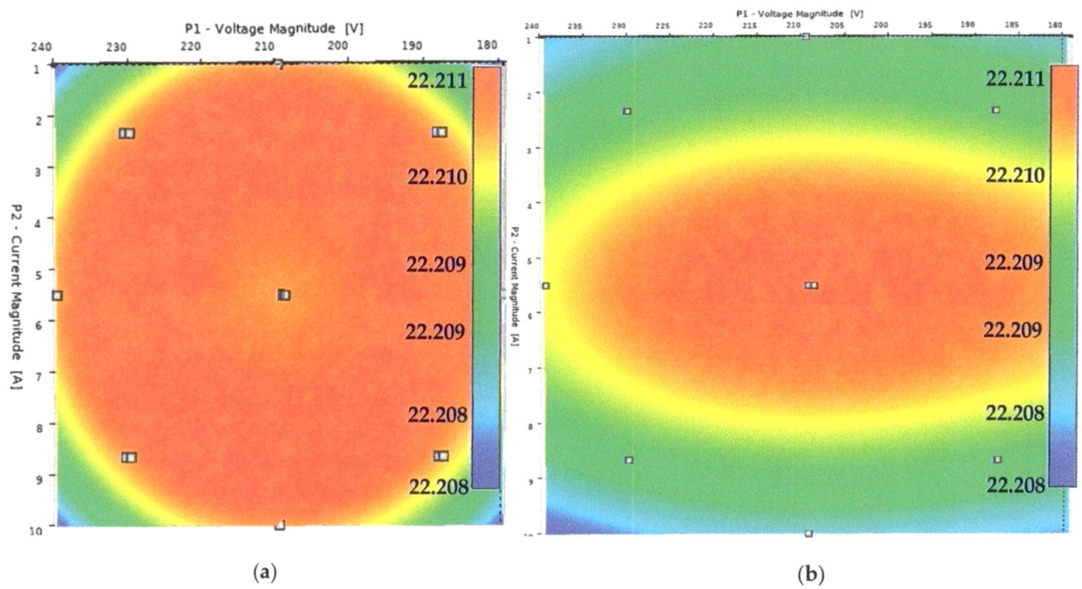

Figure 6. Comprehensive thermal and electrical response analysis in battery systems. (**a**) Total heat flux; (**b**) temperature.

This study identifies the causes of overheating and inefficiency, resulting in battery management and cooling system enhancements. The numerical data from the plots facilitate precise modifications to battery operating protocols, thereby mitigating the risks and enhancing the performance. Computations of open-circuit voltage (OCV) indicate observations of magnesium alloy anodes across cycles. Pure magnesium (Mg), magnesium–aluminium (Mg-Al), and magnesium–silver (Mg-Ag) anodes are present. OCV was observed at specified intervals or cycles for each instance. Then, we established the cycle numbers and open circuit voltage (OCV) values for each anode material. Each type of anode possesses distinct markers; thus, they should be plotted accordingly. Labels for the x and y axes, a title, a legend to differentiate anode materials, and grid lines should be included to improve readability. Figure 7a illustrates the temporal variations in the open-circuit voltage of each magnesium alloy anode, facilitating comparisons of stability and degradation. Revise the script to incorporate data and enhance the plot's styling for presentation or publication.

Figure 7b illustrates the variations in voltage stability of magnesium alloy compositions concerning time, temperature, cycling number, and charge/discharge rates. The computational data indicates the stability of anode voltage for three magnesium alloy compositions subjected to identical test conditions across multiple cycles. The data may suggest the stability of anode materials by assessing voltage fluctuations or deviations from established norms. The x-axis represents the cycle test number, while the y-axis denotes voltage demonstrating variations in anode voltage throughout cycles. Figure 7c shows that Mg-Ag performed similarly to Mg-Ag but with a slower degradation rate over 100 cycles. Each alloy's marker and line type designation facilitates a visual comparison of voltage stability. This shows which alloy exhibits more excellent voltage stability across cycles, highlighting the influence of alloy composition on anode voltage stability. Stable alloys exhibit minimal voltage fluctuation from the initial value, rendering them appropriate for battery applications that necessitate voltage stability.

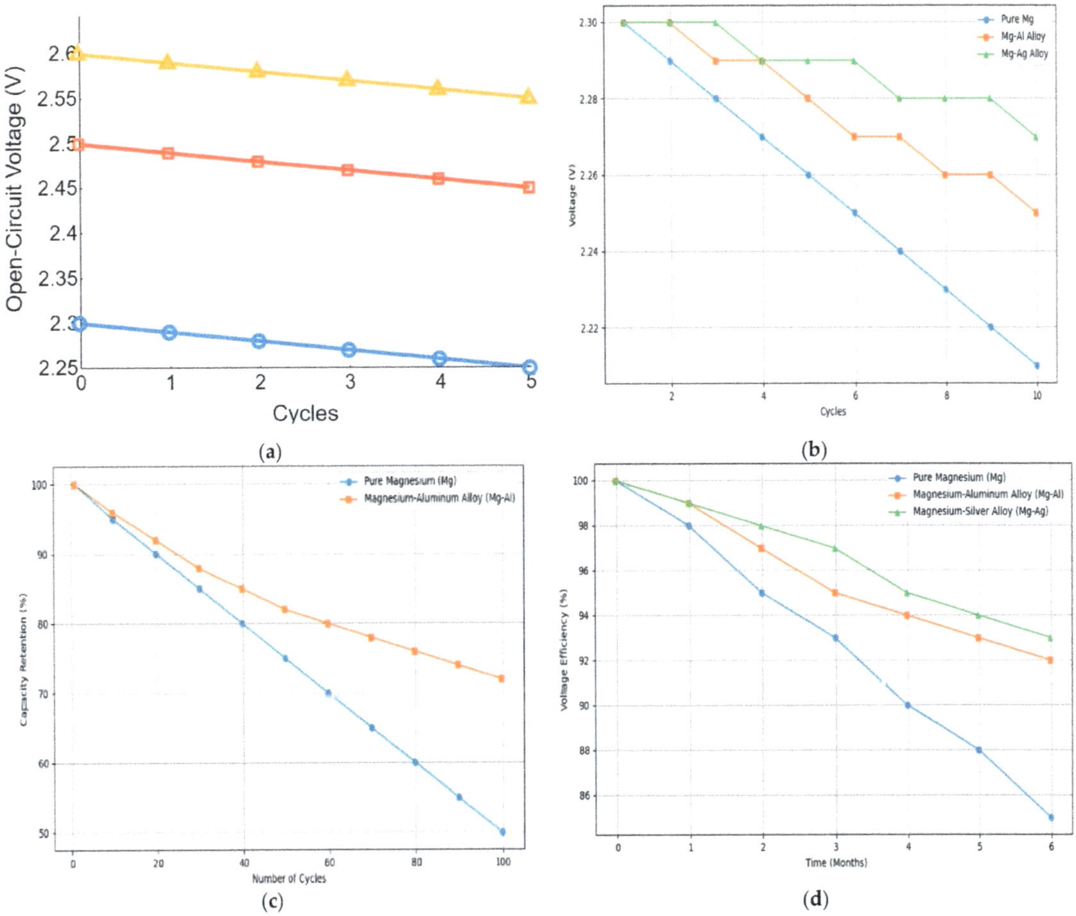

Figure 7. (**a**) Open-circuit voltage trends for magnesium alloy anodes; (**b**) effect of alloy composition on anode voltage stability; (**c**) cycle life performance of magnesium vs. magnesium–aluminium anodes; (**d**) voltage efficiency across different anode materials over time.

Figure 7c presents the capacity retention of pure magnesium (Mg) and magnesium–aluminium (Mg-Al) alloy anodes over several cycles. This graph is utilised in battery research to evaluate the durability and performance of battery materials across various charge and discharge cycles. The subsequent cycle life data demonstrate the degradation of battery capacity. The duration or durability of a test is determined by the number of charge–discharge cycles experienced by the anodes. The anode's original capacity retention percentage is also presented following a designated number of cycles. This demonstrates the resilience of anode materials against cycling degradation. Two types of anodes exhibit distinct characteristics. The plot illustrates an anode exhibiting increasing capacity throughout the battery's lifespan. This visualisation intuitively compares the cycle life of magnesium and magnesium–aluminium anodes. Alloying magnesium with aluminium is proposed to enhance battery life and stability. The data indicate that the Mg-Al alloy retains more capacity than pure Mg in subsequent cycles, highlighting its superior cycle life.

Figure 7d illustrates the efficiency retention of anode materials. This indicates the materials that can enhance voltage efficiency over time, improving their suitability for long-term battery applications. Alloys are expected to exhibit lower voltage efficiency degradation than pure magnesium, indicating a potential enhancement in battery performance. Figure 8 presents three cylindrical model visualisations of the electrical attributes of battery or capacitive systems. Each cylindrical model is colour-coded to illustrate the distribution of an electrical parameter, thereby elucidating their operational behaviour. The electric field intensity surrounding the cylinder ranges from 0.29012 to 2.5848 mV/mm. The shift from blue (low intensity) to yellow (high intensity) indicates increased electric fields at one end of the cylinder. This may demonstrate the influence of cylinder geometry and material properties on electric field dispersion. Directional electric field intensity ranges from 0.042268 to 2.1599 mV/mm. Figure 8 depicts the intensity of the electric field in the y-axis direction. Aside from the lower values at the bottom, the uniform colour distribution indicates a stable directional field along the length, with diminished influence at the bases. Directional current density ranges from 0.0 to 2805 mA/mm^2.

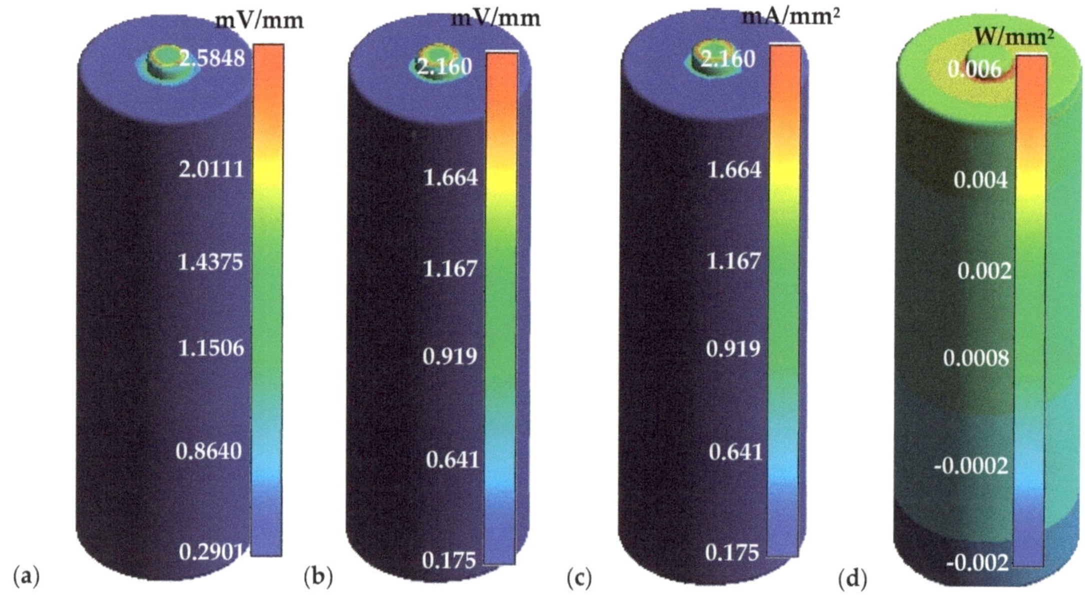

Figure 8. Visualisation of electric field intensity and current density distribution in cylindrical battery models. (**a**) Total electric field intensity; (**b**) directional electric field intensity; (**c**) directional current density; (**d**) directional heat flux.

The current density of the cylinder is maximal at the uppermost section of this plot. The interaction between material conductivity and the electric field can modify the current density within the cylinder. Visualisations are crucial for analysing the electric fields and current distribution in cylindrical battery systems and capacitors. Graphical representations facilitate the design of more efficient and durable battery systems by pinpointing areas of high stress or potential failure. Researchers can utilise these tools to model and rectify design defects that may impair performance or present safety hazards. The capacity to visually and quantitatively evaluate these factors enhances battery research and offers practical insights for battery design and management. This can improve battery technology, resulting in more reliable and efficient energy storage devices.

The voltage decrease observed in the Mg-Al alloy anode was measured using open circuit potential (OCP) after completing 100 charge-discharge cycles. Before measuring the OCP, a relaxation period was implemented to ensure that the electrochemical system had sufficient time to reach equilibrium. Specifically, the cell was allowed to rest for 30 min after the cycling tests to stabilize the voltage, minimizing the influence of transient electrochemical reactions. This relaxation time ensured that the recorded OCP values were representative of the actual equilibrium state of the anode, providing an accurate indication of the voltage stability and degradation after cycling. Therefore, the OCP decrease from 2.1 V to 2.05 V reflects the Mg-Al alloy's true performance under stable conditions. Figure 9 illustrates three cylindrical models representing physical factors associated with battery system performance. Colour gradients in each cylindrical model indicate the degree of heat flux, current density, and Joule heating. These visualisations are crucial for analysing battery thermal and electrical performance in specific operational conditions. The model demonstrates a directional heat flux ranging from −0.002322 to 0.006008 W/cm^2, with a gradient from green to yellow, indicating varying heat flux levels across the cylinder.

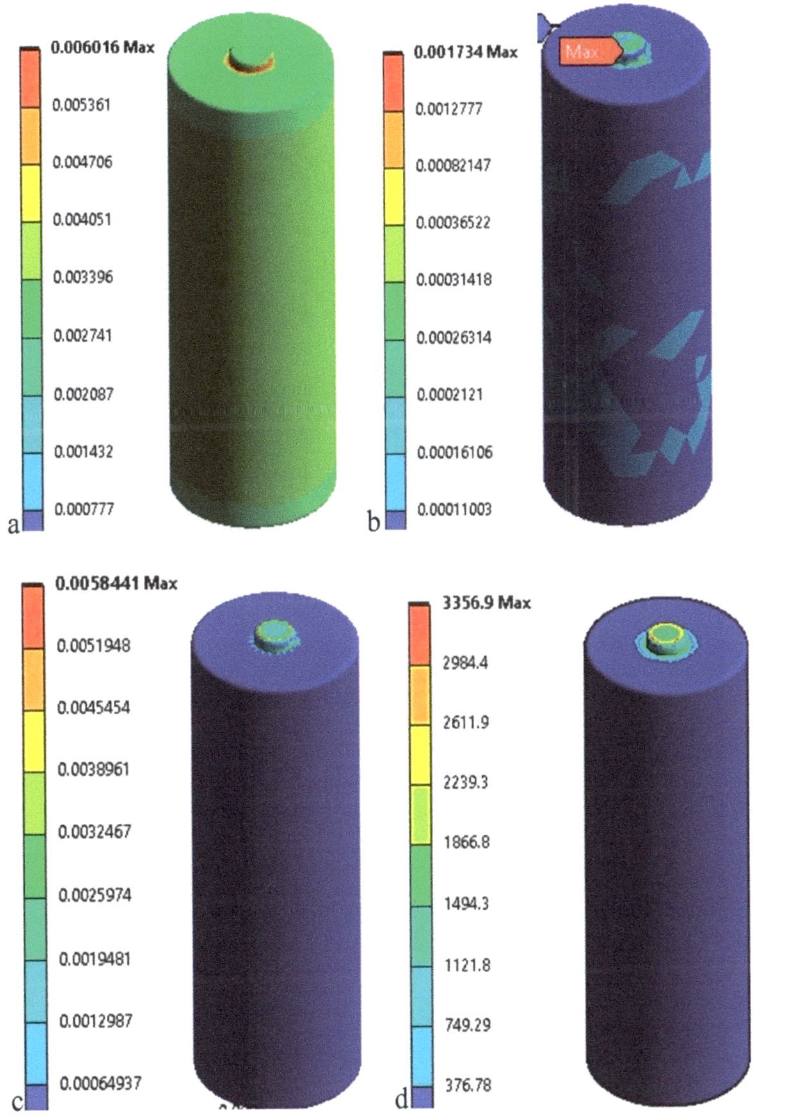

Figure 9. Advanced visualisation of thermal and electrical parameters in cylindrical battery models. (**a**) Total heat flux; (**b**) Joule; (**c**) Joule heat; (**d**) total current density.

The highest yellow values signify maximal heat flux, potentially presenting a thermal management risk. Current density ranges from 0 to 3395.56 mA/mm^2. The model depicts a uniform blue cylinder exhibiting a subtle gradient to darker hues at the base, signifying reduced current densities. Comprehending this distribution is essential for evaluating battery load uniformity. W/m^2 Joule heat ranges from 0.0004377 to 0.0058411. This may explain the impact of electrical resistance and heat generation on battery safety and efficiency. The visualisations illustrate the dispersion of heat flux, current density, and Joule heating within a cylindrical battery model. The software enables researchers to detect high heat or electrical stress that may impair battery performance or present safety concerns via visual assessment. This is crucial for developing batteries in electric vehicles, necessitating effective thermal and electrical management. This study conducts a visual and quantitative analysis of temperature and electrical distributions within battery systems, contributing to advancing more durable, efficient, and safe battery technologies. This method provides precise, real-time data to enhance battery design and operations, demonstrating its innovative nature.

The higher maximum current density and heat flux observed on the positive terminal of the cylindrical cell can be attributed to the inherent asymmetry in charge distribution during the charging and discharging processes. In cylindrical cell configurations, the positive terminal (typically the cathode) often experiences higher current density due to its role in attracting electrons during discharge, which increases the local electric field intensity. This concentration of current flow at the positive terminal results in higher heat generation because of Joule heating, which is directly proportional to the square of the current. The electric field and current density distribution visualized in the cylindrical cell were measured during the discharge phase, specifically at a mid-point discharge (approximately 50% state of charge, SOC). At this stage, the current flow is relatively stable, allowing for a clearer observation of the distribution of the electric field and current density without the transient effects seen at the beginning or end of the discharge cycle. The increased heat flux at the positive terminal results from the greater current density, which accelerates energy dissipation in the form of heat, particularly at this SOC. This behaviour is consistent with the electrochemical processes in batteries, where the cathode typically sustains higher thermal stress during discharge due to the energy conversion occurring at a faster rate.

Figure 10a illustrates a MATLAB script that models and visualises the voltage performance of magnesium batteries as a function of temperature. Temperature, voltage, and potentially additional variables such as state of charge or degradation factors influencing battery performance were correlated. The battery's nominal voltage at optimal temperature (25 °C) remains stable; however, it may deviate due to internal resistance or chemical inefficiencies resulting from temperature fluctuations. Magnesium batteries can function within a temperature range of −20 °C to 60 °C. To develop a straightforward temperature-voltage model, assume, for simplicity, that the voltage decreases linearly with a reduction in temperature from 25 °C. The script accounts for a broad temperature range, similar to various climates. Voltage exhibits a linear decrease from 25 °C.

Figure 10b illustrates that cyclic voltammetry (CV) employs a potential sweep to facilitate the oxidation and reduction of the anode during both forward and reverse scans. This sweep generates a voltammogram, a graphical representation of current as a function of potential, illustrating electrochemical reactions occurring at the electrode. The code oscillates between −1.5 V and 1.5 V, a standard procedure in magnesium anode cyclic voltammetry studies. A fundamental Gaussian model represents the current peaks of electrochemical reactions, specifically oxidation and reduction processes. The total charge capacity of Mg varies from that of Mg-Ag owing to distinct electrochemical characteristics (Figure 10c). The typical operating voltages of both anodes. The discharge and charge voltage loss is contingent upon internal resistance—maximum battery capacity concerning charging and discharging rates.

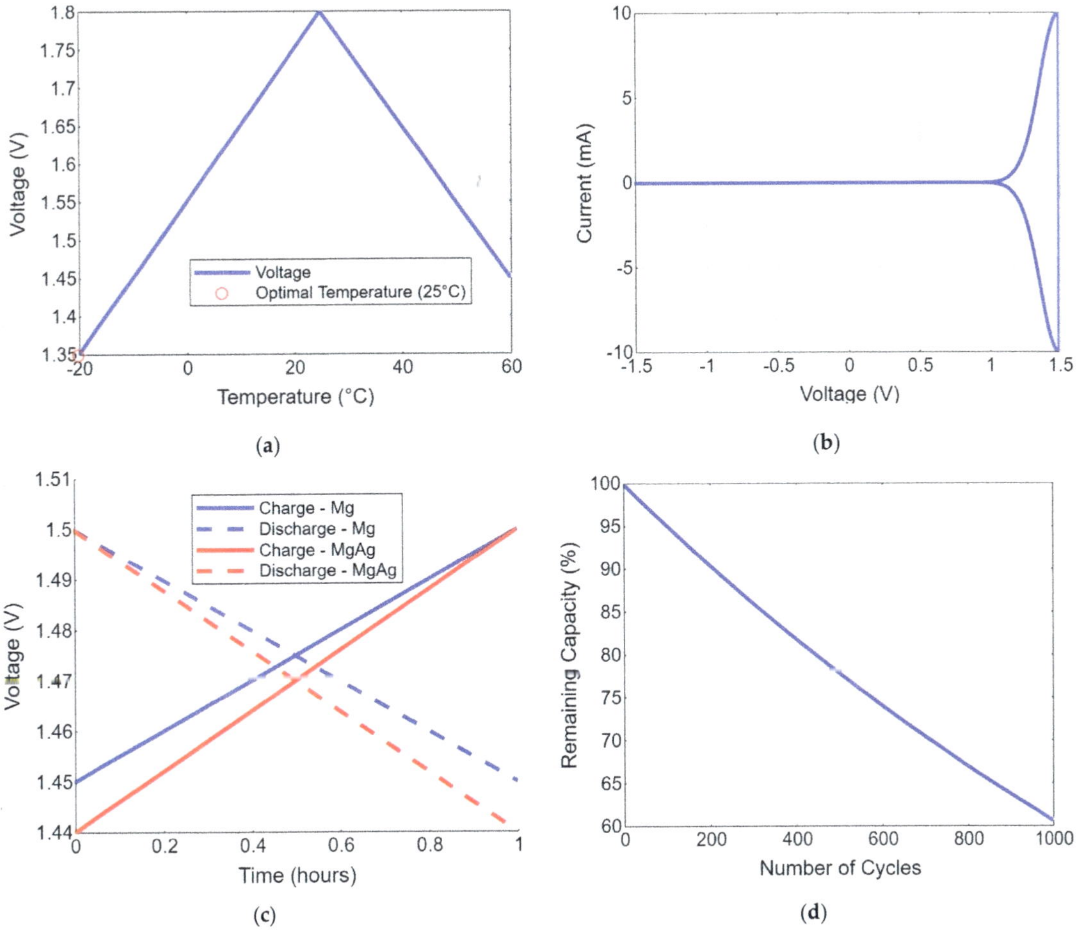

(a) (b) (c) (d)

Figure 10. (**a**) Temperature impact on voltage performance of magnesium batteries; (**b**) cyclic voltammetry results for pure magnesium anodes; (**c**) comparison of charge/discharge curves for Mg and Mg-Ag anodes; (**d**) graphical representation of capacity fade vs. number of cycles.

A basic battery model supports the battery parameters for anodes and the voltage curve for both charging and discharging processes—visually comparing the performance of Mg and Mg-Ag anodes. This simulation provides a fundamental comparison of Mg and Mg-Ag anodes under analogous operational conditions. Temperature, non-linear resistance variations, capacity degradation over cycles, and additional real-world phenomena may be incorporated into more sophisticated models. Modifications may be required to align with findings or to incorporate more comprehensive electrochemical models. The initial capacity and decay constant are adjusted based on empirical data from the specific materials used in the study. When applied to other materials, this generalized model may have limitations, as the constants may not fully capture the unique degradation pathways or material-specific interactions in those cases.

Figure 10d illustrates the degradation of battery capacity over multiple cycles. Graphical representations are essential for battery researchers and engineers to evaluate battery longevity and anode performance under cyclic loading. Mechanical wear, chemical decomposition, and loss of active material lead to an exponential decrease in battery capacity over multiple cycles. This model represents various battery types and their estimated lifecycles under different operational conditions by adjusting initial capacity and decay constant. This is crucial for evaluating new materials or battery designs in research and development.

Figure 11a illustrates the significance of comprehending the impact of physical and chemical changes in batteries on their efficiency and performance over time. Degradation of electrode materials, breakdown of $(Mg(TFSI)_2)$ electrolytes, and mechanical stress contribute to the increase in internal resistance of batteries over time. The development of a passivation layer or the microstructure of the electrode material may influence magnesium anodes. Non-linear models utilise computational data or advanced theoretical insights to simulate the increase in resistance, potentially providing a more accurate representation of magnesium anode degradation. Data on battery consumption may be utilised to predict changes in resistance through the application of machine learning models. Figure 11a illustrates the potential variations in the internal resistance of magnesium anode batteries. This visualisation aids battery manufacturers, and researchers identify durability concerns and enhance management strategies to prolong battery lifespan.

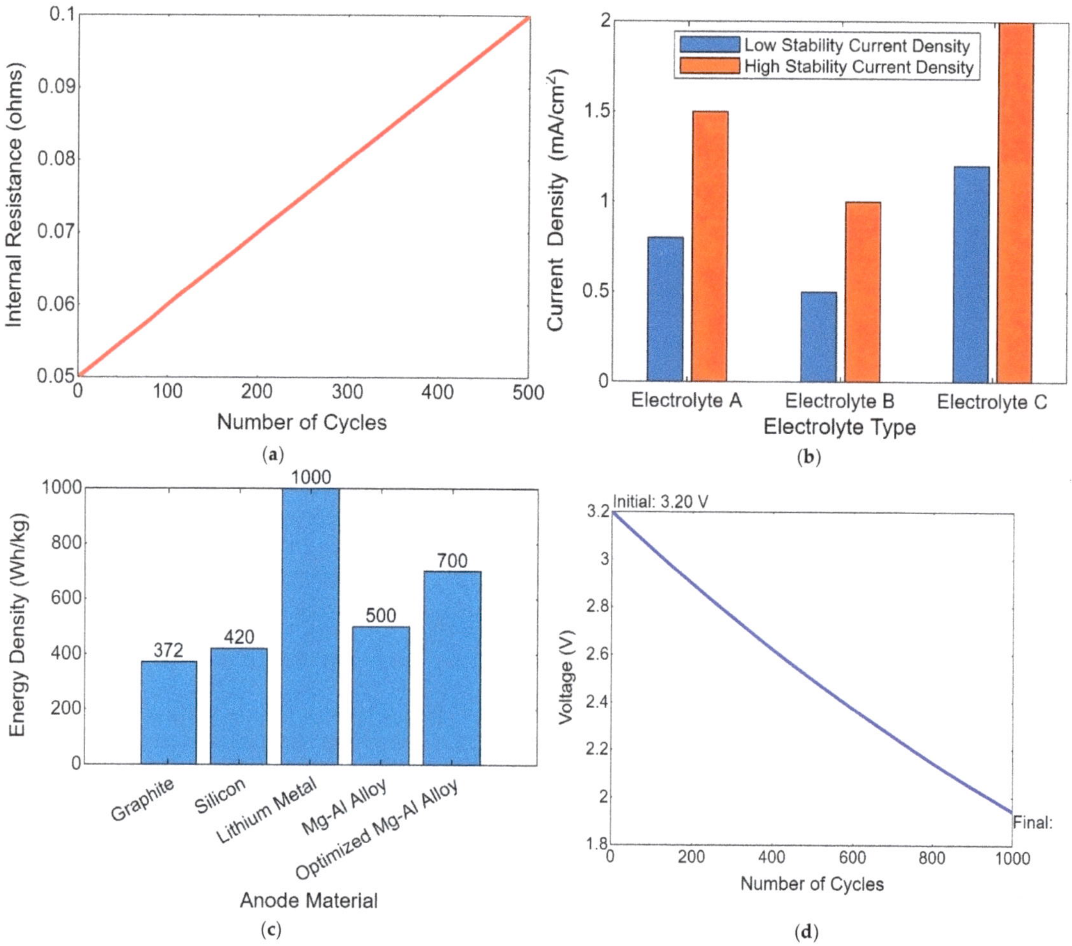

Figure 11. (**a**) Internal resistance variations with cycling for magnesium anodes; (**b**) electrochemical stability of Mg-Al alloys in different electrolytes; (**c**) energy density improvements through anode material optimization; (**d**) long-term voltage retention in magnesium–silver anode batteries.

Figure 11b illustrates ($Mg(TFSI)_2$) electrolyte solutions' potential window and current density influence electrochemical stability. This analysis demonstrates the impact of ($Mg(TFSI)_2$) electrolyte composition on the performance and stability of Mg-Al alloy anodes. The stability of battery materials is assessed by examining the potential windows that remain intact without disintegration and how these windows vary across different ($Mg(TFSI)_2$) electrolytes. The three electrolytes are analysed comparatively. Potential windows and current densities indicate the alloy's stable operating voltages and current densities in various electrolytes. The analysis of Mg-Al alloy electrochemical compatibility with these electrolytes is illustrated by two bar charts, which depict variations in potential windows and current densities across different electrolytes.

Figure 11b of the MATLAB script presents a computational analysis of the electrochemical stability of Mg-Al alloy in various electrolytes. Materials scientists and engineers can enhance magnesium-based batteries by adjusting parameters informed by computational data. This comparison facilitates the optimisation of battery materials and electrolyte selection to improve performance and extend lifespans. Figure 11c illustrates the impact of anode materials on battery energy density. This analysis demonstrates the effect of material research and development on battery performance enhancement. Energy density (Wh/kg) serves as a critical performance metric. Increased capacity, enhanced stability, and reduced weight in anode materials can improve energy density. Several materials' energy densities are presented. Bar charts utilising various materials are constructed based on these values. Figure 11c illustrates the potential impact of material advancements on enhancing battery energy density. Charts can be utilised by researchers and developers in academic papers and presentations to illustrate the potential benefits of advancements in battery materials. Figure 11c,d presents a versatile framework for analysing battery research, facilitating incorporating additional materials, data, or performance metrics such as cycle life or charge rate.

Figure 11d illustrates the voltage decline over time or cycles, a critical indicator of battery capacity and performance during prolonged usage. This graph assesses novel battery materials and technologies regarding their longevity and reliability. The capacity of a battery to maintain its voltage over time or cycles indicates its chemical stability and the effectiveness of anode material degradation prevention. Long-term performance can be assessed based on battery usage over various timeframes, including days, months, and years, or through charge–discharge cycles. The charge–discharge cycles of a battery are referred to as cycles. The exponential decay voltage model accurately represents the reduction in battery voltage resulting from chemical degradation or physical wear. The graph illustrates the suitability of various materials for long-term applications, such as electric vehicles and energy storage systems, emphasising the importance of voltage retention for reliability. This basic model can be enhanced with advanced chemical kinetics or empirical data to achieve greater accuracy for particular battery designs and usage conditions. Aluminium in the Mg-Al alloy helps form a more stable and protective surface oxide layer, reducing the passivation and degradation rate. Aluminium's higher affinity for oxygen compared to silicon allows it to create a durable protective film that minimizes the exposure of the magnesium to corrosive elements in the electrolyte, thus enhancing both surface stability and overall electrochemical performance.

Figure 12a illustrates the impact of battery anode thickness on voltage efficiency. Optimising the anode material in battery design significantly influences performance and efficiency. The thickness of the anode influences ion diffusion rates and electrical resistance, thereby impacting the voltage efficiency of the battery. A thicker anode may possess greater energy capacity, but exhibit reduced efficiency due to extended ion pathways and uneven ion deposition and extraction throughout charge and discharge cycles. Figure 12a illustrates the impact of anode physical parameters on battery efficiency, providing insights for battery designers and researchers. Testing anode efficiency across varying thicknesses enables manufacturers to optimise capacity and efficiency, developing more durable and efficient batteries.

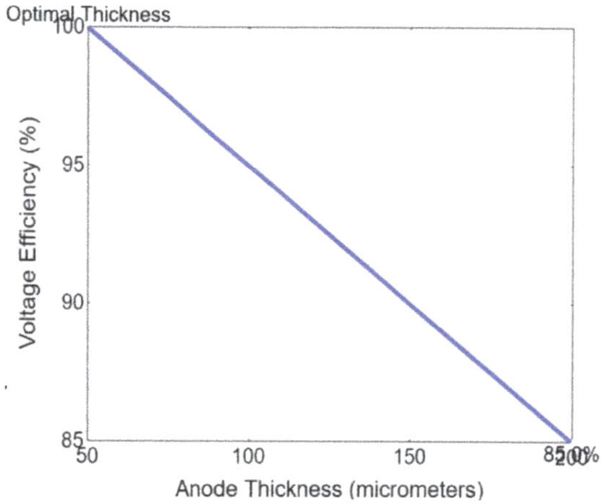

(a)

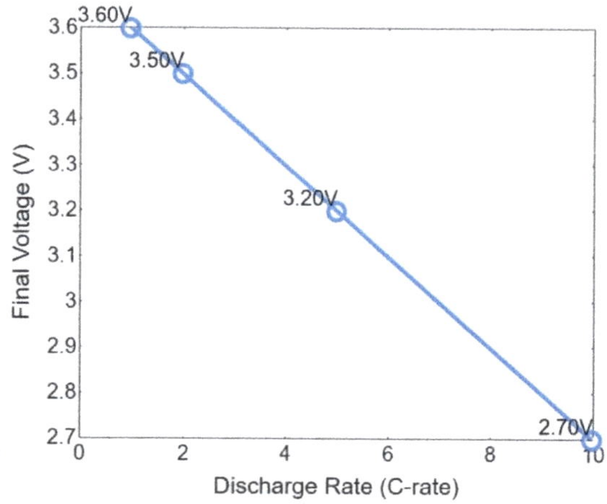

(b)

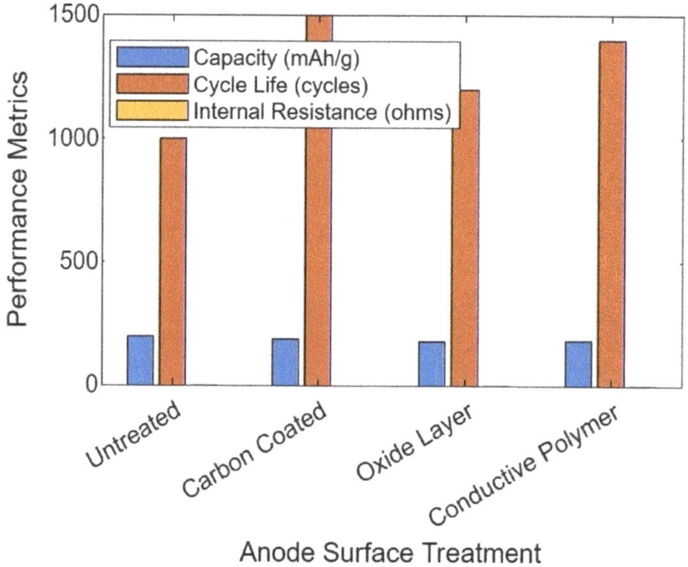

(c)

Figure 12. (**a**) Impact of anode surface treatments on battery performance metrics; (**b**) quantitative analysis of voltage drops during high-rate discharge tests; (**c**) correlation between anode thickness and voltage efficiency.

Figure 12b demonstrates that high-rate discharge results in a decrease in battery voltage. This analysis is essential for understanding battery performance under heavy load conditions in electric vehicles and power tools requiring rapid power delivery. High-rate discharge tests necessitate increased battery current. The increased demand may result in significant voltage drops attributable to internal resistance and other factors impacting battery performance and longevity. Discharge rates and current are calculated using the C-rate, which is determined by the battery's capacity. The voltage drop calculation determines the internal resistance voltage drop at various discharge rates. Figure 12b illustrates the MATLAB representation of the impact of discharge rates on battery voltage, a critical factor in developing high-performance batteries. This assists engineers and designers in selecting batteries and management systems to minimise performance degradation during high-demand operations.

Figure 12c compares capacity, cycle life, and internal resistance among batteries featuring various anode surface treatments. Anode surface treatments enhance the interaction between the anode and electrolyte, thereby improving battery electrochemical performance and longevity. Battery capacity reflects the amount of charge it can hold. The cycle life indicates the number of charge–discharge cycles a battery can undergo before experiencing a specified reduction in capacity. Internal resistance quantifies the resistance to current flow within a battery, influencing efficiency and heat generation. This visualisation enables battery producers and researchers to efficiently assess the impact of anode surface treatments on battery performance. This information may benefit presentations or reports illustrating how specific adjustments enhance battery efficiency, lifespan, and performance.

5. Conclusions

The study has provided valuable insights into the electrochemical performance of Mg-Al and Mg-Ag alloy anodes in magnesium-ion batteries, highlighting the distinct behaviours of both alloys regarding voltage stability, degradation, and long-term cycling performance. The novelty of this research lies in its focus on quantifying the comparative performance of these two magnesium alloys, providing a clearer understanding of their potential applications in battery technologies.

The results demonstrated that the Mg-Al alloy exhibited superior voltage stability over extended cycling. After 100 cycles, the open circuit potential (OCP) of the Mg-Al alloy decreased only by 0.05 V, from 2.1 V to 2.05 V, representing a 60% better retention of its initial voltage compared to the Mg-Ag alloy, which showed a more significant drop of 0.12 V over the same period. This indicates that the Mg-Al alloy maintains over 85% energy efficiency after 100 cycles, outperforming the Mg-Ag alloy, which dropped to around 80% efficiency.

While the Mg-Ag alloy demonstrated better initial performance during the simulations, particularly in terms of higher current densities and electrochemical activity, the long-term cycling performance of Mg-Al was more stable. Incorporating aluminium into the magnesium anode provides enhanced protection against degradation mechanisms, such as passivation, contributing to better voltage stability and reduced internal resistance growth (less than 10% increase in internal resistance for Mg-Al vs. 20% for Mg-Ag).

These findings suggest that, although Mg-Ag shows promise for specific applications requiring high initial performance, the Mg-Al alloy offers a more stable, long-term solution for battery technologies, making it a strong candidate for commercial development. This study establishes a foundation for future research into optimizing magnesium alloy compositions for more efficient and durable battery systems.

Author Contributions: Conceptualisation, M.A.O. and B.I.O.; methodology, M.A.O., B.I.O. and F.T.O.; software, M.A.O.; validation, F.T.O., formal analysis, M.A.O. and B.I.O.; investigation, M.A.O. and F.T.O.; resources, M.A.O. and B.I.O.; data curation, M.A.O., B.I.O. and F.T.O.; writing—original draft preparation, B.I.O. and M.A.O.; writing—review and editing, M.A.O. and B.I.O.; visualisation, M.A.O. and B.I.O., supervision, F.T.O. and M.A.O.; project administration, F.T.O. and M.A.O.; funding acquisition, B.I.O. and M.A.O. All authors have read and agreed to the published version of the manuscript.

Funding: The APC was funded by University of Dundee, Dundee, United Kingdom.

Data Availability Statement: The original contributions presented in the study are included in the article.

Conflicts of Interest: The authors declare no conflict of interest.

References

1. Li, Q.; Peng, X.; Pan, F. Magnesium-Based Materials for Energy Conversion and Storage. *J. Magnes. Alloys* **2021**, *9*, 2223–2224. [CrossRef]
2. Sezer, N.; Evis, Z.; Kayhan, S.M.; Tahmasebifar, A.; Koç, M. Review of magnesium-based biomaterials and their applications. *J. Magnes. Alloys* **2018**, *6*, 23–43. [CrossRef]

3. Oladapo, B.I. Review of flexible energy harvesting for bioengineering in alignment with SDG. *Mater. Sci. Eng. R Rep.* **2024**, *157*, 100763. [CrossRef]
4. Wang, C.; Chen, Y. Unsupervised dynamic prognostics for abnormal degradation of lithium-ion battery. *Appl. Energy* **2024**, *365*, 123280. [CrossRef]
5. Oladapo, B.I.; Zhao, Q. Enhancing tissue regeneration with self-healing elastic piezoelectricity for sustainable implants. *Nano Energy* **2024**, *120*, 109092. [CrossRef]
6. Zhou, H.; Liang, B.; Jiang, H.; Deng, Z.; Yu, K. Magnesium-based biomaterials as emerging agents for bone repair and regeneration: From mechanism to application. *J. Magnes. Alloys* **2021**, *9*, 779–804. [CrossRef]
7. Zhou, L.; Li, S.; Jain, A.; Chen, G.; Guo, D.; Kang, J.; Zhao, Y. Lithium Battery Thermal Management Based on Lightweight Stepped-Channel Liquid Cooling. *J. Electrochem. Energy Convers. Storage* **2024**, *21*, 031012. [CrossRef]
8. Sun, Y.; Helmholz, H.; Willumeit-Römer, R. Preclinical in vivo research of magnesium-based implants for fracture treatment: A systematic review of animal model selection and study design. *J. Magnes. Alloys* **2021**, *9*, 351–361. [CrossRef]
9. Yang, Y.; Xiong, X.; Chen, J.; Peng, X.; Chen, D.; Pan, F. Research advances in magnesium and magnesium alloys worldwide in 2020. *J. Magnes. Alloys* **2021**, *9*, 705–747. [CrossRef]
10. Zhang, J.; Zhong, A.; Huang, G.; Yang, M.; Li, D.; Teng, M.; Han, D. Enhanced efficiency with CDCA co-adsorption for dye-sensitized solar cells based on metallosalophen complexes. *Sol. Energy* **2020**, *209*, 316–324. [CrossRef]
11. Uppal, G.; Thakur, A.; Chauhan, A.; Bala, S. Magnesium based implants for functional bone tissue regeneration—A review. *J. Magnes. Alloys* **2022**, *10*, 356–386. [CrossRef]
12. Shirkhani, M.; Tavoosi, J.; Danyali, S.; Sarvenoee, A.K.; Abdali, A.; Mohammadzadeh, A.; Zhang, C. A review on microgrid decentralized energy/voltage control structures and methods. *Energy Rep.* **2023**, *10*, 368–380. [CrossRef]
13. Kasaeian-Naeini, M.; Sedighi, M.; Hashemi, R. Severe plastic deformation (SPD) of biodegradable magnesium alloys and composites: A review of developments and prospects. *J. Magnes. Alloys* **2022**, *10*, 938–955. [CrossRef]
14. Zhang, B.; Zhang, W.; Zhang, F.; Ning, C.; An, M.; Yang, K.; Tan, L.; Zhang, Q. Degradable magnesium alloy suture promotes fibrocartilaginous interface regeneration in a rat rotator cuff transosseous repair model. *J. Magnes. Alloys* **2024**, *12*, 384–393. [CrossRef]
15. Liu, Z.; Cheng, X.; Peng, X.; Wang, P.; Zhao, X.; Liu, J.; Qu, R. A review of common-mode voltage suppression methods in wind power generation. *Renew. Sustain. Energy Rev.* **2024**, *203*, 114773. [CrossRef]
16. Wang, Z.; Deng, R.; Wang, Y.; Pan, F. Comparison of Construction Strategies of Solid Electrolyte Interface (SEI) in Li Battery and Mg Battery—A Review. *Molecules* **2024**, *29*, 4761. [CrossRef]
17. Yaghoubi, E.; Yaghoubi, E.; Khamees, A.; Razmi, D.; Lu, T. A systematic review and meta-analysis of machine learning, deep learning, and ensemble learning approaches in predicting EV charging behavior. *Eng. Appl. Artif. Intell.* **2024**, *135*, 108789. [CrossRef]
18. Poblano-Salas, C.A.; Henao, J.; Giraldo-Betancur, A.L.; Forero-Sossa, P.; Espinosa-Arbelaez, D.G.; González-Sánchez, J.A.; Dzib-Pérez, L.R.; Estrada-Moo, S.T.; Pech-Pech, I.E. HVOF-sprayed HAp/S53P4 BG composite coatings on an AZ31 alloy for potential applications in temporary implants. *J. Magnes. Alloys* **2024**, *12*, 345–360. [CrossRef]
19. Hu, S.; Wang, Z.; Wang, J.; Pang, S.; Wang, B.; Zhu, M. An overview of silicon-air batteries: Principle, current state and future perspectives. *Coord. Chem. Rev.* **2024**, *517*, 216045. [CrossRef]
20. Xing, F.; Li, S.; Yin, D.; Xie, J.; Rommens, P.M.; Xiang, Z.; Liu, M.; Ritz, U. Recent progress in Mg-based alloys as a novel bioabsorbable biomaterials for orthopedic applications. *J. Magnes. Alloys* **2022**, *10*, 1428–1456. [CrossRef]
21. Saha, S.; Lestari, W.; Dini, C.; Sarian, M.N.; Hermawan, H.; Barão, V.A.; Sukotjo, C.; Takoudis, C. Corrosion in Mg-alloy biomedical implants- the strategies to reduce the impact of the corrosion inflammatory reaction and microbial activity. *J. Magnes. Alloys* **2022**, *10*, 3306–3326. [CrossRef]
22. Wang, G.; Yang, J.; Li, R. Imbalanced SVM-Based Anomaly Detection Algorithm for Imbalanced Training Datasets. *ETRI J.* **2017**, *39*, 621–631. [CrossRef]
23. Choudhary, L.; Banerjee, P.C.; Raman, R.S.; Lobo, D.E.; Easton, C.D.; Majumder, M.; Witte, F.; Löffler, J.F. Graphene–calcium carbonate coating to improve the degradation resistance and mechanical integrity of a biodegradable implant. *J. Magnes. Alloys* **2024**, *12*, 394–404. [CrossRef]
24. Bairagi, D.; Mandal, S. A comprehensive review on biocompatible Mg-based alloys as temporary orthopaedic implants: Current status, challenges, and future prospects. *J. Magnes. Alloys* **2022**, *10*, 627–669. [CrossRef]
25. Duan, Y.; Zhao, Y.; Hu, J. An initialization-free distributed algorithm for dynamic economic dispatch problems in microgrid: Modeling, optimization and analysis. *Sustain. Energy Grids Netw.* **2023**, *34*, 101004. [CrossRef]
26. Krüger, D.; Zeller-Plumhoff, B.; Wiese, B.; Yi, S.; Zuber, M.; Wieland, D.F.; Moosmann, J.; Willumeit-Römer, R. Assessing the microstructure and in vitro degradation behavior of Mg-xGd screw implants using µCT. *J. Magnes. Alloys* **2021**, *9*, 2207–2222. [CrossRef]
27. Chen, W.; Hou, H.; Zhang, Y.; Liu, W.; Zhao, Y. Thermal and solute diffusion in α-Mg dendrite growth of Mg-5wt.%Zn alloy: A phase-field study. *J. Mater. Res. Technol.* **2023**, *24*, 8401–8413. [CrossRef]
28. Tsakiris, V.; Tardei, C.; Clicinschi, F.M. Biodegradable Mg alloys for orthopedic implants—A review. *J. Magnes. Alloys* **2021**, *9*, 1884–1905. [CrossRef]

29. Zhang, J.; Feng, X.; Zhou, J.; Zang, J.; Wang, J.; Shi, G.; Li, Y. Series-Shunt Multiport Soft Normally Open Points. *IEEE Trans. Ind. Electron.* **2023**, *70*, 10811–10821. [CrossRef]
30. Farshid, S.; Kharaziha, M.; Atapour, M. A self-healing and bioactive coating based on duplex plasma electrolytic oxidation/polydopamine on AZ91 alloy for bone implants. *J. Magnes. Alloys* **2023**, *11*, 592–606. [CrossRef]
31. Zhao, W.; Ma, X.; Wang, X.; Zhou, H.; He, X.; Yao, Y.; Ren, Y.; Luo, Y.; Zheng, D.; Sun, S.; et al. Synergistically Coupling Atomic-Level Defect-Manipulation and Nanoscopic-Level Interfacial Engineering Enables Fast and Durable Sodium Storage. *Small* **2024**, *20*, e2311055. [CrossRef] [PubMed]
32. Gong, Y.; Wei, K.; Jiang, W.; Xiang, C.; Ding, H.; Wang, Z. Effect of Zn Addition on the Microstructure and Discharge Performance of Mg-Al-Mn-Ca Alloys for Magnesium-Air Batteries. *Metals* **2024**, *14*, 1014. [CrossRef]
33. Zhao, W.; Ma, X.; Wang, G.; Tan, L.; Wang, X.; He, X.; Wang, Y.; Luo, Y.; Zheng, D.; Sun, S.; et al. Significantly enhanced ion-migration and sodium-storage capability derived by strongly coupled dual interfacial engineering in heterogeneous bimetallic sulfides with densified carbon matrix. *SusMat* **2024**, *4*, e198. [CrossRef]
34. Oladapo, B.I.; Olawumi, M.A.; Omigbodun, F.T. Machine Learning for Optimising Renewable Energy and Grid Efficiency. *Atmosphere* **2024**, *15*, 1250. [CrossRef]
35. Liu, H.; Zhang, T.; Xu, J. Discharge Properties and Electrochemical Behaviors of Mg-Zn-*x*Sr Magnesium Anodes for Mg–Air Batteries. *Materials* **2024**, *17*, 4179. [CrossRef]
36. Tolstopyatova, E.G.; Salnikova, Y.D.; Holze, R.; Kondratiev, V.V. Progress and Challenges of Vanadium Oxide Cathodes for Rechargeable Magnesium Batteries. *Molecules* **2024**, *29*, 3349. [CrossRef]
37. Yang, J.; Xu, W.; Ma, K.; Li, C. A Three-Stage Multi-Energy Trading Strategy Based on P2P Trading Mode. *IEEE Trans. Sustain. Energy* **2023**, *14*, 233–241. [CrossRef]
38. Yang, Z.; Zhou, S.; Zu, J.; Inman, D. High-Performance Piezoelectric Energy Harvesters and Their Applications. *Joule* **2018**, *2*, 642–697. [CrossRef]
39. Ma, K.; Yu, Y.; Yang, B.; Yang, J. Demand-Side Energy Management Considering Price Oscillations for Residential Building Heating and Ventilation Systems. *IEEE Trans. Ind. Inform.* **2019**, *15*, 4742–4752. [CrossRef]
40. Huang, J.; Wang, Y.; Guo, L. Energy intensity and energy-specific technological progress: A case study in Guangdong province of China. *Renew. Energy* **2022**, *184*, 990–1001. [CrossRef]
41. Wang, Z.; Wang, S.; Wang, X.; Luo, X. Permanent Magnet-Based Superficial Flow Velometer With Ultralow Output Drift. *IEEE Trans. Instrum. Meas.* **2023**, *72*, 1–12. [CrossRef]
42. Guigon, R.; Chaillout, J.J.; Jager, T.; Despesse, G. Harvesting raindrop energy: Experimental study. *Smart Mater. Struct.* **2008**, *17*, 015039. [CrossRef]
43. Li, S.; Zhou, J.; Zhou, F.; Niu, F.; Deng, W. A Reduced Current Ripple Overmodulation Strategy for Indirect Matrix Converter. *IEEE Trans. Ind. Electron.* **2024**, 1–10. [CrossRef]
44. Wang, R.; Liu, S.; Wang, L.; Li, Q.; Zhang, S.; Chen, B.; Jiang, L.; Zhang, Y. Superior energy-saving splitter in monoethanolamine-based biphasic solvents for CO2 capture from coal-fired flue gas. *Appl. Energy* **2019**, *242*, 302–310. [CrossRef]
45. Liu, M.; Qian, F.; Mi, J.; Zuo, L. Dynamic interaction of energy-harvesting backpack and the human body to improve walking comfort. *Mech. Syst. Signal Process.* **2022**, *174*, 109101. [CrossRef]
46. Liu, F.; Zhao, X.; Zhu, Z.; Zhai, Z.; Liu, Y. Dual-microphone active noise cancellation paved with Doppler assimilation for TADS. *Mech. Syst. Signal Process.* **2023**, *184*, 109727. [CrossRef]
47. Riahi, K.; Van Vuuren, D.P.; Kriegler, E.; Edmonds, J.; O'Neill, B.C.; Fujimori, S.; Bauer, N.; Calvin, K.; Dellink, R.; Fricko, O.; et al. The Shared Socioeconomic Pathways and their energy, land use, and greenhouse gas emissions implications: An overview. *Glob. Environ. Chang.* **2017**, *42*, 153–168. [CrossRef]
48. Sun, Q.; Lyu, G.; Liu, X.; Niu, F.; Gan, C. Virtual Current Compensation-Based Quasi-Sinusoidal-Wave Excitation Scheme for Switched Reluctance Motor Drives. *IEEE Trans. Ind. Electron.* **2024**, *71*, 10162–10172. [CrossRef]
49. Abbas, D.; Current, D.; Phillips, M.; Rossman, R.; Hoganson, H.; Brooks, K.N. Guidelines for harvesting forest biomass for energy: A synthesis of environmental considerations. *Biomass Bioenergy* **2011**, *35*, 4538–4546. [CrossRef]
50. Wang, S.; Wang, Z.; Dong, H.; Chen, Y. Recursive state estimation for stochastic non-linear non-Gaussian systems using energy-harvesting sensors: A quadratic estimation approach. *Automatica* **2023**, *147*, 110671. [CrossRef]
51. Meng, Q.; Tong, X.; Hussain, S.; Luo, F.; Zhou, F.; He, Y.; Liu, L.; Sun, B.; Li, B. Enhancing Distribution System Stability and Efficiency through Multi-Power Supply Startup Optimization for New Energy Integration. *IET Gener. Transm. Distrib.* **2024**, *early access*. [CrossRef]
52. Collier, S.H.C.; House, J.I.; Connor, P.M.; Harris, R. Distributed local energy: Assessing the determinants of domestic-scale solar photovoltaic uptake at the local level across England and Wales. *Renew. Sustain. Energy Rev.* **2023**, *171*, 113036. [CrossRef]
53. Viet, N.V.; Xie, X.D.; Liew, K.M.; Banthia, N.; Wang, Q. Energy harvesting from ocean waves by a floating energy harvester. *Energy* **2016**, *112*, 1219–1226. [CrossRef]
54. Beeby, S.P.; O'Donnell, T. Electromagnetic energy harvesting. In *Energy Harvesting Technologies*; Springer: Berlin/Heidelberg, Germany, 2009; pp. 129–161. [CrossRef]
55. Li, W.; Guo, X.; Zhang, Y.; Wang, K. A Comprehensive Review on the Synergetic Effect of Physicochemical Properties of Carbon Materials in Polysulfide Adsorption and Electrocatalysis for Advanced Li–S Batteries. *Chem. Eng. J.* **2023**, *459*, 141629. [CrossRef]

56. Guan, Z.; Li, P.; Wen, Y.; Du, Y.; Wang, Y. Efficient bubble energy harvesting by promoting pressure potential energy release using helix flow channel. *Appl. Energy* **2022**, *328*, 120159. [CrossRef]
57. Wang, W.; Cionca, V.; Wang, N.; Hayes, M.; O'Flynn, B.; O'Mathuna, C. Thermoelectric energy harvesting for building energy management wireless sensor networks. *Int. J. Distrib. Sens. Netw.* **2013**, *9*, 232438. [CrossRef]
58. Wang, Y.; Liu, X.; Tan, G.; Han, B.; Zhu, H.; Yuan, G. Electrocatalytic Water Splitting over Metal–Organic Framework-Derived Nanostructures. *Nano Res. Energy* **2022**, *1*, e9120015. [CrossRef]
59. Kassan, S.; Gaber, J.; Lorenz, P. Autonomous Energy Management System Achieving Piezoelectric Energy Harvesting in Wireless Sensors. *Mob. Netw. Appl.* **2020**, *25*, 794–805. [CrossRef]
60. Igliński, B.; Pietrzak, M.B.; Kiełkowska, U.; Skrzatek, M.; Kumar, G.; Piechota, G. The assessment of renewable energy in Poland on the background of the world renewable energy sector. *Energy* **2022**, *261*, 125319. [CrossRef]
61. Luo, R.; Yu, Z.; Wu, P.; Hou, Z. Analytical solutions of the energy harvesting potential from vehicle vertical vibration based on statistical energy conservation. *Energy* **2023**, *264*, 126111. [CrossRef]
62. Østergaard, P.A.; Duic, N.; Noorollahi, Y.; Mikulcic, H.; Kalogirou, S. Sustainable development using renewable energy technology. *Renew Energy* **2020**, *146*, 2430–2437. [CrossRef]
63. Zhao, W.; Ma, X.; Gao, L.; Wang, X.; Luo, Y.; Wang, Y.; Li, T.; Ying, B.; Zheng, D.; Sun, S.; et al. Hierarchical Architecture Engineering of Branch-Leaf-Shaped Cobalt Phosphosulfide Quantum Dots: Enabling Multi-Dimensional Ion-Transport Channels for High-Efficiency Sodium Storage. *Adv Mater.* **2024**, *36*, e2305190. [CrossRef] [PubMed]
64. Legerstee, W.J.; Kiriinya, L.; Kwakernaak, M.; Kelder, E.M. Magnesium Transfer between Atomic Force Microscopy Probes and Metal Electrodes in Aqueous Alginate Electrolytes. *Polymers* **2024**, *16*, 1615. [CrossRef] [PubMed]
65. Xia, J.; Liang, F.; Zhou, S.; Zhang, Y.; Wang, X.; Cao, J.; Cai, S.; Zhang, J.; Wang, L.; Zhu, Y.; et al. Recent Advances in Metal–Organic Frameworks and Their Derivatives for Electrocatalytic Water Splitting. *Chem. Eng. J.* **2022**, *433*, 134477. [CrossRef]
66. Li, C.; Liu, B.; Jiang, N.; Ding, Y. Elucidating the charge-transfer and Li-ion-migration mechanisms in commercial lithium-ion batteries with advanced electron microscopy. *Nano Res. Energy* **2022**, *1*, 9120031. [CrossRef]
67. Xu, Y.; Zhou, Y.; Li, Y.; Hao, Y.; Wu, P.; Ding, Z. Magnesium-Based Hydrogen Storage Alloys: Advances, Strategies, and Future Outlook for Clean Energy Applications. *Molecules* **2024**, *29*, 2525. [CrossRef]
68. Xie, X.D.; Wang, Q. Energy harvesting from a vehicle suspension system. *Energy* **2015**, *86*, 382–395. [CrossRef]
69. Zhang, H.; Li, P.; Jin, H.; Bi, R.; Xu, D. Non-linear wave energy dissipator with wave attenuation and energy harvesting at low frequencies. *Ocean. Eng.* **2022**, *266*, 112935. [CrossRef]

batteries

MDPI

Article

Multi-Step Ageing Prediction of NMC Lithium-Ion Batteries Based on Temperature Characteristics

Abdelilah Hammou, Boubekeur Tala-Ighil, Philippe Makany and Hamid Gualous *

LUSAC Laboratory, University of Caen Normandy, 50130 Cherbourg-Octeville, France; abdelilah.hammou@unicaen.fr (A.H.)

* Correspondence: hamid.gualous@unicaen.fr

Abstract: The performance of lithium-ion batteries depends strongly on their ageing state; therefore, the monitoring and the prediction of the battery state of health (SoH) is necessary for an optimized and secured functioning of battery systems. This paper evaluates and compares three artificial neural network architectures for multi-step ageing prediction of lithium-ion cells: Recurrent Neural Network (RNN), Gated Recurrent Unit (GRU) and Long short-term memory (LSTM). These models use the features extracted from the cell's temperature to predict the cell's capacity. The features are extracted from experimental measurements of the cell's surface temperature and selected based on Spearman correlation analysis. The prediction results were evaluated and compared considering three different percentages of the training dataset: 60%, 70%, and 80%. Training and testing data were generated experimentally based on accelerated ageing cycling tests. During these experiments, four Nickel Manganese Cobalt/Graphite (NMC) cells were cycled under a controlled temperature environment based on a dynamic current profile extracted from the Worldwide Harmonized Light Vehicles Test Cycles.

Keywords: lithium-ion batteries; ageing; prediction; state of health; artificial neural networks

1. Introduction

Lithium-ion batteries present the main energy storage system used in transport applications [1,2], thanks to their greater advantages in comparison with other energy storage systems: high power and energy density, fast response time, and wide operating temperature range [3,4]. However, these characteristics depend strongly on the battery ageing stage [5,6]. The degradation of lithium-ion batteries generally leads to three degradation modes: the loss of lithium inventory, the loss of active material, and the increase in the internal resistance [7]. These ageing modes lead to the fade of the battery capacity and power [8]. Therefore, to ensure the proper functioning of the battery and to anticipate failure, several studies have been carried out to predict the battery capacity for its ageing prognosis [9]. The state-of-health prediction methods in the literature are classified in two categories: model-based methods and data-driven methods [10,11].

Model-based methods exploit mathematical/physical models of battery ageing to predict the future evolution of a battery's capacity. These models are first calibrated and fitted based on the historical data of battery capacity. The prediction is obtained by the extrapolation of the model to future cycles [12]. The main advantage of the model-based approaches is the ability to learn even when the available data are limited in quantity [13]. In [14], the authors combined the Particle Filter (PF) with the Genetic Algorithms (GA) for the prediction of the battery capacity; the modelling of the capacity fade is performed based on the multiphase sigmoidal model. The results evaluation showed that the Root Mean Squared Percentage Error (RMSPE) of the capacity prediction is less than 1.2% when the model is fitted based on 60% of cycles. In addition, the study in [15] used the Particle Swarm Optimization (PSO) for the prediction of the battery capacity based on the dual exponential model, and the performance of PSO was enhanced by PF. The method was verified using two public datasets: NASA and CALSE datasets. Moreover, in [16] the authors combined two empirical models for the prediction of the battery capacity: the double-exponential model and the polynomial model; the aim of this combination is to enhance the adaptability of the perdition approach to different degradation trends. The model parameters are estimated using the PF. The evaluation of the proposed method confirms the accuracy of these methods in comparison with one single model.

Data-driven methods, on the other hand, are based on designing and fitting forecasting models based on the historical data to predict the battery ageing evolution [17]. The main advantage of these methods is their flexibility, since they are model-free approaches; consequently, they can be adapted to several scenarios and cycling conditions [18]. In [19], the authors combined a CNN-LSTM-Attention neural network with transfer learning for the capacity prediction; the input features of the proposed model were extracted using Complete Ensemble Empirical Mode Decomposition with Adaptive Noise (CEEDMAN). The experimental evaluation of the proposed approach shows that the Mean Absolute Percentage Error (*MAPE*) of the predictions is less than 2%. The study in [20] compared two deep LSTM network architectures for the prediction of the battery capacity: the one-step prediction architecture, and the multi-step prediction architecture. Furthermore, the work presented in [21] combines the Convolutional Neural Network (CNN) with the bidirectional LSTM neural network for the prediction of the battery capacity. The method was trained and tested based on two public datasets: NASA and CALCE. The Root Mean Squared Error (*RMSE*) of the prediction is less than 0.025 when the model is trained with 50% of cycles.

The discussed prognosis methods are interesting and presented accurate prediction results. However, in vehicular applications, the working conditions of batteries depend on driver behavior and also climatic conditions [22]. As a result, the ageing trajectory of the battery systems varies with the variation of working conditions [23]. Therefore, the prediction algorithms should be trained based on the historical data of each vehicle battery. Additionally, the computational power for embedded applications is limited; thus, the prediction algorithms should be less computationally intensive [24]. This paper presents and compares three variants of recurrent neural networks for multi-step prediction of battery capacity based on the cell's thermal behavior. The models' assessment aims to select the suitable method for vehicular conditions. The ageing data used for model training and testing were generated experimentally based on accelerated ageing tests of four NMC cells at controlled temperature. During these tests, the cells were cycled under a dynamic current profile extracted from the Worldwide Harmonized Light Vehicles Test Cycles (WLTC) [25]. Prediction results were analyzed and compared considering three percentages of the cell cycles in order to evaluate the prediction models against training size and cell ageing stage. The evaluation also takes into consideration the model accuracy and complexity.

The rest of the paper is organized as follows: Section 2 presents the experimental data used in this study and details the prediction structure and the prognosis models. Section 3 presents the prediction results and their analysis. A conclusion closes the paper.

2. Materials and Methods

2.1. Experimental Data

The experimental data used for model training and validation are obtained from accelerated cycling tests of four cells of Nickel Manganese Cobalt (NMC) at a controlled temperature of 35 °C [26]. The cells are charged using Constant-Current Constant Voltage mode (CC-CV) and then discharged using a dynamic current profile adapted from the Worldwide Harmonized Light Vehicles Test Cycles (WLTC) [25]. Figure 1 displays the current profile applied during the cycling process (charge and discharge). This cycling operation is repeated during five days, and then a check-up test is carried out to measure the cell's performance at 25 °C. The check-up tests aim to measure the battery capacity, energy, and direct current internal resistance (DCIR) [26].

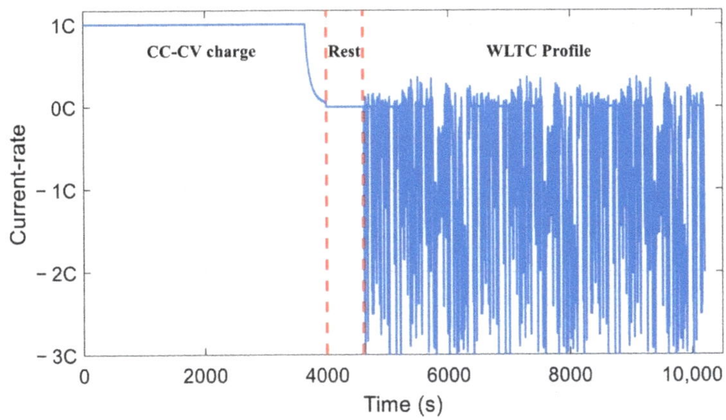

Figure 1. Applied current during cell cycling.

These experiments are carried out using a CHROMA 17020® battery tester (Chroma ATE Europe B.V., AH Ede, The Netherlands) that allows the cycling of eight cells at the same time. This device enables the acquisition of the charging/discharging current, the cell voltage and the cell surface temperature using a type T thermocouple. The temperature is conditioned using a Weiss Technik® climate chamber (Weiss Technik France S.A.S, Eragny, France). In these tests, four cells of Nickel Manganese Cobalt/Graphite (NMC) with a rated capacity of 3 Ah are cycled [27].

The temperature is measured at three points on the cell's surface: near the anode, near the cathode, and in the middle of the cell, as presented in Figure 2. The sampling period for data acquisition is set at $T_s = 1$ s. Figure 3 shows the evolution of capacity as a function of the number of cycles for the four cells in each chemistry.

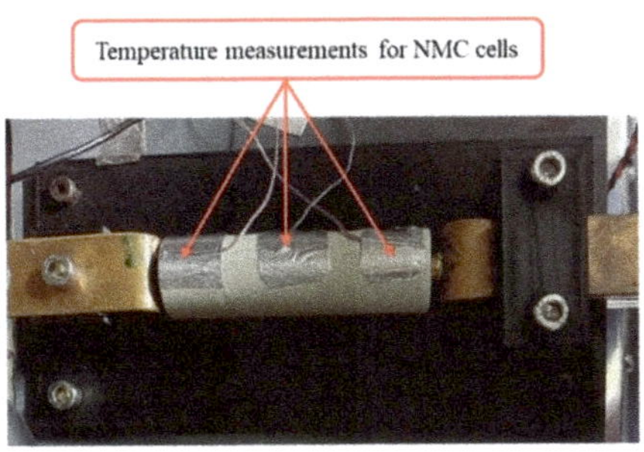

Figure 2. Temperature measurement points at the cell surface.

The experimental tests validate the effect of cycling on the loss of cell performance; the results show the fade of battery capacity and energy and the increase in the DCIR with cycles. The cell's capacity and energy decrease with 20% and 27%, respectively, while the DCIR growth is greater than 200%. The results also show that the energy decline is higher than the capacity decline; this difference is related to the dependence of the energy on both cell capacity and internal resistance.

The comparison between cells' ageing behavior reveals the variation of ageing evolution between cells, which is related to the cell's inconsistency [28]. Thus, prediction models should consider cell inconsistency for improving prediction results.

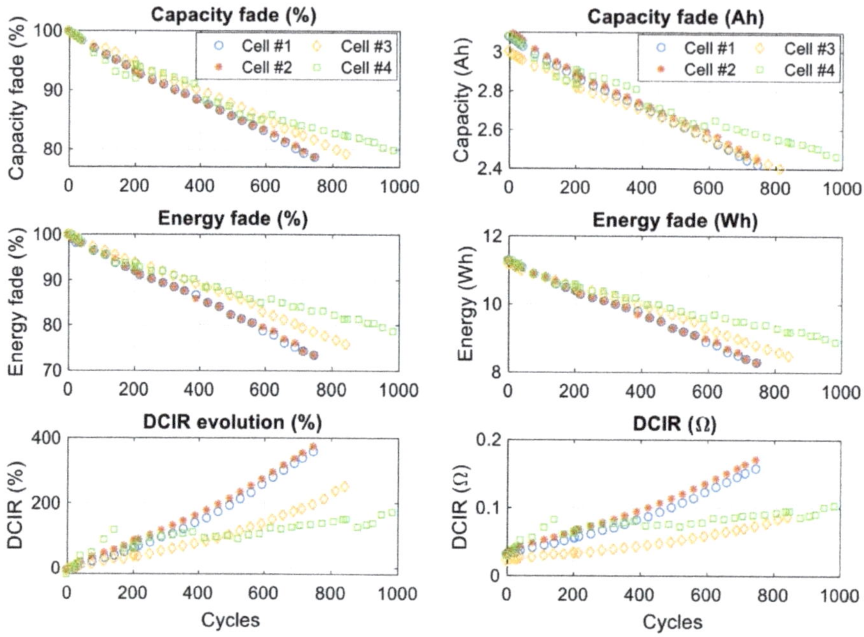

Figure 3. Cycling characterization results.

The generated experimental data allow for the extraction and the analysis of thermal features during battery discharging with the dynamic current profile, which allows the evaluation of the prediction models under conditions close to vehicular conditions. However, in order to design generalized prediction models, the future work will focus on the extension of the experimental tests to different temperatures and state-of-charge ranges.

2.2. Features Extraction

The cell ageing rate is highly dependent on the battery temperature, as reported in several studies [29], and at the same time the temperature behavior of a battery varies with ageing [30]. Therefore, exploring the temperature characteristics of the batteries will provide information about the battery SoH and rate. The experimental results presented in Figure 4 confirm the effect of ageing on the temperature behavior of the cells. The temperature measured on the cell surface shows that the cell's temperature increases with ageing; the maximum temperature reaches 38 °C at the beginning of the life, while it increases with 1 °C after 350 cycles; at the end of life, the temperature reaches 42 °C.

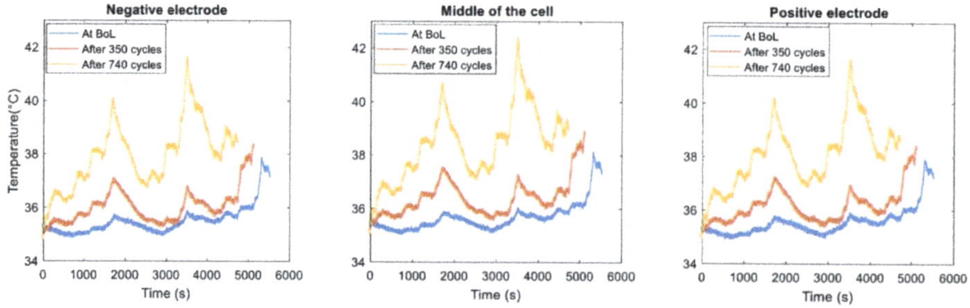

Figure 4. Temperature profile measured at the cell's surface at three different ageing stages.

In order to exploit the information contained in the temperature curves, several features are calculated from the cell's surface temperature during its discharging: the mean, the variance, and the maximum of the temperature curve under the WLTC profile. Figure 5 shows the evolution of the features extracted from temperature measurements as a function of cycles for one cell (Cell 1).

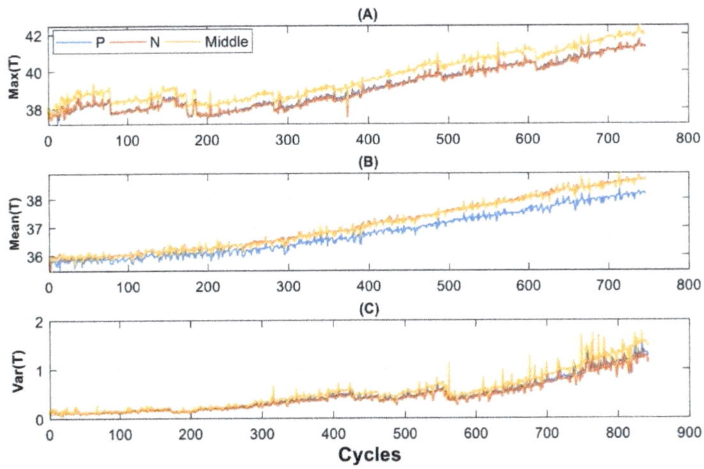

Figure 5. Evolution of the extracted features as a function of the performed cycles: (**A**) the evolution of the maximum temperature, (**B**) the evolution of the mean temperature, and (**C**) the evolution of the temperature variance.

The results obtained show that the extracted features increase with the cycles. The comparison between the temperatures at three different points on the surface of the cell shows that the three points present the same trends. In order to evaluate and analyze the correlation between the extracted features and the cell's SoH, the Spearman correlation is used to assess their correlation (Equation (1)). The Spearman correlation coefficient [31] evaluates the existence of a monotonic linear or non-linear relationship between two variables, X and Y. Figure 6 presents the heat map of the Spearman coefficient correlation between the extracted features and the cell capacity.

$$R_{Spearman} = 1 - \frac{6\sum_{i=1}^{n} d_i^2}{n^3 - n} \quad (1)$$

where n is the number of observations, and d_i is the difference between $rank(X_i)$ and $rank(Y_i)$.

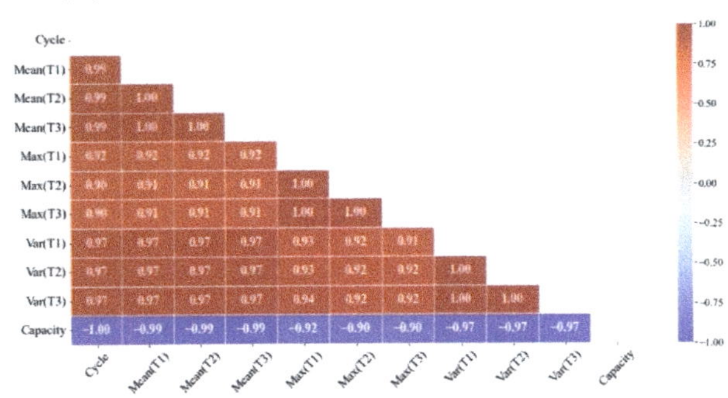

Figure 6. Heatmap of the Spearman coefficient correlation between the extracted features and the capacity for cell 1.

The analysis of the correlation results showed that the cycle number and the mean value of the surface temperature are the most correlated features with battery capacity; the Spearman coefficient correlation is greater than 0.99. In addition, based on the results in Figure 5, it is noted that the evolution of the mean value is more regular, unlike the maximum value and the variance, which present some fluctuations and disturbances. The comparison of the evolution of the means of the three measurement points shows that they have the same correlation coefficient and follow almost the same trend with the degradation (Figure 5). Therefore, to simplify the architecture of neural networks, instead of using the three temperature measurements, only the temperature measured in the middle of the cell will be used, since it is the least sensitive point to external temperature variation [32]. Consequently, the mean value of the temperature and the cycle number are used as inputs for the neural networks.

2.3. Prognosis Models

In this study, three variants of recurrent neural networks have been used for the prediction of the cell capacity: Recurrent Neural Network (RNN), Long Short-Term Memory (LSTM), and Gated Recurrent Unit network (GRU). The use of recurrent networks is related to their ability to handle time series data, as discussed in several studies [17–21,33], which is the case for this study, where the objective is to predict the temporal evolution of the cell's capacity. Figure 7 presents the architectures of the RNN, LSTM, and GRU units.

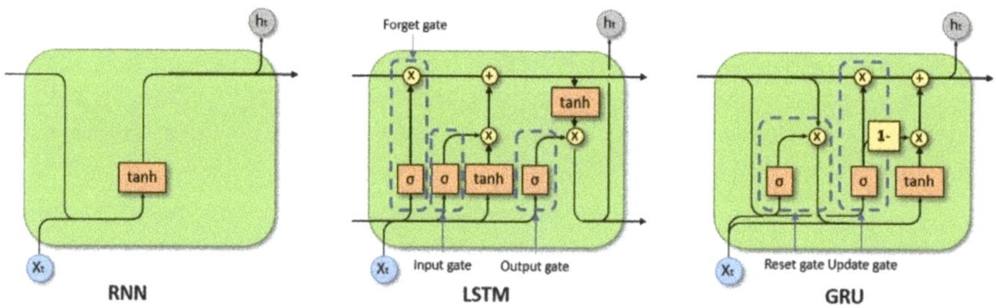

Figure 7. Architecture of RNN, LSTM, and GRU units [34].

(a) Recurrent Neural Networks

A Recurrent Neural Network (RNN) is designed to work with sequential data [35], such as word sequences or time series. In classical neural networks, inputs are processed independently of each other, whereas RNNs are able to take into account the history of the input data to predict the next output value. This is possible thanks to feedback loops that allow information to flow through the network and be stored in an internal memory called a hidden state. The hidden state of the RNN units, h_t, is updated based on the previous state h_{t-1} and the input at the time step t (Equation (2)):

$$h_t = tanh(w_i[h_{t-1}, x_t] + b_h) \tag{2}$$

where w_i is the weight matrix, b_h is the bias vector, and $tanh$ is the hyperbolic tangent function. Therefore, the unit output is calculated based on the hidden state h_t (Equation (3)):

$$y_t = W_o * h_t + b_y \tag{3}$$

where W_o and b_y are the weight matrix and the bias vector, respectively. Table 1 summarizes the RNN configuration (architecture and hyperparameters) used for the multi-step capacity prediction.

Table 1. RNN neural network configuration and hyperparameters.

Parameter	Configurations
Input layer	300 RNN units
Hidden layer-1	200 RNN units
Hidden layer-2	100 RNN units
Output layer	Dense unit with Leaky Rectified Linear Unit activation function [36]
Optimizer	'ADAM' optimizer [37] with Mean Squared Error (MSE) as a loss function

(b) Long Short-Term Memory

Long Short-Term Memory (LSTM) neural networks are a class of recurrent neural networks. They were introduced to solve the vanishing gradient problem in the back-propagation of traditional RNN [38]. LSTM takes into account long-term dependencies in temporal data by using an internal memory to store important contextual information. The LSTM architecture consists of three gates: the input gate, the forget gate, and the output gate described by equations (Equation (4)), (Equation (5)) and (Equation (6)), respectively [38]. These gates play a role in determining which data to keep and which to discard, which helps to keep only the relevant data, to facilitate prediction by the sequence chain.

$$i_t = \sigma(w_i[h_{t-1}, x_k] + b_i) \tag{4}$$

$$f_t = \sigma\left(w_f[h_{t-1}, x_t] + b_f\right) \tag{5}$$

$$o_t = \sigma(w_o[h_{t-1}, x_t] + b_o) \tag{6}$$

where w_X is the weight matrix, b_X the bias, and σ the sigmoid function for gate X. The candidate state cell $\tilde{c}_k$ is calculated using (Equation (7)):

$$\tilde{c}_t = tanh(w_c[h_{t-1}, x_t] + b_c) \tag{7}$$

where w_c and b_c are the weight and bias for the candidate state cell $\tilde{c}_t$, and $tanh$ is the hyperbolic tangent function.

Subsequently, the state cell c_t is updated based on the previous state c_{t-1} and the candidate state $\tilde{c}_t$ (Equation (8)). The final output h_t is calculated using (Equation (9)).

$$c_t = f_t * c_{t-1} + i_t * \tilde{c}_t \tag{8}$$

$$h_t = o_t * tanh(c_t) \tag{9}$$

Table 2 sums up the LSTM architecture and hyperparameters used for the multi-step capacity prediction.

Table 2. LSTM neural network configuration and hyperparameters.

Parameter	Configurations
Input layer	300 LSTM units
Hidden layer-1	200 LSTM units
Hidden layer-2	100 LSTM units
Output layer	Dense unit with Leaky Rectified Linear Unit activation function [36]
Optimizer	'ADAM' optimizer [37] with Mean Squared Error (MSE) as a loss function

(c) Gated Recurrent Unit network

The Gated Recurrent Unit (GRU) network aims to keep the advantages of the gating mechanism in the LSTM unit while reducing the number of parameters, thus reducing the network complexity [39]. The GRU unit is composed of two gates: update and reset gates. The update gate u_k evaluates the significance of the features in the time series and catches long-term dependencies. The reset gate r_k determines the amount of past information to be forgotten. The update and reset gates are calculated based on the current input vector x_k and the previous hidden state h_{k-1}, as described by the following equations (Equations (10) and (11)):

$$u_t = \sigma(w_u[h_{t-1}, x_t] + b_u) \tag{10}$$

$$r_t = \sigma(w_r[h_{t-1}, x_t] + b_t) \tag{11}$$

where w_u, w_r are the weight matrices of the update and reset gates, b_u and b_k are the biases, and σ denotes the sigmoid function. The candidate hidden state $\tilde{h}_t$ can be expressed using (Equation (12)):

$$\tilde{h}_t = tanh(w_h[r_t * h_{t-1}, x_t] + b_h) \tag{12}$$

where w_h and b_h are the weight matrix and bias of the candidate hidden state, respectively, and $tanh$ refers to the hyperbolic tangent function. Therefore, the current hidden state h_k is obtained based on the candidate state and the previous hidden state (Equation (13)):

$$h_t = (1 - u_t) * h_{t-1} + u_t * \tilde{h}_t \tag{13}$$

Table 3 sums up the GRU architecture and hyperparameters used for the capacity prediction.

Table 3. GRU neural network configurations and hyperparameters.

Parameter	Configurations
Input layer	100 GRU units
Hidden layer-1	50 GRU units
Hidden layer-2	25 GRU units
Output layer	Dense unit with Leaky Rectified Linear Unit activation function [36]
Optimizer	'ADAM' optimizer [37] with Mean Squared Error (MSE) as a loss function

(d) Parameter initialization and regularization

- He Initialization

The weights of the neural network models are initialized using the He initialization method [40]. The objective of this initialization is to provide an appropriate initialization of the weights that avoids the saturation and vanishing gradient problems during the training of the network.

The He initialization method takes into account the size of the previous layers when initializing the weights. The weights are initialized with a normal distribution centered on zero and a variance adapted to the size of the previous layer. More precisely, the weights are randomly drawn according to a normal distribution with a variance of 2/N, where N represents the number of neurons in the previous layer. This initialization promotes a more efficient propagation of information through the network, by avoiding too low or too high values of the weights.

- L2 regularization

L2 regularization is a technique used in model training to prevent overfitting of neural network models [41]. It consists of penalizing high values of the model's weights, which allows controlling their relative importance during training.

In this case, the loss function is Mean Squared Error (MSE) [42] (Equation (14)):

$$L = \frac{1}{N_s} * \sum_{i=1}^{N_s} \left(\hat{Y}_i - Y_{exp}\right)^2 \tag{14}$$

where N_s represents the number of samples, $\hat{Y}_i$ represents the output of the model, and Y_{exp} is the actual value. The training process of the model consist of adjusting the model parameters (weights and biases) to minimize the loss function L.

L2 regularization consists of adding the following term to the cost function (Equation (15)) [41]:

$$L = \frac{1}{N_s} * \sum_{i=1}^{N_s} \left(\hat{X}_i - X_{exp}\right)^2 + \lambda \sum_{i=1}^{P} w_i^2 \tag{15}$$

where w_i are the model weights and λ is the regularization parameter. L2 regularization adds a penalty term to the loss function that applies strong penalties to larger weights, which favorizes smaller ones.

3. Results

3.1. Model Training and Evaluation

The neural network models are trained with three different percentages of data (60%, 70%, and 80% of data). This percentage refers to the ratio of the number of cycles used for model training to the number of cycles performed by the cells during their lifetime. This procedure allows the evaluation at different ageing stages of the cells and also considering different amounts of data.

The prediction errors are evaluated using the *MAPE* and the *RMSE* described by equations (Equation (16)) and (Equation (17)), respectively.

$$MAPE(\%) = 100 * \frac{1}{N} * \sum_{1}^{N} \left|\hat{X} - X_{exp}\right| / X_{exp} \tag{16}$$

$$RMSE(\text{Ah}) = \frac{1}{N} * \sum_{1}^{N} \sqrt{\left(\hat{X} - X_{exp}\right)^2} \tag{17}$$

3.2. Prediction Results and Discussion

With the aim to maximize the number of capacity values that can be predicted in one shot (multi-step prediction), Figure 8 presents the evolution of the average *MAPE* of all cells as a function of the prediction steps (the capacity values predicted simultaneously). These results are obtained when the neural network models are trained with 60% of cycles.

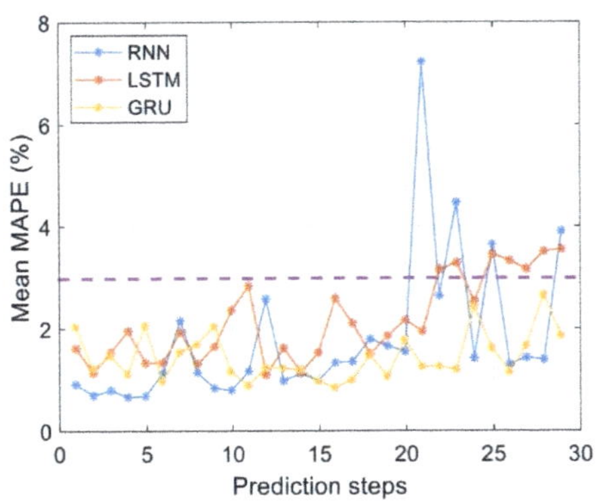

Figure 8. Evolution of the mean *MAPE* error as a function of the prediction steps for the three prognosis models (RNN, LSTM, and GRU).

The analysis of the evolution (Figure 8) of the prediction errors as a function of the predicted steps shows that the RNN model presents the lowest error when predicting five capacity values. However, as the number of predicted steps increases, the prediction errors of the three models are very close and stay lower than 3% until 20 predicted steps. When predicting more than 20 capacity values, the mean *MAPE* for the RNN reaches the value of 7%, fluctuates and becomes unstable, pointing to its weakness to capture long-term dependencies due to its simple architecture [35]. Concerning the LSTM model, the *MAPE* continues to increase progressively with the cycles to exceed the value of 3%. The *MAPE* for the GRU model is always lower than 3%. Based on the LSTM and the GRU architectures discussed in Section 2.3, the LSTM model is expected to show high accuracy compared to the GRU model due to its gating mechanisms. However, the results show the opposite; this could be related to the insufficiency of the training data for the LSTM network, since the number of parameters to optimize for LSTM is higher than that of the GRU [43].

In order to compare and analyze the results of the three neural networks for each cell, the prediction steps are fixed at 20 in the rest of this study. Figures 9–11 present the prediction results of the three prediction models in comparison with the measured capacity, for the tree training set 60%, 70%, and 80%, respectively. Table 4 presents the calculation of the *MAPE* and the *RMSE* of the prediction results of the three prognosis models; the mean value of errors of the four cells is also calculated.

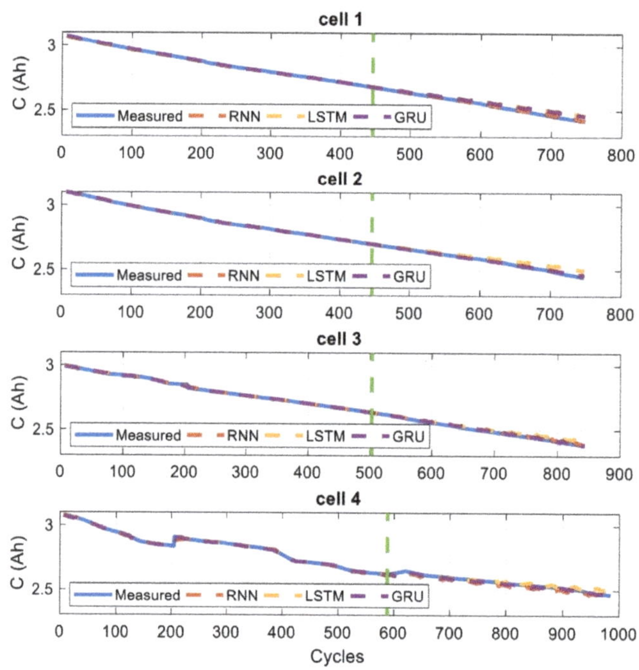

Figure 9. Prediction results of the three prognosis models in comparison with the measured capacity when the models are trained with 60% of the cycles.

The results obtained in Figures 9–11 show that the three prognosis models present an accurate prediction result for the three training percentages: the *MAPE* and the *RMSE* errors are less than 2.5% and 0.054 Ah, respectively, for the four cells.

The analysis of the evolution of the prediction accuracy (Table 4) shows that, for the three neural networks, the prediction accuracy increases with the increasing of training data: the increase in the percentage of the training cycles from 60% to 80% allowed the mean *MAPE* to be reduced from 2% to 5% for the RNN model, from 1.616% to 0.732%

for the LSTM model, and from 1.183% to 0.169% for the GRU model. However, when comparing the error evolution for individual cells, the analysis shows that the error does not decrease with the training data increasing for all cells: for cells 1, 2 and 3, the results show that in some cases the error increases with the increase in training data. Concerning cell 4, the prediction accuracy increases with the increasing training set. This difference can be related to the overfitting of the prediction models on the data of cells 1, 2 and 3, where the capacity evolution is quasi-linear in comparison with cell 4, where the evolution is nonlinear.

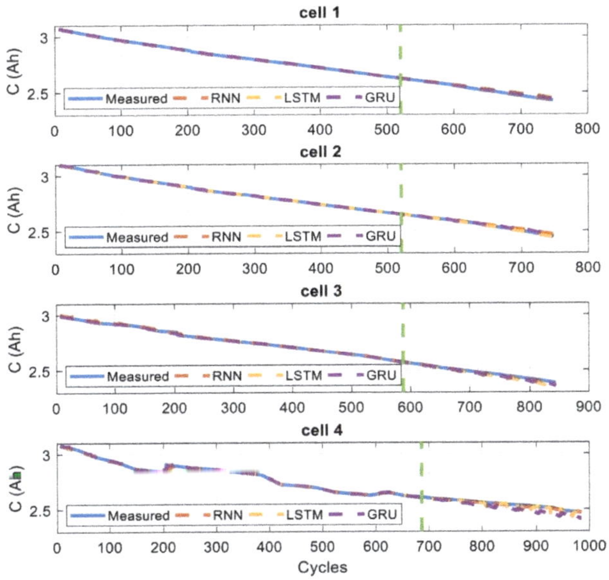

Figure 10. Prediction results of the three prognosis models in comparison with the measured capacity when the models are trained with 70% of the cycles.

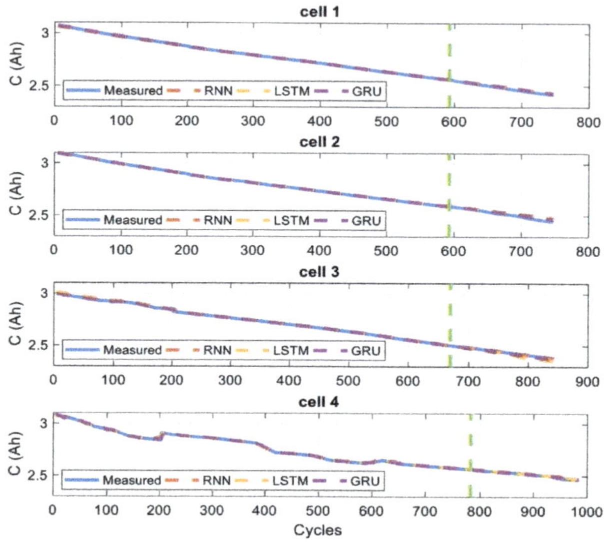

Figure 11. Prediction results of the three prognosis models in comparison with the measured capacity when the models are trained with 80% of the cycles.

Table 4. Parameters of GRU neural networks.

Prognosis Models	Battery Cells	60% of Cycles		70% of Cycles		80% of Cycles	
		MAPE (%)	*RMSE* (Ah)	*MAPE* (%)	*RMSE* (Ah)	*MAPE* (%)	*RMSE* (Ah)
RNN	Cell 1	0.555	0.014	0.486	0.012	0.566	0.014
	Cell 2	0.515	0.013	0.676	0.017	0.809	0.020
	Cell 3	0.560	0.014	0.642	0.016	0.228	0.006
	Cell 4	1.552	0.041	0.279	0.007	0.442	0.012
	Mean	0.795	0.020	0.521	0.013	0.511	0.013
LSTM	Cell 1	2.043	0.050	1.197	0.029	0.509	0.012
	Cell 2	2.171	0.054	0.500	0.013	1.586	0.039
	Cell 3	1.043	0.026	1.071	0.026	0.645	0.016
	Cell 4	1.206	0.032	0.529	0.014	0.187	0.005
	Mean	1.616	0.040	0.824	0.020	0.732	0.018
GRU	Cell 1	1.786	0.044	0.731	0.018	0.572	0.014
	Cell 2	0.644	0.016	0.737	0.018	0.938	0.023
	Cell 3	0.670	0.017	1.000	0.025	0.605	0.015
	Cell 4	1.183	0.031	0.729	0.019	0.169	0.004
	Mean	1.071	0.027	0.799	0.020	0.571	0.014

The comparisons between the prognosis models show that the RNN model presents accurate prediction results for the three training percentage sets: in the case of 60% of training cycles, the mean *MAPE* error is less than 0.8% for the RNN model, less than 1.1% for the GRU model, and less than 1.62% for the LSTM model. The superiority of the RNN model is related to its simplicity, which allows it to achieve high accuracy even with a small amount of data; however, as the training set increases, we observe that the prediction errors of the LSTM and the GRU decrease and the differences between the prediction errors of the three models decreases. As presented in Table 4, with 80% of training cycles, the GRU and the RNN present the same mean error (about 0.7%), while the LSTM prediction mean error is less than 0.75%.

In order to compare the prediction models in term of complexity, the computational time of the training process is presented in Table 5. The three recurrent networks (RNN, LTSM, and GRU) are developed using a Python library, TensorFlow. The algorithms are trained and tested in the Windows 10® environment. The computer used is equipped with an i-5 processor 2.50 (GHz) and 16 GB of RAM.

Table 5. Training time of three prediction models for the three training sets: 60%, 70%, and 80% of the cycles.

Prediction Model	Training Time (s)		
	60%	**70%**	**80%**
RNN	32.7	35.1	44.6
LSTM	101.6	128.4	146.6
GRU	38.2	45.0	50.9

The results presented in the Table 5 show, as expected, that increasing the training data leads to increasing the training time for all models. The results obtained are in agreement with the description of the architectures of the prediction models in Section 2.3: the comparison shows that the LSTM neural network has the highest training time (higher than 100 s) followed by the GRU neural networks with a training time less than 51 s, where the RNN has a training time less than 44.6 s.

Finally, the precision of the prediction of the three prognosis algorithms is compared with the methods proposed in the literature in Table 6. The comparison shows that the simple architectures proposed in this study have prediction accuracies similar to those of other studies that proposed complex and hybrid models for the prognosis of batteries.

Table 6. Results comparison with the literature.

Prediction Model	Input Features	Dataset	*MAPE*
Variational modal decomposition with entropy detection-GRU [44]	Capacity	NASA and CALCE datasets, laboratory tests	<0.64%
CEEDMAN algorithm—CNN-LSTM [21]	Capacity	NASA datasets and laboratory tests	<2%
LSTM with Explainable IA-Optimal features selection [45]	Cycles, Current, Voltage, Temperature	NASA dataset	<2%
Multi-kernel RVM-Particle swarm optimization [46]	Capacity	Laboratory tests	<1.34%
Multi-Layer Perceptron deep neural network [47]	Impedance measurements features	Cavendish Laboratory University of Cambridge	<0.97%
RNN this study	Cycle number and mean temperature	Experimental tests	<1.55%
LSTM this study			<2.17%
GRU this study			<1.78%

4. Conclusions

Lithium-ion battery health prognosis is necessary for an optimal and reliable battery management of an electrical vehicle. This paper compares and evaluates three recurrent neural network models for a multi-step capacity prediction of NMC lithium-ion batteries. These prognosis models allow the prediction of the capacity based on the thermal features of the battery cells. The features are extracted and selected based on the correlation analysis between the experimental data and the battery capacity. The model training and evaluation were performed based on experimental data generated from accelerated ageing tests. During these tests, four NMC cells were cycled using dynamic current profile extracted from the WLTC procedure under a controlled temperature of 35 °C. The predictions were evaluated and compared considering three different percentages of the training sets: 60%, 70% and 80% of the cells cycles. The analysis and comparison of the results show that the GRU model is the most suitable prognosis model for vehicular applications, as it has shown good performance for multi-step prediction and offers a good compromise between prediction accuracy and complexity.

Future work will focus on the extension and the evaluation of the prediction models at different temperatures and state-of-charge cycling windows, in order to generalize their performance to the different working conditions of vehicle batteries.

Author Contributions: A.H.: Methodology, Formal analysis, Data curation, Writing original draft preparation. B.T.-I.: Review, Data curation, Investigation, Visualization, Supervision. P.M.: Methodology, Review, Investigation, Visualization, Supervision. H.G.: Methodology, Investigation, Review, Visualization, Supervision. All authors have read and agreed to the published version of the manuscript.

Funding: This research received no external funding.

Data Availability Statement: The data presented in this study are available on request from the corresponding author. The data are not publicly available due to [the privacy of the laboratory tests conducted].

Conflicts of Interest: The authors declare no conflicts of interest.

References

1. Veza, I.; Syaifuddin, M.; Idris, M.; Herawan, S.G.; Yusuf, A.A.; Fattah, I.M.R. Electric Vehicle (EV) Review: Bibliometric Analysis of Electric Vehicle Trend, Policy, Lithium-Ion Battery, Battery Management, Charging Infrastructure, Smart Charging, and Electric Vehicle-to-Everything (V2X). *Energies* **2024**, *17*, 3786. [CrossRef]
2. Gong, X.; Xiong, R.; Mi, C.C. Study of the Characteristics of Battery Packs in Electric Vehicles With Parallel-Connected Lithium-Ion Battery Cells. *IEEE Trans. Ind. Appl.* **2014**, *51*, 1872–1879. [CrossRef]
3. Chaoui, H.; Ibe-Ekeocha, C.C. State of Charge and State of Health Estimation for Lithium Batteries Using Recurrent Neural Networks. *IEEE Trans. Veh. Technol.* **2017**, *66*, 8773–8783. [CrossRef]
4. Xia, Z.; Abu Qahouq, J.A. State-of-Charge Balancing of Lithium-Ion Batteries With State-of-Health Awareness Capability. *IEEE Trans. Ind. Appl.* **2020**, *57*, 673–684. [CrossRef]
5. Rahman, T.; Alharbi, T. Exploring Lithium-Ion Battery Degradation: A Concise Review of Critical Factors, Impacts, Data-Driven Degradation Estimation Techniques, and Sustainable Directions for Energy Storage Systems. *Batteries* **2024**, *10*, 220. [CrossRef]
6. Cai, J.; Zhang, H.; Jin, X. Aging-aware predictive control of PV-battery assets in buildings. *Appl. Energy* **2018**, *236*, 478–488. [CrossRef]
7. Shan, C.; Chin, C.S.; Mohan, V.; Zhang, C. Review of Various Machine Learning Approaches for Predicting Parameters of Lithium-Ion Batteries in Electric Vehicles. *Batteries* **2024**, *10*, 181. [CrossRef]
8. Xiong, R.; Pan, Y.; Shen, W.; Li, H.; Sun, F. Lithium-ion battery aging mechanisms and diagnosis method for automotive applications: Recent advances and perspectives. *Renew. Sustain. Energy Rev.* **2020**, *131*, 110048. [CrossRef]
9. Lipu, M.H.; Hannan, M.; Hussain, A.; Hoque, M.; Ker, P.J.; Saad, M.; Ayob, A. A review of state of health and remaining useful life estimation methods for lithium-ion battery in electric vehicles: Challenges and recommendations. *J. Clean. Prod.* **2018**, *205*, 115–133. [CrossRef]
10. Zhang, Y.; Li, Y.-F. Prognostics and health management of Lithium-ion battery using deep learning methods: A review. *Renew. Sustain. Energy Rev.* **2022**, *161*, 112282. [CrossRef]
11. Toughzaoui, Y.; Toosi, S.B.; Chaoui, H.; Louahlia, H.; Petrone, R.; Le Masson, S.; Gualous, H. State of health estimation and remaining useful life assessment of lithium-ion batteries: A comparative study. *J. Energy Storage* **2022**, *51*, 104520. [CrossRef]
12. El Mejdoubi, A.; Chaoui, H.; Gualous, H.; Bossche, P.V.D.; Omar, N.; Van Mierlo, J. Lithium-Ion Batteries Health Prognosis Considering Aging Conditions. *IEEE Trans. Power Electron.* **2018**, *34*, 6834–6844. [CrossRef]
13. Hu, X.; Xu, L.; Lin, X.; Pecht, M. Battery Lifetime Prognostics. *Joule* **2020**, *4*, 310–346. [CrossRef]
14. Ossai, C.I. Prognosis and Remaining Useful Life Estimation of Lithium-Ion Battery with Optimal Multi-Level Particle Filter and Genetic Algorithm. *Batteries* **2018**, *4*, 15. [CrossRef]

15. Pang, H.; Chen, K.; Geng, Y.; Wu, L.; Wang, F.; Liu, J. Accurate capacity and remaining useful life prediction of lithium-ion batteries based on improved particle swarm optimization and particle filter. *Energy* **2024**, *293*, 130555. [CrossRef]
16. Lin, Y.-H.; Tian, L.-L.; Ding, Z.-Q. Ensemble Remaining Useful Life Prediction for Lithium-Ion Batteries With the Fusion of Historical and Real-Time Degradation Data. *IEEE Trans. Veh. Technol.* **2023**, *72*, 5934–5947. [CrossRef]
17. Hu, X.; Che, Y.; Lin, X.; Deng, Z. Health Prognosis for Electric Vehicle Battery Packs: A Data-Driven Approach. *IEEE/ASME Trans. Mechatron.* **2020**, *25*, 2622–2632. [CrossRef]
18. Li, J.; Peng, Y.; Wang, Q.; Liu, H. Status and Prospects of Research on Lithium-Ion Battery Parameter Identification. *Batteries* **2024**, *10*, 194. [CrossRef]
19. Sun, C.; Lu, T.; Li, Q.; Liu, Y.; Yang, W.; Xie, J. Predicting the Future Capacity and Remaining Useful Life of Lithium-Ion Batteries Based on Deep Transfer Learning. *Batteries* **2024**, *10*, 303. [CrossRef]
20. Bellomo, M.; Giazitzis, S.; Badha, S.; Rosetti, F.; Dolara, A.; Ogliari, E. Deep Learning Regression with Sequences of Different Length: An Application for State of Health Trajectory Prediction and Remaining Useful Life Estimation in Lithium-Ion Batteries. *Batteries* **2024**, *10*, 292. [CrossRef]
21. Jia, Z.; Jia, Z.; Li, Z.; Li, Z.; Zhao, K.; Zhao, K.; Wang, K.; Wang, K.; Wang, S.; Wang, S.; et al. CNN-DBLSTM: A long-term remaining life prediction framework for lithium-ion battery with small number of samples. *J. Energy Storage* **2024**, *97 Pt B*, 112947. [CrossRef]
22. Al-Wreikat, Y.; Serrano, C.; Sodré, J.R. Driving behaviour and trip condition effects on the energy consumption of an electric vehicle under real-world driving. *Appl. Energy* **2021**, *297*, 117096. [CrossRef]
23. Chen, X.; Hu, Y.; Li, S.; Wang, Y.; Li, D.; Luo, C.; Xue, X.; Xu, F.; Zhang, Z.; Gong, Z.; et al. State of health (SoH) estimation and degradation modes analysis of pouch NMC532/graphite Li-ion battery. *J. Power Sources* **2021**, *498*, 229884. [CrossRef]
24. Vespaiboon, S.; Pora, W. Analyzing Computational Complexity and Footprint: A Comparative Study of State of Charge Estimation across Various Equivalent Circuit Models for Batteries. In Proceedings of the 2024 21st International Conference on Electrical Engineering/Electronics, Computer, Telecommunications and Information Technology (ECTI-CON), Khon Kaen, Thailand, 27–30 May 2024; pp. 1–5. [CrossRef]
25. Sadil, J.; Kekula, F.; Majera, J.; Pisharodi, V. Comparison of Capacity Fade for the Constant Current and WLTC Drive Cycle Discharge Modes for Commercial $LiFeYPO_4$ Cells Used in xEV Vehicles. *Batteries* **2022**, *8*, 282. [CrossRef]
26. Hammou, A.; Petrone, R.; Diallo, D.; Gualous, H. Estimating the Health Status of Li-ion NMC Batteries from Energy Characteristics for EV Applications. *IEEE Trans. Energy Convers.* **2023**, *38*, 2160–2168. [CrossRef]
27. Hammou, A.; Petrone, R.; Gualous, H.; Diallo, D. Deep learning framework for state of health estimation of NMC and LFP Li-ion batteries for vehicular applications. *J. Energy Storage* **2023**, *70*, 108083. [CrossRef]
28. Tian, J.; Fan, Y.; Pan, T.; Zhang, X.; Yin, J.; Zhang, Q. A critical review on inconsistency mechanism, evaluation methods and improvement measures for lithium-ion battery energy storage systems. *Renew. Sustain. Energy Rev.* **2024**, *189 Pt B*, 113978. [CrossRef]
29. Leng, F.; Tan, C.M.; Pecht, M. Effect of Temperature on the Aging rate of Li Ion Battery Operating above Room Temperature. *Sci. Rep.* **2015**, *5*, 12967. [CrossRef] [PubMed]
30. Shen, W.; Wang, N.; Zhang, J.; Wang, F.; Zhang, G. Heat Generation and Degradation Mechanism of Lithium-Ion Batteries during High-Temperature Aging. *ACS Omega* **2022**, *7*, 44733–44742. [CrossRef]
31. Kampik, M.; Bodzek, K.; Piaskowy, A.; Pilśniak, A.; Fice, M. An Analysis of Energy Consumption in Railway Signal Boxes. *Energies* **2023**, *16*, 7985. [CrossRef]
32. Azizighalehsari, S.; Venugopal, P.; Singh, D.P.; Rietveld, G. Performance Evaluation of Retired Lithium-ion Batteries for Echelon Utilization. In Proceedings of the IECON 2022—48th Annual Conference of the IEEE Industrial Electronics Society, Brussels, Belgium, 17–20 October 2022; pp. 1–6. [CrossRef]
33. Hewamalage, H.; Bergmeir, C.; Bandara, K. Recurrent Neural Networks for Time Series Forecasting: Current status and future directions. *Int. J. Forecast.* **2021**, *37*, 388–427. [CrossRef]
34. Runge, J.; Zmeureanu, R. A Review of Deep Learning Techniques for Forecasting Energy Use in Buildings. *Energies* **2021**, *14*, 608. [CrossRef]
35. Chen, Y.; Cheng, Q.; Cheng, Y.; Yang, H.; Yu, H. Applications of recurrent neural networks in environmental factor forecasting: A review. *Neural Comput.* **2018**, *30*, 2855–2881. [CrossRef]
36. Liu, Y.; Wang, X.; Wang, L.; Liu, D. A modified leaky ReLU scheme (MLRS) for topology optimization with multiple materials. *Appl. Math. Comput.* **2019**, *352*, 188–204. [CrossRef]
37. Kingma, D.P.; Ba, J. Adam: A Method for Stochastic Optimization. In Proceedings of the 3rd International Conference on Learning Representations, ICLR, Conference Track Proceedings, Diego, CA, USA, 7–9 May 2015.
38. Hochreiter, S.; Schmidhuber, J. Long Short-Term Memory. *Neural Comput.* **1997**, *9*, 1735–1780. [CrossRef] [PubMed]
39. Cho, K.; Van Merrienboer, B.; Bahdanau, D.; Bengio, Y. On the properties of neural machine translation: Encoder-decoder approaches. *arXiv* **2014**, arXiv:1409.1259. [CrossRef]
40. He, K.; Zhang, X.; Ren, S.; Sun, J. Delving deep into rectifiers: Surpassing human-level performance on imagenet classification. In Proceedings of the IEEE International Conference on Computer Vision, Las Condes, Chile, 11–18 December 2015; pp. 1026–1034.
41. Safavi, V.; Vaniar, A.M.; Bazmohammadi, N.; Vasquez, J.C.; Keysan, O.; Guerrero, J.M. Early prediction of battery remaining useful life using CNN-XGBoost model and Coati optimization algorithm. *J. Energy Storage* **2024**, *98 Pt B*, 113176. [CrossRef]

42. Jadon, A.; Patil, A.; Jadon, S. A Comprehensive Survey of Regression-Based Loss Functions for Time Series Forecasting. In *Data Management, Analytics and Innovation. ICDMAI 2024. Lecture Notes in Networks and Systems*; Sharma, N., Goje, A.C., Chakrabarti, A., Bruckstein, A.M., Eds.; Springer: Singapore, 2024; Volume 998. [CrossRef]
43. Chen, X.; Yang, L.; Xue, H.; Li, L.; Yu, Y. A Machine Learning Model Based on GRU and LSTM to Predict the Environmental Parameters in a Layer House, Taking CO_2 Concentration as an Example. *Sensors* **2024**, *24*, 244. [CrossRef]
44. Liu, H.; Li, Y.; Luo, L.; Zhang, C. A Lithium-Ion Battery Capacity and RUL Prediction Fusion Method Based on Decomposition Strategy and GRU. *Batteries* **2023**, *9*, 323. [CrossRef]
45. Vakharia, V.; Shah, M.; Nair, P.; Borade, H.; Sahlot, P.; Wankhede, V. Estimation of Lithium-ion Battery Discharge Capacity by Integrating Optimized Explainable-AI and Stacked LSTM Model. *Batteries* **2023**, *9*, 125. [CrossRef]
46. Jiang, B.; Dai, H.; Wei, X.; Jiang, Z. Multi-Kernel Relevance Vector Machine With Parameter Optimization for Cycling Aging Prediction of Lithium-Ion Batteries. *IEEE J. Emerg. Sel. Top. Power Electron.* **2021**, *11*, 175–186. [CrossRef]
47. Bao, M.; Liu, D.; Wu, Y.; Wang, Z.; Yang, J.; Lan, L.; Ru, Q. Interpretable machine learning prediction for li-ion battery's state of health based on electrochemical impedance spectroscopy and temporal features. *Electrochim. Acta* **2024**, *494*, 144449. [CrossRef]

batteries

MDPI

Article

Li Chemical and Tracer Diffusivities in $LiCoO_2$ Sintered Pellets

Erwin Hüger [1,2,*] and Harald Schmidt [1,2]

1 Solid State Kinetics Group, Institute of Metallurgy, Clausthal University of Technology, 38678 Clausthal-Zellerfeld, Germany; harald.schmidt@tu-clausthal.de
2 Clausthal Center for Materials Technology, 38678 Clausthal-Zellerfeld, Germany
* Correspondence: erwin.hueger@tu-clausthal.de

Abstract: $LiCoO_2$ (LCO) is a crucial active material for positive electrodes of commercial lithium-ion batteries. It is typically present in the form of micrometer-sized LCO particles, which are surrounded by binders and conductive agents with a thickness of tens of microns. In order to determine the intrinsic Li transport parameters of pure crystalline LCO, it is necessary to measure the Li diffusivity at room temperature in sintered LCO pellets free of additives. The LCO sintered bulk material consists of interconnected, about 3 µm clusters, composed of grains of about 70 nanometers in size. The Li chemical and tracer diffusivities are determined using electrochemical impedance spectroscopy (EIS) and potentiostatic intermittent titration technique (PITT), while the latter ones are in the range between 10^{-9} and 10^{-28} m^2s^{-1}, depending on the application of different relevant formulas and characteristic parameters. Consequently, it is essential to apply a classical non-electrochemical and Li selective method of tracer diffusion determination like ^6Li depth profiling and secondary ion mass spectrometry (SIMS) for comparison. Li tracer diffusivities of about 10^{-22} m^2s^{-1} at room temperature are obtained by the extrapolation of the SIMS results from higher temperatures. This significantly narrows the range of reliable electrochemically determined Li tracer diffusivities to a more limited range, between 10^{-21} and 10^{-22} m^2s^{-1}.

Keywords: $LiCoO_2$; diffusion; thermodynamic factor; EIS; PITT; SIMS; LIB; cathode materials

1. Introduction

In order to accurately describe lithium-ion battery (LIB) performance, it is essential to understand and quantify not only the charge/discharge behaviour, but also the process of solid-state diffusion in electrodes that regulates the kinetics of this behaviour [1]. The diffusion of Li has a direct impact on the charge/discharge times (and consequently on the power density), the maximum capacity, the generation of stress (electrode stability), and the occurrence of side reactions. This was demonstrated by experiments on lithium cobalt oxide ($LiCoO_2$), referred to as LCO, which is a well-established cathode (positive electrode) active material in commercial LIBs [2,3].

LCO is a layered material comprising lithium oxide (001) atomic layers and cobalt oxide (001) atomic layers stacked sequentially along the crystallographic c-axis [2,3]. It is assumed that rapid Li diffusion will occur within the lithium oxide layers (i.e., in the ab-plane) where Li vacancies are present. In the direction perpendicular to the lithium oxide layers, i.e., in the c-axis direction, Li has to permeate through the cobalt oxide layers, which produces a low Li flux (low Li^+ current) due to the lack of Li vacancies in the cobalt oxide layers. This may be considered as an almost blocked Li diffusion perpendicular to the cobalt oxide layer (in the c-axis direction) [4]. Recent Li tracer diffusion experiments on LCO single crystals based on secondary ion mass spectrometry (SIMS) have corroborated this hypothesis [5]. The fast Li diffusivity path perpendicular to the c-axis may facilitate the extraction and insertion of Li^+ into LCO, which is of interest for the operation of LIBs. The impact of the Li diffusivity anisotropy in LCO on LIB performance has also been demonstrated by electrochemical experiments. In films of LCO oriented along the c-axis (where the c-axis is perpendicular to the electrolyte/film interface), the capacity and rate capability were observed to be inferior to those of LCO films where the c-axis is parallel to the electrolyte/film interface [6]. The high-rate capacity is dependent on the presence of fast Li diffusion paths that are perpendicular to the electrolyte/cathode interface. This has been demonstrated in (104) (ab-plane) oriented LCO films. In the case of polycrystalline LCO, the Li diffusivity was determined in sintered polycrystalline bulk LCO pellets [5,7] and in polycrystalline LCO thin films [4,6,8]. The Li diffusivity in polycrystalline LCO was found to be faster than in c-axis oriented LCO films and single crystals, and similarly faster than in ab-plane oriented LCO single crystals [5].

A highly convenient methodology for the determination of the Li diffusivity in LIB electrodes employs electrochemical-based measurement techniques, including electrochemical impedance spectroscopy (EIS) [9–32] and pulse methods such as potentiostatic intermittent titration technique (PITT) [4,9,23,33,34] and galvanostatic intermittent titration technique (GITT) [23,35]. All these methods offer the advantage of being applied in situ, without the need for cell disassembly or electrode washing procedures. At least for PITT and GITT techniques, measurements can be performed during LIB operation, a process referred to as operando measurements. A recent review article provides an overview of how these electrochemically based methodologies are applied to determine the Li diffusivity [9]. Using SIMS [9,36–38] the diffusivities are directly determined by measuring the diffusion induced modification of a ^{6}Li tracer distribution during annealing [9]. In contrast, electrochemical methods are more indirect, involving the application of different models resulting often in contradictory results. In electrochemical experiments, the observed signal may not be exclusively attributable to Li diffusion within the active material. Rather, it may also be influenced by other processes, including electrolyte/surface polarization. Accordingly, the Li diffusivities derived in this study may not fully reflect the diffusion process occurring in the cathode active material. See reference [9] for a detailed discussion of this topic. The objective of the present study is to use the diffusivity data acquired by SIMS depth profiling to evaluate the results obtained from electrochemical methods.

The SIMS method has the following advantages for Li diffusivity determination. First, Li easily loses its valence electron (high ionisation cross section), which results in Li^{+} SIMS signals of relatively high intensity. A further advantage is that it can discriminate between the two stable Li isotopes, which enables Li selective tracer studies. A third reason is that, in contrast to nuclear magnetic resonance (NMR) spectroscopy, which is also employed to determine diffusivities, SIMS is not influenced by the magnetic element cobalt. The fourth and fifth reasons pertain to a comparison of SIMS with the established electrochemical methodologies of GITT/PITT and EIS. These methods are typically only applicable at or near room temperature, thus providing no insight into activation energies and the underlying defect structure. Furthermore, they lack selectivity for lithium. Electrochemical measurements of diffusivities may involve interfacial processes, (parasitic) electronic currents, ohmic resistances, and side reactions that frequently interfere with the analysis. In contrast, tracer methods provide a direct measurement of the Li tracer diffusion coefficient, eliminating the need for model-dependent correction factors.

Table 1 of reference [4] presents a literature overview of the diffusivities of Li in LCO electrodes, as determined via EIS, PITT, and GITT at room temperature. These electrodes were prepared by two methods: deposition of thin films and drying slurries of an ensemble of LCO powder particles with binder and conductive additives to form a surrounding layer with a thickness in the tens of microns. The diffusivities span a range from 10^{-11} to 10^{-15} m^2s^{-1}, which is significantly higher than the tracer diffusivities at room temperature (Figure 1). The discrepancy can be attributed to the following factors: (i) the application of different models (e.g., different equations) to extract diffusivities and (ii) the difference between chemical and Li tracer diffusivities by the thermodynamic factor (TF), (iii) differing preparation procedures (e.g., the use of different electrode additives), and (iv) different state-of charge (SOC) of the electrode.

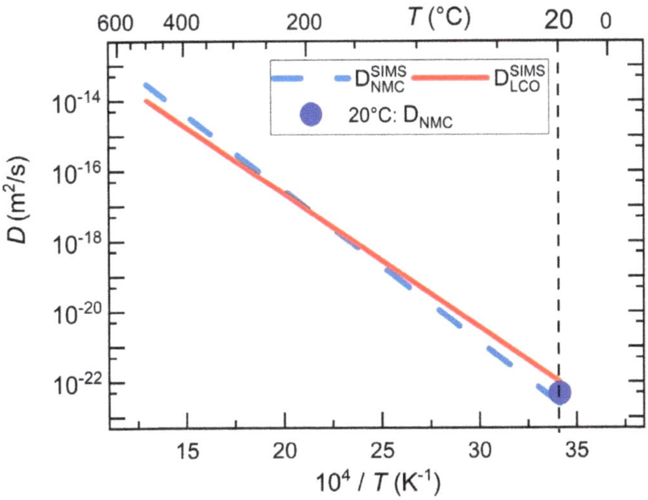

Figure 1. Arrhenius plot of the Li tracer diffusivities of LCO and NMC111 as determined from SIMS depth profiling on sintered pellets. The plots were obtained as described in the main text. The Li diffusivities were experimentally determined between 100 °C and 400 °C for NMC and between 200 °C and 500 °C for LCO. The values below 100 °C for NMC and below 200 °C for LCO are extrapolated values derived in accordance with the Arrhenius law. The blue dot at 20 °C represents the Li diffusivity in the sintered NMC111 pellet, which was determined to be $(4.7 \pm 1.9) \times 10^{-23}$ m^2s^{-1} by SIMS in a long-term (800 days) experiment performed at room temperature.

LCO represents one of the corners of the class of layered positive electrode materials, which includes NMC ($LiNi_xMn_yCo_zO_2$, where x + y + z = 1). All of these materials are of interest as cathode materials for LIBs [2,9]. Of the corner materials, i.e., $LiNiO_2$ (x = 1) [39], $LiMnO_2$ (y = 1), and LCO (z = 1) [5], LCO is the most stable material during LIB operation and thus one of the well-established materials in commercial LIBs [2,3]. LCO is a suitable cathode material due to the high electrochemical potential of 4.2 V versus Li metal reference. Furthermore, LCO possess a relatively high theoretical capacity of 270 $mAhg^{-1}$ for cathode materials. The disadvantages of using $LiCoO_2$ are safety reasons, difficult access to cobalt resources, capacity fading at high potentials, poor rate performance, Co toxicity, and a low practical capacity of only 148 $mAhg^{-1}$ [1,39,40]. For $LiNiO_2$, capacities higher than 240 $mAhg^{-1}$ are obtained in practice for the first cycle at an average voltage of 3.8 V [1,39]. Unfortunately, $LiNiO_2$ does not show a stable cycling behaviour. Therefore, one way to increase the capacity of the cathode material is to replace Co partly with the less expensive metal Ni in $LiCoO_2$ [9,39]. This has been successfully done by adding a third metal, i.e., Mn, to form NMC. The strong bonds of Mn to oxygen atoms stabilize the material during cycling, while Mn is not involved in the Li^+ release and uptake processes [1]. The theoretical capacity for NMC is reported to be 273 $mAhg^{-1}$ [3]. The practical capacity of Ni-rich and of Ni-poor NMC (including $LiNi_{0.33}Mn_{0.33}Co_{0.33}O_2$ named NMC111) is over 200 $mAhg^{-1}$ and up to 165 $mAhg^{-1}$, respectively [3]. Consequently, LCO and NMC are often used in commercial LIBs.

An ab-initio study on Li diffusion in Li_xCoO_2 [40] suggests Li migration via a monovacancy or divacancy mechanism. However, experiments indicate that a monovacancy mechanism is more probable [5]. Van der Ven and Ceder [40] concluded that although the results of their theoretical work were calculated for Li_xCoO_2, the same conclusions are likely to be valid for other NMC layered materials. Consequently, the Li diffusion in LCO should behave similarly to that in NMC111, as also experimentally shown [7].

The temperature dependence of the Li tracer diffusivities in pressed and sintered LCO [5,7] and NMC111 [41] pellets was determined using SIMS experiments. The diffusivities were determined over the temperature range of 100–500 °C. The diffusivities of NMC111 and LCO are the same within experimental error limits (Figure 1). An activation enthalpy of $\Delta H = (0.85 \pm 0.03)$ eV and a pre-exponential factor (D_0) of $\ln(D_0/m^2s^{-1}) = -18.42 \pm 0.74$ was determined for NMC111. A pre-exponential factor of $\ln(D_0/m^2s^{-1}) = -20.96 \pm 1.4$ and a activation enthalpy of $\Delta H = (0.75 \pm 0.03)$ eV was found for LCO. Figure 1 shows diffusivities of LCO extrapolated to room temperature according to the Arrhenius law.

$$D = D_0 \cdot exp(-\Delta H/kT) \tag{1}$$

In this equation, D is the diffusion coefficient, D_0 is the pre-exponential factor, k is the Boltzmann constant and T is the temperature. Extrapolation from higher temperatures represents a possibility to obtain the diffusivity at room temperature, under the assumption that the diffusion mechanism remains unchanged. This has been demonstrated experimentally for the case of NMC111 [9].

Figure 1 shows that the Li tracer diffusivity of LCO is expected to be by extrapolation even below 10^{-22} m^2s^{-1} at room temperature. Such a low diffusivity may need a considerable long annealing time in the order of years in form of storage at room temperature to be measurable by SIMS. Such a long-time experiment was not realized in the case of LCO due to time limitations, but for the case of NMC111 [9]. The exact Li tracer diffusivity was determined to be $(4.7 \pm 1.9) \times 10^{-23}$ m^2s^{-1}, which is indicated by a blue dot in Figure 1. This is a low diffusivity, but in agreement with the value extrapolated from the diffusivities at higher temperatures. This suggests that the extrapolation is a valid assumption for NMC111, which is likely also applicable for LCO. The error range of Li tracer diffusivities in NMC111 extrapolated to room temperature is between 3×10^{-23} m^2s^{-1} and 8×10^{-23} m^2s^{-1}, taking into account the measured error in activation energy and pre-exponential factor. In the same way extrapolated Li tracer diffusivities between 1×10^{-22} m^2s^{-1} and 3×10^{-22} m^2s^{-1} are found for LCO. The observed diffusivities for LCO are only marginally larger, which may be attributed to experimental error. Consequently, diffusivities varying only an order of magnitude can be estimated at room temperature for LCO due to experimental error and extrapolation. This should also be a reasonable approximation for the comparison of diffusivities determined by different methods and obtained from materials produced by different techniques.

The objective of the present study is to provide an improved understanding of Li diffusion in LCO at room temperature. To this end, the chemical diffusivity will be measured in the same LCO material where Li tracer diffusivity was previously determined using SIMS (Figure 1). Chemical diffusivities are measured in pressed and sintered LCO pellets free of additives for $x \approx 1$, employing electrochemically based techniques (EIS, PITT). These methods will be applied in a way that the Li content in the LCO pellet is unmodified to a significant extent. This approach includes the assessment of various models used to determine Li chemical diffusivities in LCO. This paper is structured as follows: The following section provides a description of the materials under study and the measurement techniques employed. The third section is dedicated to the presentation of the data obtained from EIS and PITT. The fourth section comprises a critical evaluation of the results, which are compared to the diffusivities obtained from SIMS depth profiling experiments. The last section summarises these results.

2. Materials and Methods

The procedure for preparing and characterizing the sintered LCO samples was described in detail in reference [5]. First, commercial LCO powders (Sigma Aldrich, Taufkirchen, Germany) were high-energy ball milled. Afterwards, cylindrical bulk samples with 12 mm diameter were produced by uniaxial pressing at 100 MPa. The pellets were then subjected to a 24 h sintering process at 800 °C in air. The results of the inductively coupled plasma-optical emission spectroscopy (ICP-OES) (Analytik Jena PlasmaQuant PQ 9000 Elite, Jena, Germany) analysis indicated that the relative element concentrations are Li (24.64 ± 0.13) at.%, Co (24.94 ± 0.21) at.%, and O (50.42 ± 0.34) at.%. This gives a lithium-to-metal ratio of x = 0.99 ± 0.01 [5], which is identical within error limits to the value of 1. Therefore, a relative Li content of x = 1 is assigned to the fully lithiated sintered LCO pellets. Scanning electron microscopy (ZEISS Evo 15, Oberkochen, Germany) revealed that the pellet is composed of interconnected fused LCO clusters of about 3 μm diameter. X-ray diffraction (XRD) measurements (Bruker D8 Discover, Karlsruhe, Germany) revealed polycrystalline LCO with grain size of about 70 nm.

The EIS and PITT measurements were conducted using a custom-built three-electrode electrochemical cell. The electrolyte was propylene carbonate (PC, Sigma Aldrich, Taufkirchen, Germany, anhydrous, 99.7%) with 1 M lithium perchlorate (Sigma Aldrich, Taufkirchen, Germany, battery grade). The assembly and disassembly of the cell was done within a glove box containing argon gas (MBraun, Garching bei München, Germany), with a concentration of oxygen and water vapour of less than 1 ppm.

The current collector of the LCO electrode, which is the working electrode designated as "we", is composed of a polished nickel disk with 1 mm in thickness and 14 mm in diameter. The LCO pellet with a thickness of 250 μm was fixed to the nickel disk by conductive agents. The counter and reference electrodes were lithium plates with a thickness of 1.5 mm and a purity of 99.9%, obtained from Alfa Aesar (Kandel, Germany). All reported LCO potentials (E_{we}) are referenced to the Li metal reference electrode. The LCO electrode was then inserted into the electrolyte-filled electrochemical cell. The process of Li^+ extraction (delithiation) from the LCO electrode is equivalent to cell charging. The electrochemical studies were conducted on a Biologic SP150 potentiostat using EC-lab software version V11.43 (Biologic, Seyssinet-Pariset, France).

Potentiostatic EIS was carried out with an amplitude of 10 mV. This study focuses on ionic diffusion, which is slow at room temperature. Consequently, the mass transfer impedance response manifests in the extremely low frequency range, and EIS was measured down to 1 mHz with an acquisition time of 7 h per spectrum.

PITT measurements were performed down to low Li^+ current densities over long time intervals as required for PITT analysis [9]. Further discussion of PITT is given in detail in reference [9]. The measurements were done at room temperature.

3. Results

3.1. Determination of the Chemical Diffusivity with EIS

An EIS spectrum measured at an open circuit voltage (OCV) of 1.4 V corresponding to the state right after sample preparation is plotted in Figure 2a and of 3.51 V after slight delithiation in Figure 2b.

The analysis of EIS data typically employs the use of equivalent circuits [9,13,14]. The data obtained in the present study can be accurately represented by the equivalent circuit schematised in Figure 2, which is given by the following expression (R1 + Q2/R2 + Q3/(R3 + Ma5)). This circuit represents a physically plausible model that provides a comprehensive description of the observed phenomena. The resistance R1 is ascribed to the cable, current collector, and electrolyte resistance. The constant phase element (CPE) Q2 and the resistor R2 can be attributed to the electrochemical double layer formed at the surface of the LCO electrode. The CPE Q3 and the resistor R3 can be attributed to the charge transfer process occurring at the LCO electrode. A linear modified restricted diffusion element (Ma5) is used for the simulation of the Li diffusion process in the LCO electrode.

It should be noted that direct diffusion at the interface may occur from the electrochemical double-layer, which tends to establish equilibrium between the electrochemical potentials of the electrolyte and electrode through a dynamic process. In that case, an additional Warburg element has to be connected in series with R2 and in parallel with Q2. However, this is omitted here due to the general requirement of EIS data fitting to restrict the number of elements in the equivalent circuit and to reduce the number of fitting parameters as much as possible.

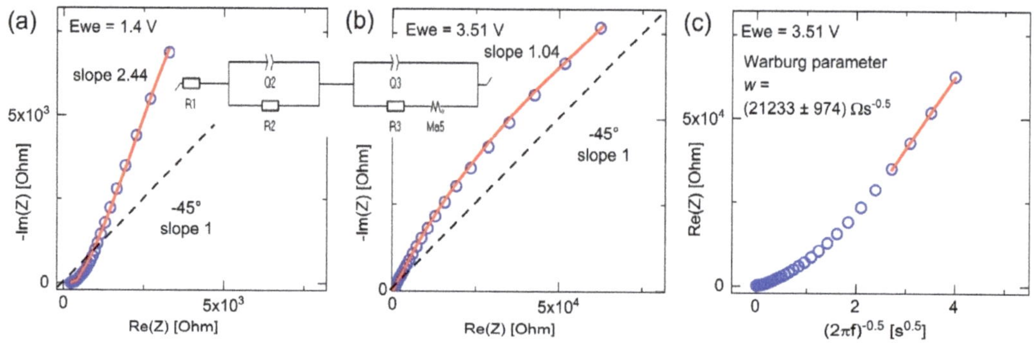

Figure 2. Results of EIS measurements on the LCO electrode marked with blue circles. The Nyquist plots were measured at (**a**) 1.40 V and (**b**) 3.51 V, respectively. The fits (red curves) were obtained using the equivalent electrical circuit [R1 + Q2/R2 + Q3/(R3 + Ma5)] schematised in (**a**,**b**). The dotted line represents the behavior of a Warburg impedance. (**c**) Diffusion-based analysis of the EIS data from (**b**) by plotting the real part of the impedance as a function of the reciprocal square of the frequency.

The CPE is dependent on two parameters, Q and the exponent a. The diffusion element Ma5 is dependent on three parameters, Rd5, a5, and t5. Rd5 represents the resistance associated with modified restricted linear diffusion, a5 is an exponent, and t5 denotes the diffusion time constant (for further details, please see reference [9] and references therein). The diffusivity is directly related to t5 by the equation $D = L^2(2\ t5)^{-1}$, where L is the thickness of the region where the restricted diffusion takes place. The fit to the measured data was performed using the randomized simplex method with 5000 iterations, and the following values were obtained with regard to Figure 2a: R1 = 226 Ω, R2 = 208 Ω, Q2 = 153 × 10^{-6} $Fs^{-0.5}$, a2 = 0.5, R3 = 0. Ω, Q3 = 500 ×10^{-6} $Fs^{-0.1}$, a3 = 0.9, Rd5 = 2300 Ω, a5 = 0.73, and t5 = 2.1 s. In the case of Figure 2b, we get R1 = 223 Ω, R2 = 428 Ω, Q2 = 301 ×10^{-6} $Fsa^{-0.1}$, a2 = 0.96, R3 = 89,100 Ω, and Q3 = 60 × 10^{-6} $Fs^{-0.25}$, a3 = 0.75, Rd5 = 831853 Ω, a5 = 0.46, and t5 = 8943 s. The value of a2 = 0.96 is here near unity, which is indicative for an ideal electrochemical double layer capacity, in contrast to that at 1.4 V. This demonstrates that the double layer is more developed at 3.5 V than at 1.4 V. Further differences are observed in the charge transfer resistance R3 and the values of the diffusion element, particularly in Rd5 and t5. The calculated Li diffusivities from t5, considering the entire pellet thickness of L = 250 μm, yield D(1.4 V) = 1.6 × 10^{-9} $m^2\ s^{-1}$ and D(3.51 V) = 4 × 10^{-12} $m^2\ s^{-1}$. These diffusivities are relatively high.

It is important to note that, as discussed in detail in reference [9] and the supporting information of that reference, the methodology to obtain results from EIS data by performing fittings with equivalent circuits is ambiguous [9,13]. This is because although different equivalent circuits can produce a similar good fits to the measured EIS data, they may give different results. Furthermore, it should be noted that the precise thickness of the region where the Li diffusion occurs is currently unknown. Consequently, in the present study, an older, less frequently employed, but robust method of diffusivity determination from EIS measurements independent of film thickness is also employed, as discussed in detail in reference [9].

In the low-frequency range, the measured impedance is defined by a linear relationship with a slope of $|d(ImZ)/d(ReZ)|$ = 2.44 in the Nyquist plot of Figure 2a. In reference [9] it was shown that for such slopes the Li diffusivity in the pellet cannot be determined unambiguously. This is different for slopes approaching the value 1, which is the case for the EIS spectra obtained at 3.51 V (Figure 2b). In such cases, a reasonable Li diffusivity can be obtained using the concept of the Warburg impedance [9] as shown in Figure 2c.

In a plot of Re(Z) versus $f^{-0.5}$, where f is the frequency, the Warburg impedance produces a straight line. The diffusivity is determined from the slope of the line in Figure 2c according to [9]

$$D = \left(\frac{R \cdot T}{\left(\sqrt{2}\right) \cdot n^2 \cdot F^2 \cdot a_e \cdot c} \cdot \frac{1}{c} \cdot \frac{1}{w} \right)^2 \quad (2)$$

In this equation, R represents the universal gas constant, T denotes a temperature of 300 K, a_e signifies the lithiated area, n is the valence number, F is the Faraday constant, and c is the Li^+ concentration, which is expressed as a function of the molar mass (M_{LCO}) and the mass density (ρ_{LCO} = 4.8 gcm^{-3}) of LCO. The SOC is defined as the percentage of Li ions that are present in the LCO material. The SOC is equal to 1 when the material is fully lithiated. The Warburg parameter (w) is a measure of Li ion diffusion through the material. It is calculated by fitting a linear function to the data in Figure 2c and determining the slope.

In the case of high frequencies and high diffusivities, or in the context of thick films (i.e., in the absence of any restrictions on diffusion), the following equation can be applied [9]:

$$D = \left(\frac{\frac{V_M}{SOC}}{\left(\sqrt{2}\right) \cdot n \cdot F \cdot a_e} \cdot \frac{dE}{dy} \cdot \frac{1}{w} \right)^2 = \left(\frac{1}{\left(\sqrt{2}\right) \cdot n \cdot F \cdot a_e} \cdot \frac{dE}{dy} \cdot \frac{1}{c} \cdot \frac{1}{w} \right)^2 \quad (3)$$

The slope of the open circuit voltage versus composition curve represented by dE/dy is calculated using the Li content y. The mole volume of Li in LCO is calculated to $V_M = (M_{LCO}/\rho_{LCO})$. Consequently, the ratio of V_M to SOC is equal to $1/c$. In the present study, the dE/dy value is determined by examining the relationship between the SOC and the electrode potential in LCO (Figure 3d).

The fit in Figure 2c gives a Warburg parameter of (21233 ± 974) $\Omega s^{-0.5}$. Diffusivities of (5.5 ± 3) × 10^{-24} m^2s^{-1} and (3.4 ± 2) × 10^{-25} m^2s^{-1} are obtained using Equations (2) and (3), respectively. These are low values compared to the results of equivalent circuit fitting. The Li diffusivity determined by Equation (3) at the potential of 3.51 V for NMC111 is (1.6 ± 0.6) × 10^{-12} m^2s^{-1} which is thirteen orders of magnitude higher than in this work measured for LCO. This large discrepancy may be due to different SOCs of the two materials. For LCO and NMC111, the SOC at a potential of 3.51 V is about 99.95% and 98.00%, respectively. This means that at 3.51 V, 40 times more Li is extracted from NMC111 than from LCO. Reference [9] has shown for NMC111 that there is an exponential increase in the Li diffusivity with the amount of Li extracted close to 3.5 V. The SOC of 99.95% obtained for LCO at 3.51 V is obtained for NMC111 at a potential of 2.8 V. The literature also shows that the onset of Li extraction from LCO [4] occurs at higher potentials than from NMC111 [9]. For NMC111, delithiation starts at 3.5 V [9], whereas for LCO, the delithiation plateau in the Ewe vs. Li composition curve does not start before 3.9 V [4]. Since the Li diffusivity increases sharply with decreasing Li content at the onset of charging in both LCO [4] and NMC111 [9], the higher Li content in LCO at 3.51 V compared to the lower Li content in NMC111 at 3.51 V results in a lower Li diffusivity in LCO at 3.51 V, closer to the completely lithiated state.

3.2. Determination of the Chemical Diffusivity and of the Thermodynamic Factor with PITT

PITT is a potential-step chronoamperometric method that measures the current response versus time as successive small voltage steps are applied. It is a widely used method to determine diffusion coefficients in electrochemical materials, especially Li diffusion in LIBs [9]. The voltage steps and current response of typical measured PITT curves are shown in Figures 3a and 3b, respectively. Figure 3c illustrates the plot of the Li^+ charge extracted from the LCO electrode during each voltage step as a function of potential. It can be observed that the extracted charge is in the range of $\mu Ahcm^{-2}$, which is a low value in comparison to the total Li^+ content of the LCO electrode, which is calculated to be

≈30,000 μAhcm^{-2}. The low amount of extracted charge in Figure 3c fulfills the general requirement of diffusivity determination via the pulse method [9]. The cumulative Li^+ charge, expressed in terms of the SOC of the LCO electrode at each potential (after each PITT step), is plotted in Figure 3d.

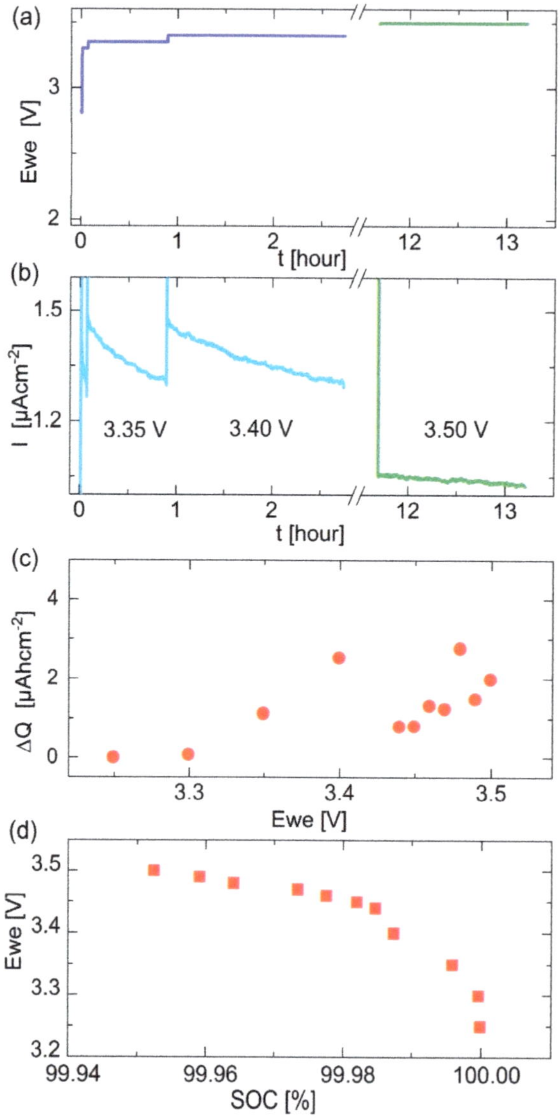

Figure 3. Results of PITT measurements. (**a**) Voltage steps between 3.2 and 3.4 (blue lines) and at 3.5 V (green line) versus time. (**b**) Current response between 3.2 and 3.4 (blue lines) and at 3.5 V (green line). versus time. (**c**) Charge (red circles) extracted during each voltage step (**d**) The SOC (red squares) present in the electrode during the subsequent voltage step that is indicative for the degree of lithiation (SOC = 100% represents the fully lithiated state). Voltage steps of 50 mV were used up to 3.4 V and 10 mV steps above.

Figure 4 illustrates the measured PITT data at 3.50 V, presented in a series of distinct plots. The Li diffusivity can be calculated from a plot of current versus the square root of the inverse of time (I vs. $t^{-0.5}$, Figure 4b) and from a logarithmic plot of current (ln(I) vs. t, Figure 4c). It should be noted that the analysis of PITT data depends on the dimensions of the electrode, particularly the thickness of the electrode or the radius of the particles present in the case of powders or agglomerates. In the Equation (4), L represents the characteristic length of the electrode material. The LCO electrode used in the present study is composed of approximately 3 μm clusters of sintered particles with grain sizes of approximately 70 nm. There are no well-separated particles, but rather a bulk material. Consequently, we first employ the LCO electrode thickness (L = 250 μm) for PITT analysis.

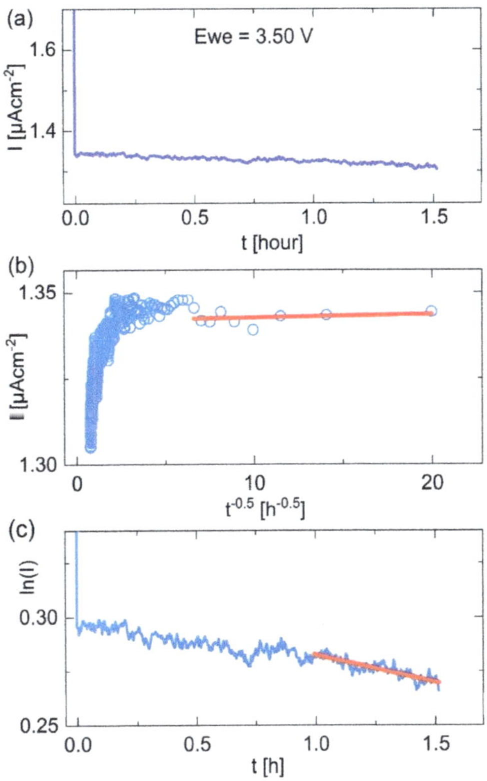

Figure 4. (**a–c**) The current response (blue curves) as a function of time during a typical PITT voltage step recorded at an LCO electrode potential of 3.50 V on different scales. The fits using Equations (4) and (6) are represented in (**b**,**c**) as the red line.

For short times (Figure 4b), the following equation is used to determine diffusivities by PITT [9]

$$I = a_{short} + b_{short} \cdot \frac{1}{\sqrt{t}} \text{ with } a_{short} = 0 \Longrightarrow D_{short} = \frac{\pi \cdot L^2}{Q^2} \cdot {b_{short}}^2 \quad (4)$$

where D_{short} is the corresponding diffusivity and Q is the charge extracted or inserted during the PITT step. Furthermore, b_{short} is the slope (red line in Figure 4b).

In addition to the requirement a_{short} = 0, the time interval Δt_{short}, from which the diffusivity is determined via Equation (4), have to satisfy the following inequality [9]

$$\Delta t_{short} < 0.14 \cdot t_D = 0.14 \cdot \frac{L^2}{2 \cdot D_{short}} = 0.14 \cdot \frac{Q^2}{2 \cdot \pi \cdot {b_{short}}^2} \tag{5}$$

where $t_D = L^2(2D_{short})^{-1}$ is the time required for Li to diffuse across the characteristic length of the material under investigation (e.g., electrode thickness L). The last term shows that the requirement (5) is formally independent of the characteristic length of the electrode material.

In the case of long-term domain PITT (Figure 4c), the following equation is employed in order to ascertain the diffusivities [9].

$$\ln(I) = a + b \cdot t \Longrightarrow D_{a-long} = \frac{L^2}{2 \cdot Q} \cdot e^a, D_{b-long} = \frac{4 \cdot L^2}{\pi^2} \cdot b \tag{6}$$

where $D_{a\text{-long}}$ and $D_{b\text{-long}}$ are the diffusivities obtained in this analysis (Figure 4c), $D_{a\text{-long}}$ from the intercept of the y-axis (parameter a) and $D_{b\text{-long}}$ from the slope (parameter b). The time interval Δt_{long} (necessary for diffusivity determination) has to fulfill the following conditions [9].

$$\Delta t_{a-long} > 0.14 \cdot t_D = 0.14 \cdot \frac{L^2}{2 \cdot D_{a-long}} = 0.14 \cdot \frac{Q}{\pi \cdot e^a} \tag{7}$$

$$\Delta t_{b-long} > 0.14 \cdot t_D = 0.14 \cdot \frac{L^2}{2 \cdot D_{b-long}} = 0.14 \cdot \frac{\pi^2}{8 \cdot b} \tag{8}$$

The last term of Equations (7) and (8) shows that the requirements (7) and (8) are independent of the characteristic length of the electrode material (L).

As obvious, linear segments were detected in the I vs. $t^{-0.5}$ (Figure 4b) and ln(I) vs t (Figure 4c) plots. These were fitted with Equations (4) and (6), respectively. Figure 5a shows the diffusivities D_{short}, $D_{a\text{-long}}$ and $D_{b\text{-long}}$ from the PITT, and also those from EIS measurements.

EIS and PITT give chemical diffusivities (D_{chem}). In case of the electronic conductivity is higher than ion conductivity they are correlated to the tracer diffusivities (D_{tracer}), (e.g., obtained by SIMS) by the thermodynamic factor (TF) [4,34] according to Equation (9)

$$D^{PITT,GITT,EIS} = D_{chem} = \mathrm{TF} \cdot D_{tracer} \tag{9}$$

PITT measurements [4,34] can be used to determine the TF according to

$$\mathrm{TF} = \frac{F}{R \cdot T} \cdot \frac{Q}{\Delta Q} \cdot \Delta \mathrm{E}_{we} \tag{10}$$

In this equation, F represents the Faraday constant, R is the universal gas constant, T = 300 K is the temperature, Q is the total number of Li^+ ions in the LCO electrode at the current SOC, and ΔQ is the Li^+ charge extracted from the LCO electrode during a potential step. Additionally, ΔE_{we} is the potential step. The thermodynamic factor calculated to Equation (10) is plotted in Figure 6a.

The chemical and tracer diffusivities plotted in Figures 5 and 6 will be further discussed in the next section by taken into account constraints and corrective considerations. Here, we briefly mention that the chemical diffusivities in Figure 5a decrease with higher LCO potentials (lower SOC, see Figure 3d). The trend of diffusivity decrease is not only observed for PITT experiments, but it can also be evidenced from the diffusivities obtained by fitting the EIS data with equivalent circuits (the black crosses and the violet stars in Figure 5). On the other hand, the tracer diffusivities (Figure 6b) are almost constant versus the SOC with the exception of some PITT data very close to 100% SOC. However, an unsatisfying result of Figures 5 and 6 is the high scatter of the diffusivity data over more than 15 orders of magnitudes, a result which will be discussed in more detail in the next section.

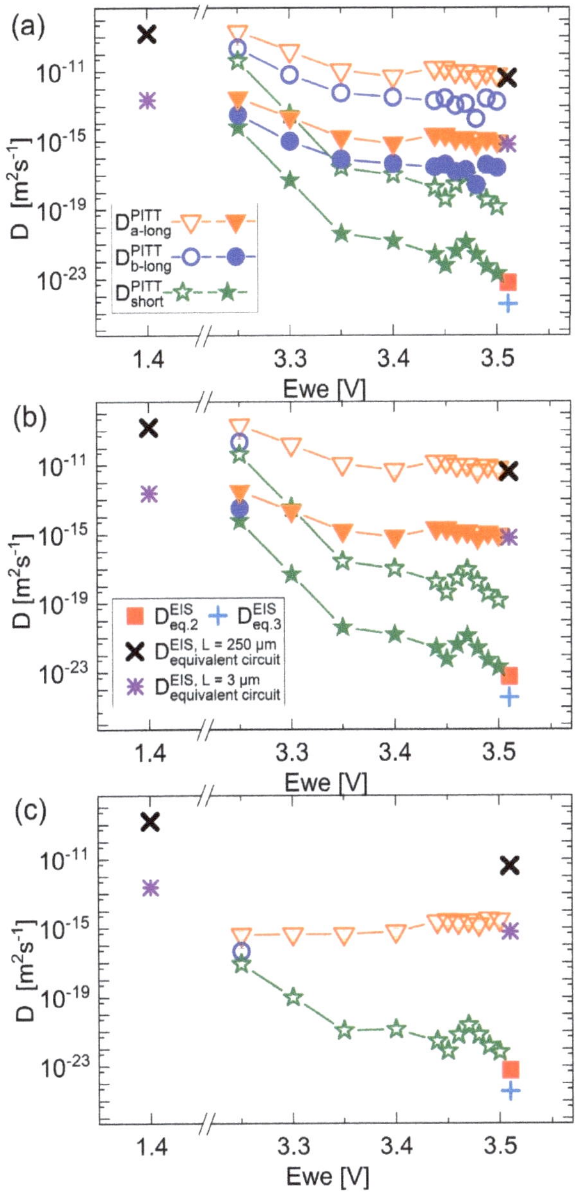

Figure 5. (**a–c**) Chemical diffusivities versus the LCO electrode potential: (**a**) As obtained from PITT and EIS. (**b**) All diffusivities that do not satisfy the requirements (4), (5), (7) and (8) are excluded from the plot. (**c**) The corrected Li diffusivities (D^{PITT}) were calculated under consideration of a moving phase boundary mechanism during delithiation by Equation (11). D^{PITT} values shown with unfilled and solid symbols correspond to values calculated using L = 250 μm and 3 μm, respectively. The black crosses and the violet stars correspond to the diffusivities obtained from equivalent circuit fitting of EIS data measured at an LCO potential of 1.4 V and 3.51 V, using L = 250 μm and 3 μm, respectively.

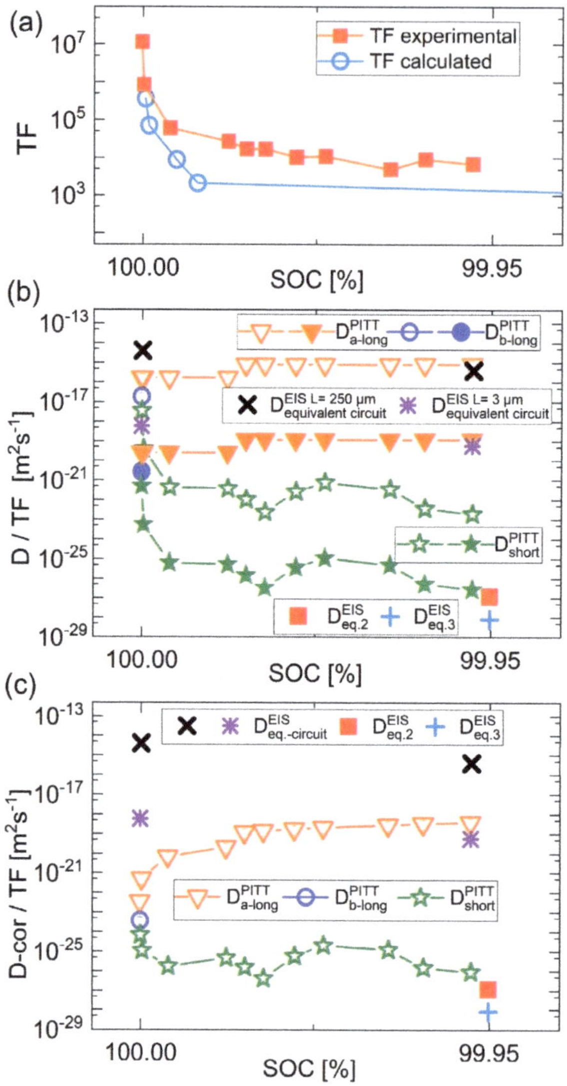

Figure 6. (**a**) The thermodynamic factor (TF) plotted as a function of SOC, as determined from PITT experiments on LCO. Additionally, the TF is theoretically obtained from first-principle calculations on LCO, as presented in the literature [40]. (**b**) The SOC dependence of the Li tracer diffusivity in LCO. (**c**) The corrected Li diffusivities, taking into account a presumed two phase delithiation process. D^{PITT} values shown with unfilled and solid symbols correspond to values calculated using L = 250 µm and 3 µm, respectively. D^{EIS} values shown with black crosses and violet stars correspond to diffusivities obtained from equivalent circuit fitting using L = 250 µm and 3 µm, respectively.

4. Discussion

First, note that the diffusivities under discussion represent average values due to the anisotropy of the LCO structure and pellet morphology. It should be noted that the formation of an equilibrium between the electrochemical potentials of the electrolyte and electrode may be associated with a diffusion process from the electrochemical double-layer capacitor. Our results indicate that Li-ion diffusion may occur in regions with a reduced thickness, probably also in proximity to the electrochemical double layer. This might be associated with the tunneling of electrons that neutralize the Li ions in the electrolyte, thereby reducing the influence of coulombic forces on short-range diffusion. Subsequently, the neutral Li atom may lose an electron while migrating within the cathode.

The diffusivities determined from the EIS measurements with the help of equivalent circuit fitting, and those according to Equations (2) and (3) are shown in Figure 5a–c as black crosses, violet stars, red-filled squares and blue plus signs, respectively. Figure 5a also gives the diffusivities obtained from PITT by the three distinct methods, using Equations (4) and (6). The PITT data indicate a similar qualitative behaviour as a function of potential. As previously mentioned, the potential increases during delithiation and the diffusivities decrease. In quantitative terms, the diffusivities $D_{a\text{-}long}^{PITT}$ and $D_{b\text{-}long}^{PITT}$, obtained by applying Equation (6) to the long-time domain of the PITT signal, are more similar to each other than to D_{short}^{PITT}. These values are up to seven orders of magnitude higher than those obtained from the short-time domain of the PITT signal (D_{short}^{PITT}). The considerable discrepancy between the diffusivities obtained from the short-time and long-time domains remains unexplained, underlining the inherent limitations of deriving reliable diffusivities from electrochemical methods with different models. An adjustment can be made when these diffusivities are examined for their validity in accordance with the conditions given in Equations (4), (5), (7) and (8). Data points that do not satisfy these conditions are excluded, and the remaining data are presented in Figure 5b. Further details can be found in reference [9]. $D_{b\text{-}long}^{PITT}$ does not fulfill condition (8) for nearly all potentials and is not considered further (Figure 5b). The discrepancy between the values of $D_{a\text{-}long}^{PITT}$ and D_{short}^{PITT} persists.

A further issue arises when the diffusivities obtained from the EIS measurements are compared to those obtained by PITT (Figure 5). The diffusivities determined from the EIS measurements by Equations (2) or (3), are found to be lower than those obtained from the PITT experiments. This suggests that either the Equations (2)–(4) and (6) or the values of the parameters used in those formulas are incorrect. The diffusivities extracted using the equivalent circuit on EIS data are in agreement to the D_{long}^{PITT} data.

With respect to the parameters used in the equations, two possible errors may exist: the characteristic length (L) inserted in the PITT Formulas (4) and (6) and the Li concentration (c) in the EIS Formulas (2) and (3). Concerning the characteristic length L, it should be noted that this quantity may differ from the overall thickness of the sintered LCO pellet of about 250 μm. Due to the low Li diffusivities present, the Li^+ current delivered by the PITT step on the electrode surface has insufficient time to cross the whole thickness of the LCO pellet during the relevant time scale. Since the quantity L is a variable in the PITT Formulas (4) and (6), it is not possible to derive the true diffusion length during the PITT experiment. The much lower diffusivity determined by EIS using Equations (2) or (3) (Figure 5b) indicates that the value of L to be used in the PITT Equations (4) and (6) has to be much smaller than 250 μm.

Using the assumption that the characteristic length L is the cluster size of 3 μm as visible by SEM measurements, instead, we recalculated the D^{PITT} values correspondingly. This gives the diffusivities shown in Figure 5 with full symbols. However, the D_{long}^{PITT} values are still higher than those obtained from the EIS measurements using the Equations (2) and (3). On the other hand, the D_{short}^{PITT} values now agree much better with that D^{EIS} values (Figure 5a,b).

An alternative procedure is to consider a specific delithiation mechanism at the onset of LCO charging, as has been done for sintered NMC111 pellets in reference [9]. Assuming that a two-phase delithiation mechanism is at work during the onset of LCO charging, it is possible to calculate the thicknesses in Equations (4) and (6) using the following formula

$$L = L_0 \cdot \frac{100\% - SOC}{100\%} \quad (11)$$

where L_0 = 250 μm is the thickness of the electrode, and the SOC is given in percentage. Since the potential remains approximately constant during the onset of charging (see, for example, Figure 2a of reference [4]), a two-phase lithiation process might be justified. A more detailed explanation can be found in the supporting information of reference [9].

Combining Equations (4) and (6) with Equation (11), the resulting corrected D^{PITT} values are presented in Figure 5c. There is now also a good agreement between the diffusivities determined in the present study from EIS (using Equations (2) and (3)) and PITT experiments for D_{short}^{PITT}. As the electrode potential decreases, the corrected D_{short}^{PITT} exhibit a corresponding decline, a phenomenon that is analogous to the observations made in the case of NMC111 [9]. Close to a SOC of 100%, all diffusivities obtained from PITT show approximately the same value within about an order of magnitude.

It is important to note that the assumption of a planar two-phase delithiation mechanism, as recognized in amorphous silicon thin films [35,42,43], requires experimental verification for the case of LCO. For LCO [s53,s55,s56], Lu et al. [44,45] used high-resolution transmission electron microscopy with a resolution of 0.04 nm to gain insight into the atomic structure of delithiated LCO nanoparticles. They constructed a phase diagram for the delithiation of LCO nanoparticles as a function of the LCO electrode potential and the amount of Li extracted from LCO, for instance, as a function of y in $Li_{1\text{-}y}CoO_2$ up to y ≈ 0.5 (Figure 5 in reference [44] and Figure 6c in reference [45]). For Li extraction corresponding to y between 0.06 and 0.23, there is a two-phase situation, but not for lower values of extracted Li, which is the range of the current study. It is noteworthy that the aforementioned correction, as given in Equation (11), was not applied to the EIS results obtained with Equations (2) and (3) because they are independent of the thickness, *L*. We are uncertain whether to apply the correction to EIS results from equivalent circuit fitting.

The second uncertainty relates to the use of the appropriate value for the Li concentration in the EIS Formulas (2) and (3). It is unclear whether the total Li concentration of LCO should be used in these equations. The hypothesis is that not all Li ions in LCO migrate during the EIS experiment, but only a fraction of the total Li amount. The precise quantity of Li participating in the EIS measurement remains unknown. Therefore, we propose an alternative approach, where the diffusivities derived from EIS and the uncorrected PITT values are assumed to be equal at 3.5 V, allowing us to assess the Li concentration that is mobile during the EIS measurement. Assuming $D^{EIS} = D_{short}^{PITT} = 1.3 \times 10^{-19}$ m^2s^{-1} and 2×10^{-23} m^2s^{-1}, and $D^{EIS} = D_{a\text{-}long}^{PITT} = 6 \times 10^{-12}$ m^2s^{-1} and 8×10^{-16} m^2s^{-1}, for the case of L = 250 μm and 3 μm, respectively, the partial Li concentration that is mobile during the EIS experiments is determined from Equation (12), and the resulting values are listed in Table 1.

$$\frac{c_{Li}^{EIS}}{c_{Li}^{total}} = \sqrt{\frac{D_{eq2,eq3}^{EIS}}{D_{short,\ a-long}^{PITT}}} \tag{12}$$

Table 1. The partial Li concentrations calculated from Equation (12), which is likely to be mobile during the EIS experiments. The partial Li concentrations obtained using L = 250 μm and 3 μm in the D^{PITT} calculation are listed in the second and third rows, respectively.

L		D_{eq2}^{EIS}, D_{short}^{PITT}	D_{eq2}^{EIS}, $D_{a\text{-}long}^{PITT}$	D_{eq3}^{EIS}, D_{short}^{PITT}	D_{eq3}^{EIS}, $D_{a\text{-}long}^{PITT}$
250 μm	$c_{Li}^{EIS}/c_{Li}^{total}$	0.6%	0.00009%	0.16%	0.000024%
3 μm	$c_{Li}^{EIS}/c_{Li}^{total}$	50%	0.075%	14%	0.002%

The values are relatively low, ranging from 0.6% to less than 0.00003% of the total Li concentration in LCO using L = 250 μm for PITT analysis. Using L = 3 μm, the values are, obviously, higher, reaching even 50% (Table 1). However, it is unclear which of these values are relevant.

It can be concluded that the true Li diffusivity cannot be identified in a consistent way from the widely varying diffusivities obtained by the indirect diffusion determination methods based on electrochemical measurements. To solve this problem, it is necessary to determine the Li diffusivity independently using tracer diffusion methods with SIMS depth profiling experiments for comparison.

The Li diffusivities determined via EIS and PITT represent chemical diffusivities (D_{chem}), which are different from the tracer diffusivities (D_{tracer}) by the thermodynamic factor (TF) [4,34], as illustrated by Equation (10). The TF obtained from the PITT measurements is plotted in Figure 6a. It decreases significantly with decreasing SOC, corresponding to an increasing deviation from stoichiometry. These results are in good agreement with first-principles calculations of Li diffusion in layered Li_xCoO_2 [40]. Calculations show that if x approaches a value of 1, the Li chemical potential deviates strongly from ideal behavior. The current regime exhibits a substantial amplifying effect of a slight variation in Li concentration on the chemical potential gradient, resulting in a large TF.

Equation (9) was used to calculate the tracer diffusivities from the chemical diffusivities as illustrated in Figure 6b and with consideration of the two-phase delithiation mechanism in Figure 6c. The tracer diffusivities demonstrate a relatively independent behavior with respect to the SOC, below the exact stoichiometric composition (SOC = 100%).

Figure 7 presents a comparison of Li tracer diffusivities obtained from electrochemical methods using diverse models and appropriate assumptions for diffusivity determination (i.e., varying equations and parameters) as discussed above (positions (1) to (15)). These values are compared to the tracer diffusivities obtained from SIMS depth profiling experiments (position (16)). This is undertaken in order to provide a reliable basis for the determination of correct diffusion coefficients. As previously discussed, the Li tracer diffusivities determined from electrochemically based measurements exhibit a considerable degree of scatter with values ranging between 10^{-15} and 10^{-28} m^2s^{-1} (Figure 7).

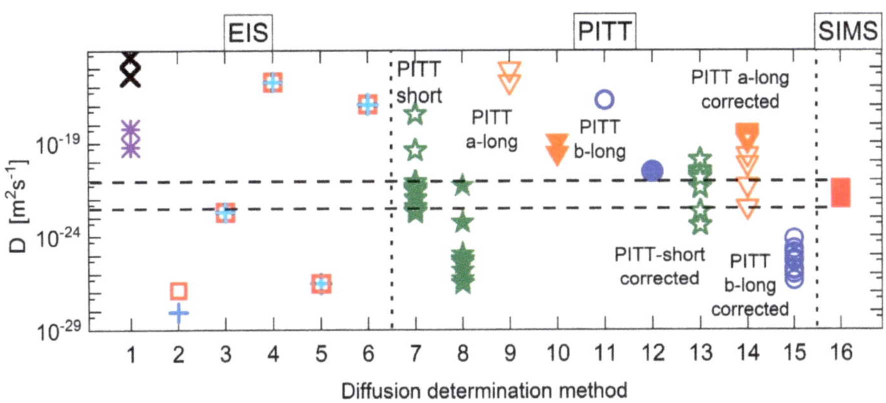

Figure 7. Li tracer diffusivities at room temperature derived from various electrochemical diffusion experiments on fully lithiated and slightly delithiated sintered LCO pellets (numbers (1) to (15)). The data at position (1) to (6) were obtained from EIS, while the data at positions (7) to (15) were obtained from PITT measurements, using different models, including different equations and different parameters. D^{PITT} values shown with unfilled and solid symbols correspond to values calculated using L = 250 μm and 3 μm, respectively. Position (16) refers to values extrapolated to room temperature from the SIMS data at higher temperatures. The diffusivity region within the two horizontal dashed lines comprises the range of diffusivities that is acceptable from the perspective of the SIMS results. For further details the reader is directed to the main text.

As previously stated in the introduction section, the extrapolation of SIMS data obtained at elevated temperatures to room temperature may be a valuable method for the class of NMC materials, which includes LCO (Figure 1). The two horizontal dashed lines in Figure 7 indicate the range of Li tracer diffusivities that are reliable from the perspective of the SIMS result (position (16)). The initial six positions in Figure 7 pertain to the diffusivities derived from EIS measurements. The black crosses (×) and violet stars were obtained from fitting the EIS data with an equivalent circuit (Figure 2) by using L = 250 μm and L = 3 μm, respectively, in the diffusivity calculation. The red-filled squares and the blue plus signs indicate the data obtained by EIS, which were obtained using Equations (2) and (3), respectively. Position (2) pertains to diffusivities for the maximum concentration of mobile Li. Positions (3) and (4) refer to a partial Li concentration from the third and fourth as well as fifth and sixth columns of the second row of Table 1. Finally, the data at positions (5) and (6) refer to the partial Li concentration of the third row of Table 1. It can be observed that the EIS data at position (3) fall within the range of acceptable diffusivities, which would indicate that not all Li in the sintered LCO pellet are mobile during the EIS measurement, but only 1.6%, which is considerable low.

The diffusivities obtained from PITT (positions (7) to (15)) exhibit also a high degree of scatter. The values at the different positions are marked in the plot. The D^{PITT} values plotted with unfilled symbols (at positions (7), (9), (11), (13), (14), (15)) and solid symbols (at positions (8), (10), (12)) correspond to values calculated using L = 250 μm and 3 μm, respectively. Equation (11) was used only for the values at positions (13), (14) and (15). The SIMS results are in good agreement with some of the PITT result in positions (7), (8), (13), and (14) close to a SOC of 100%. However, the problem is that at those positions the diffusivities were determined by different approaches. This means that in a comparative study like the present one, it is not possible to determine which of the various electrochemical methods of diffusivity determination is the most appropriate.

To summarize, the determination of Li diffusivities from electrochemical-based measurements gives values spanning a range of 21 orders of magnitude also close to 100% of SOC. The application of a well-established Li selective and direct tracer diffusion method, such as SIMS depth profiling, can effectively narrow the range of diffusivities to approximately one order of magnitude. This demonstrates that to ensure greater reliability in diffusivity determination, diffusivities obtained from electrochemical-based methods should be compared to those obtained by tracer diffusion methods. Another key finding of the current study is the existence of low tracer diffusivities and large TFs prior to the onset of the electrochemical delithiation process of sintered LCO pellets, which is similar to the case of NMC111 [9].

5. Conclusions

Li tracer diffusivity, Li chemical diffusivity, and thermodynamic factor were determined through experiments on pressed and sintered LCO bulk electrodes at room temperature. The material was free of binders and conductive agents in order to enable a measurement of the intrinsic Li diffusivity. The measurements were conducted using PITT and EIS on samples that were fully lithiated and only slightly delithiated (SOC > 99.95%). The chemical diffusivity was determined from EIS experiments to range between 1.5×10^{-9} m^2s^{-1} and 3.4×10^{-25} m^2s^{-1}, depending on the applied diffusivity determination formula. The chemical diffusivities obtained from PITT also span a large range, between 10^{-9} m^2s^{-1} and 10^{-23} m^2s^{-1}. The values depend on the applied formula of diffusivity determination and on the so-called characteristic length of the electrode material, to which diffusion is restricted during the PITT experiment, e.g., considering the whole thickness of the LCO pellet (≈250 μm) or only the size of ≈3 μm of the clusters of interconnected LCO particles of the LCO pellet. The agreement between the diffusivities obtained by PITT and those obtained by EIS (which were obtained without the need of a characteristic diffusion length) is improved by two factors: (i) a presumed two-phase delithiation mechanism at the onset of LCO charging for the PITT experiments, and (ii) a reduced mobile Li ion concentration for the EIS experiments. It must be acknowledged that both assumptions have yet to be validated through experiments. It is therefore imperative that future experiments are designed to address these shortcomings.

In order to identify the most reliable Li diffusivities, a comprehensive analysis was carried out on electrochemically determined values using diverse suited models and assumptions. These values were then compared with the diffusivities derived from the established ^{6}Li tracer diffusion experiments by SIMS. A comparison was done between the chemical diffusivities obtained by PITT and EIS and the Li tracer diffusivities estimated from SIMS experiments, with the thermodynamic factor measured by PITT. The Li tracer diffusivities obtained from PITT and EIS fall within the range of 10^{-15} and 10^{-28} m^2s^{-1}. Extrapolation of tracer diffusivities measured by SIMS to room temperature yielded a value between 1×10^{-22} m^2s^{-1} and 3×10^{-22} m^2s^{-1}, confirming the indication of a very low Li tracer diffusivity. Therefore, the SIMS results help to restrict the range of possible Li tracer diffusivities determined electrochemically to a narrow band within one order of magnitude around 10^{-22} m^2s^{-1}. A further significant outcome of the present study is the identification of low diffusivities and elevated thermodynamic factors at room temperature preceding the initiation of electrochemical delithiation in sintered LCO pellets, exhibiting a similar trend to that observed in the other layered cathode material, $LiNi_{0.33}Mn_{0.33}Co_{0.33}O_2$.

Author Contributions: Conceptualization, H.S. and E.H.; methodology, H.S. and E.H.; validation, H.S. and E.H.; formal analysis, E.H.; investigation, E.H.; resources, H.S.; data curation, H.S. and E.H.; writing—original draft preparation, E.H.; writing—review and editing, H.S. and E.H.; visualization, E.H.; supervision, H.S.; project administration, H.S.; funding acquisition, H.S. All authors have read and agreed to the published version of the manuscript.

Funding: This research was funded by the Deutsche Forschungsgemeinschaft (DFG, German Research Foundation) under the contract SCHM 1569/33-2 (413672097). The financial support is gratefully acknowledged.

Data Availability Statement: Dataset available on request from the authors.

Acknowledgments: We thank Daniel Uxa for preparing the sintered LCO pellet and Gerrit Zander (TU Clausthal) for determining the stoichiometry of the LCO pellet using ICP-OES.

Conflicts of Interest: The authors declare no conflicts of interest.

References

1. Booth, S.G.; Nedoma, A.J.; Anthonisamy, N.N.; Baker, P.J.; Boston, R.; Bronstein, H.; Clarke, S.J.; Cussen, E.J.; Daramalla, V.; De Volder, M.; et al. Perspectives for next generation lithium-ion battery cathode materials. *APL Mater.* **2021**, *9*, 109201. [CrossRef]
2. Hüger, E.; Riedel, L.; Zhu, J.; Stahn, J.; Heitjans, P.; Schmidt, H. Lithium niobate for fast cycling in Li-ion batteries: Review and new experimental data. *Batteries* **2023**, *9*, 244. [CrossRef]
3. Tang, Z.; Feng, D.; Xu, Y.; Chen, L.; Zhang, X.; Ma, Q. Perspectives for next generation lithium-ion battery cathode materials. *Batteries* **2023**, *9*, 156. [CrossRef]
4. Hasagawa, G.; Kuwata, N.; Tanaka, Y.; Miyazaki, T.; Ishigaki, N.; Takada, K.; Kawamura, J. Tracer diffusion coefficients of Li^+ ions in *c*-axis oriented Li_xCoO_2 thin films measured by secondary ion mass spectrometry. *Phys. Chem. Chem. Phys.* **2021**, *23*, 2438–2448. [CrossRef]
5. Uxa, D.; Hüger, E.; Meyer, K.; Dörrer, L.; Schmidt, H. Lithium-ion diffusion in near-stoichiometric polycrystalline and monocrystalline $LiCoO_2$. *Chem. Mater.* **2023**, *35*, 3307–3315. [CrossRef]
6. Kawashima, K.; Ohnishi, T.; Takada, K. High-Rate Capability of $LiCoO_2$ Cathodes. *ACS Appl. Energy Mater.* **2020**, *3*, 11803–11810. [CrossRef]
7. Uxa, D.; Holmes, H.J.; Meyer, K.; Dörrer, L.; Schmidt, H. Lithium Tracer Diffusion in in Sub-Stoichiometric Lithium-Metal-Oxide Compounds. *Defect Deffus. Forum* **2021**, *413*, 125–135. [CrossRef]
8. Bumberger, A.E.; Boehme, C.; Ring, J.; Raznjevic, S.; Zhang, Z.; Kubicek, M.; Fleig, J. Mass and Charge Transport in $Li1-\delta CoO_2$ Thin Films—A Complete Set of Properties and Its Defect Chemical Interpretation. *Chem. Mater.* **2022**, *34*, 10548–10560. [CrossRef] [PubMed]
9. Hüger, E.; Uxa, D.; Schmidt, H. Electrochemical (PITT, EIS) and Analytical (SIMS) Determination of Li Diffusivities at the Onset of Charging $LiNi_{0.33}Mn_{0.33}Co_{0.33}O_2$ Electrodes. *J. Phys. Chem. C* **2024**, *128*, 7408–7423. [CrossRef]
10. Randles, J.E.B. Kinetics of rapid electrode reactions. *Discuss. Faraday Soc.* **1947**, *1*, 11–19. [CrossRef]
11. Ho, C.; Raistrick, I.D.; Huggins, R.A. Application of A−C techniques to the study of lithium: Diffusion in tungsten trioxide thin films. *J. Electrochem. Soc.* **1980**, *127*, 343–350. [CrossRef]

12. Dawson, J.L.; John, D.G. Diffusion impedance—An extended general analysis. *J. Electroanal. Chem.* **1980**, *110*, 37–47. [CrossRef]
13. Macdonald, J.R. *Impedance Spectroscopy: Emphasizing Solid Materials and Systems*; John Wiley & Sons: Chichester, UK, 1987; ISBN 0-471-83122-0.
14. Taylor, S.R.; Gileadi, E. Physical interpretation of the Warburg impedance. *Corrosion* **1995**, *51*, 664–671. [CrossRef]
15. Muralidharan, V.S. Warburg impedance—Basics revisited. *Anti-Corros. Methods M.* **1997**, *44*, 26–29. [CrossRef]
16. Jamnik, J.; Maier, J.; Pejovnik, S. A powerful electrical network model for the impedance of mixed conductors. *Electrochim. Acta* **1999**, *44*, 4139–4145. [CrossRef]
17. Jamnik, J.; Maier, J. Generalised equivalent circuits for mass and charge transport: Chemical capacitance and its implications. *Phys. Chem. Chem. Phys.* **2001**, *3*, 1668–1678. [CrossRef]
18. Lai, W.; Haile, S.M. Impedance spectroscopy as a tool for chemical and electrochemical analysis of mixed conductors: A case study of ceria. *J. Am. Ceram. Soc.* **2005**, *88*, 2979–2997. [CrossRef]
19. Go, J.; Pyun, S. A review of anomalous diffusion phenomena at fractal interface for diffusion-controlled and non-diffusion-controlled transfer processes. *J. Solid State Electrochem.* **2007**, *11*, 323–334. [CrossRef]
20. Barbero, G.; Lelidis, I. Analysis of Warburg's impedance and its equivalent electric circuits. *Phys. Chem. Chem. Phys.* **2017**, *19*, 24934. [CrossRef]
21. Nguyen, T.Q.; Breitkopf, C. Determination of diffusion coefficients using impedance spectroscopy data. *J. Electrochem. Soc.* **2018**, *165*, E826–E831. [CrossRef]
22. BioLogic. How to Interpret Lower Frequencies Impedance in Batteries? EC-Lab—Application Note 61. 2017. Available online: https://www.biologic.net/documents/eis-low-frequencies-diffusion-battery-application-note-61/ (accessed on 20 February 2024).
23. BioLogic. Determination of the Diffusion Coefficient of an Inserted Species in a Host Electrode with EIS, PITT and GITT Techniques Battery—Application Note 70. 2024. Available online: https://www.biologic.net/documents/determination-diffusion-coefficient-inserted-species-host-electrode-eis-pitt-gitt-techniques-an70/ (accessed on 29 October 2024).
24. Diard, J.P.; Le Gotrech, B.; Montella, C. *Handbook of Electrochemical Impedance Spectroscopy. Diffusion Impedances*; Technical Report; ER@SE/LEPMI: Grenoble, France, 2020; pp. 1–39. [CrossRef]
25. Meddings, N.; Heinrich, M.; Overney, F.; Lee, J.; Ruiz, V.; Napolitano, E.; Seitz, S.; Hinds, G.; Raccichini, R.; Gaberscek, M.; et al. Application of electrochemical impedance spectroscopy to commercial Li-ion cells: A review. *J. Power Sources* **2020**, *480*, 228742. [CrossRef]
26. Charbonneau, V.; Lasia, A.; Brisard, G. Impedance studies of Li+ diffusion in nickel manganese cobalt oxide (NMC) during charge/discharge cycles. *J. Electroanal. Chem.* **2020**, *875*, 113944. [CrossRef]
27. Janssen, M.; Bisquert, J. Locating the frequency of turnover in thin-film diffusion impedance. *J. Phys. Chem. C* **2021**, *125*, 15737–15741. [CrossRef]
28. Bumberger, A.E.; Boehme, C.; Ring, J.; Raznjevic, S.; Zhang, Z.; Kubicek, M.; Fleig, J. Defect chemistry of spinel cathode materials—A case study of epitaxial LiMn2O4 thin films. *Chem. Mater.* **2023**, *35*, 5135–5149. [CrossRef]
29. Lazanas, A.C.; Prodromidis, M.I. Electrochemical impedance spectroscopy: A tutorial. *ACS Meas. Sci. Au* **2023**, *3*, 162–193. [CrossRef] [PubMed]
30. Bisquert, J.; Garcia-Belmonte, G.; Bueno, P.; Longo, E.; Bulhoes, L.O.S. Impedance of constant phase element (CPE)-blocked diffusion in film electrodes. *J. Electroanal. Chem.* **1998**, *452*, 229–234. [CrossRef]
31. Bisquert, J.; Garcia-Belmonte, G.; Fabregat-Santiago, F.; Bueno, P.R. Theoretical models for ac impedance of finite diffusion layers exhibiting low frequency dispersion. *J. Electroanal. Chem.* **1999**, *475*, 152–163. [CrossRef]
32. Moya, A.A. Comparing the characteristic frequencies of the transmissive and reflexive finite-length Warburg diffusion processes. *J. Electroanal. Chem.* **2023**, *929*, 117132. [CrossRef]
33. Montella, C. Discussion of the potential step method for the determination of the diffusion coefficients of guest species in host materials Part I. Influence of charge transfer kinetics and ohmic potential drop. *J. Electroanal. Chem.* **2002**, *518*, 61–83. [CrossRef]
34. Weppner, W.; Huggins, R.A. Determination of the kinetic parameters of mixed-conducting electrodes and application to the system Li_3Sb. *J. Electrochem. Soc.* **1977**, *124*, 1569–1578. [CrossRef]
35. Uxa, D.; Hüger, E.; Schmidt, H. Li Diffusion in Thin Film LiySi Electrodes: Galvanostatic intermittent titration technique and tracer diffusion experiments. *J. Phys. Chem. C* **2020**, *124*, 27894–27899. [CrossRef]
36. McPhail, D.S. Applications of Secondary Ion Mass Spectrometry (SIMS) in Materials Science. *J. Mater. Sci.* **2006**, *41*, 873–903. [CrossRef]
37. Wilson, R.G. SIMS quantification in Si, GaAs, and diamond—An update. *Int. J. Mass Spectrom. Ion Process.* **1995**, *143*, 43–49. [CrossRef]
38. Stevie, F.A.; Griffis, D.P. Quantification in dynamic SIMS: Current status and future needs. *Appl. Surf. Sci.* **2008**, *255*, 1364–1367. [CrossRef]
39. Bianchini, M.; Roca-Ayats, M.; Hartmann, P.; Brezesinski, T.; Janek, J. There and back again—The journey of $LiNiO_2$ as a cathode active material. *Angew. Chem. Int. Ed.* **2019**, *58*, 10434–10458. [CrossRef] [PubMed]
40. Van der Ven, A.; Ceder, G. Lithium diffusion in layered $LixCoO_2$. *Electrochem. Solid-State Lett.* **2000**, *3*, 301. [CrossRef]
41. Uxa, D.; Holmes, H.; Meyer, K.; Dörrer, L.; Schmidt, H. Lithium tracer diffusion in $LiNi_{0.33}Mn_{0.33}Co_{0.33}O_2$ cathode material for lithium-ion batteries. *Phys. Chem. Chem. Phys.* **2021**, *23*, 5992. [CrossRef] [PubMed]

42. Wang, J.W.; He, Y.; Fan, F.; Liu, X.H.; Xia, S.; Liu, Y.; Harris, C.T.; Li, H.; Huang, J.Y.; Mao, S.X.; et al. Two-Phase Electrochemical Lithiation in Amorphous Silicon. *Nano Lett.* **2013**, *13*, 709–715. [CrossRef]
43. Uxa, D.; Jerliu, B.; Hüger, E.; Dörrer, L.; Horisberger, M.; Stahn, J.; Schmidt, H. On the lithiation mechanism of amorphous silicon electrodes in Li-ion batteries. *J. Phys. Chem. C* **2019**, *123*, 22027–22039. [CrossRef]
44. Lu, X.; Sun, Y.; Jian, Z.; He, X.; Gu, L.; Hu, Y.; Li, H.; Wang, Z.; Chen, W.; Duan, X.; et al. New Insight into the Atomic Structure of Electrochemically Delithiated O_3-$Li_{(1-x)}CoO_2$ ($0 \leq x \leq 0.5$) Nanoparticles. *Nano Lett.* **2012**, *12*, 6192–6197. [CrossRef]
45. Lin, C.; Li, J.; Yin, Z.; Huang, W.; Zhao, Q.; Weng, Q.; Liu, Q.; Sun, J.; Chen, G.; Pan, F. Structural Understanding for High-Voltage Stabilization of Lithium Cobalt Oxide. *Adv. Mater.* **2024**, *36*, 2307404. [CrossRef]

batteries

MDPI

Review

Different Metal–Air Batteries as Range Extenders for the Electric Vehicle Market: A Comparative Study

Yasmin Shabeer [1], Seyed Saeed Madani [1], Satyam Panchal [1], Mahboubeh Mousavi [2] and Michael Fowler [1,*]

[1] Chemical Engineering, University of Waterloo, Waterloo, ON N2L3G1, Canada; yshabeer@uwaterloo.ca (Y.S.)
[2] AlumaPower Corporation, Kitchener, ON N2R1G1, Canada
* Correspondence: mfowler@uwaterloo.ca

Abstract: Metal–air batteries represent a category of energy storage system that leverages the reaction between metal and oxygen from the atmosphere to produce electricity. These batteries, known for their high energy density, have attracted considerable attention as potential solutions for extending the range of electric vehicles. Understanding the capabilities and limitations of metal-air batteries as range extenders is crucial for advancing electric vehicle technology, as these batteries could offer the additional energy needed to overcome current range limitations. This review paper provides a detailed overview of various metal-air battery technologies, delving into their design, functionality, and inherent challenges. By analyzing key theoretical and practical parameters, the study highlights how these factors influence overall battery performance. Additionally, the review addresses critical cost considerations, particularly the relationship between vehicle cost and driving range, uncovering the significant trade-offs involved in adopting metal-air batteries. Through an examination of nearly all the existing metal-air batteries, this paper sheds light on their potential to serve as effective range extenders, thereby facilitating the transition to a cleaner, more sustainable transportation landscape.

Keywords: metal-air batteries; range extenders; lithium-air battery; magnesium-air battery; sodium-air battery; aluminum-air battery; zinc-air battery; potassium-air battery; iron-air battery; calcium-air battery; silicon-air battery; germanium-air battery; electric vehicles

1. Introduction

The extensive use of fossil fuels has been the primary driver of long-term climate change for the past 150 years, responsible for a staggering 75% of global greenhouse gas emissions and 90% of global carbon dioxide (CO_2) emissions. As fossil fuel reserves continue to dwindle and geopolitical tensions heighten over their supply, the imperative to develop alternative energy sources is becoming increasingly urgent in modern civilizations. Solar, wind, wave, biomass, and other renewable energy sources hold promise as alternatives to reduce dependence on fossil fuels. However, their widespread adoption faces significant challenges related to efficiency and regulatory hurdles. There is still a significant distance to cover in addressing the ecological damage brought on by the exploitation of fossil fuels by transitioning to renewable energy sources [1–12].

Motivated to decrease carbon emissions and dependence on fossil energy, a global transition towards electric vehicles (EVs) signifies a critical juncture in the automotive sector. Evidently, the ambitious goal of significantly reducing CO_2 emissions necessitates a multifaceted approach that includes the widespread deployment of EVs alongside a strategic focus on integrating renewable energy sources into the grid [13–15]. Although there have been notable improvements in EV technology, the restricted range of existing battery systems hinders widespread adoption. This limitation translates into a critical concern for the EV industry in the form of range anxiety. Consumers remain apprehensive about the driving distance achievable on a single charge, ultimately dampening consumer confidence and hindering the overall growth of the EV market [16–21]. The rapid progress of energy storage technologies has been crucial in promoting the widespread use of battery-powered solutions in the automotive industry. Batteries are the core component of EVs and are essential in determining the driving range and overall performance of these environmentally friendly battery electric vehicles (BEVs). Advancements in battery technology are crucial for overcoming range limitations while alleviating concerns like range anxiety. This will demonstrably improve consumer acceptance of EVs, thereby expediting their universal adoption as a sustainable transportation solution [22–24].

Lithium-ion (Li-ion) systems currently dominate the battery industry, powering not only EVs but also a myriad of consumer electronics. They have experienced a significant advancement in energy density, rising from around 90 Wh kg^{-1} to a contemporary range of ~100 to ~300 Wh kg^{-1} between the 1990s and now. Though this advancement has permitted larger vehicles to cover typical driving patterns, this is insufficient for BEVs striving to outperform internal combustion engine vehicles (ICEVs), which typically achieve a mileage equivalent of about 7 to 11 kilometers per liter of gasoline, varying based on the model [25–31]. Contrariwise, lithium-powered systems are unable to meet the escalating demand for energy storage, particularly within the automotive sector. In the wake of recent regulations and the fact that lithium comprises only 0.0017% of the Earth's crust the demand for electric vehicle batteries is anticipated to increase substantially over the next decade [32,33]. To accommodate the driving patterns mentioned earlier, BEV manufacturers are compelled to include large and bulky Li-ion battery packs, which are both heavy and expensive, thereby limiting the ability of EVs to compete with standard ICEVs. Furthermore, safety issues related to Li-ion systems are of primary concern due to the significant potential risks to users resulting from battery malfunctions. This failure can occur due to an electrical shortage caused by the formation of dendritic lithium during rapid charging, or thermal runaway resulting from overcharging/discharging. These failures could potentially cause fires due to the highly combustible electrolyte used in Li-ion cells, consequently leading to a multitude of accidents involving these batteries, as analyzed by Yu et al. [34] in Figure 1. A further significant concern is the negative environmental impact of Li-ion systems. The extraction, production, and indiscriminate disposal of these batteries contribute to environmental issues, primarily because of the existence of hazardous substances and specifically, heavy metals classified as such, including nickel and cobalt. The recycling of the Li-ion system continues to pose challenges, primarily stemming from economic feasibility issues and the complexity of the disassembly process. Consequently, the combination of lithium scarcity, limited energy capacities, extended charging durations, adverse environmental effects, and safety apprehensions surrounding their components collectively hinder the widespread adoption and commercialization of Li-ion systems, particularly within the realm of the EV industry [35–43].

To address the challenges posed by Li-ion systems and meet the growing demands in the energy sector, researchers are focusing on the advancements of metal–air/oxygen batteries. Metal–air batteries (MABs), consisting of electrolytes and bifunctional air electrodes, typically function by extracting atmospheric oxygen and simultaneously converting metal anodes into their oxide forms. The batteries utilize a range of metal electrodes, including alkali metals like potassium (K), lithium (Li), and sodium (Na), metalloids like germanium (Ge) and silicon (Si), as well as transition metals, and post-transition metals such as aluminum (Al), calcium (Ca), iron (Fe), magnesium (Mg), tin (Sn), zinc (Zn), and vanadium (V). MABs have a remarkable theoretical energy density that falls within the range of 1353 to 11,429 Wh kg^{-1} (per unit weight) based on calculations for common MABs including those utilizing Li, Zn, Al, and Mg [44,45].

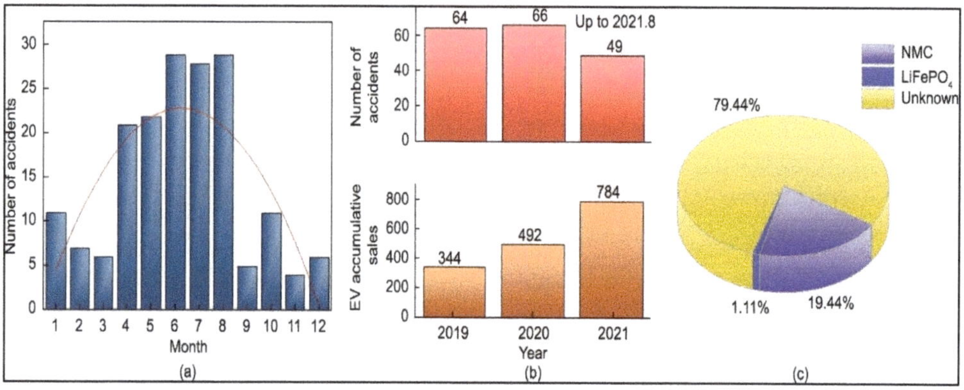

Figure 1. Compilation of EV fire occurrences from January 2019 to August 2021: (**a**) Monthly count of EV fire incidents. (**b**) Comparison of annual EV fire incidents and the total number of EVs sold. (**c**) Distribution of accidents attributed to various types of Li-ion battery [34].

The development of MABs has a rich history rooted in the quest for high-energy-density storage systems. Originating in the mid-20th century, early research focused on zinc–air batteries due to their simplicity and practical energy potential. However, advancements gained momentum with the rising demand for sustainable energy solutions, particularly for electric vehicles and renewable energy integration. Innovations in electrode design, catalysts, and electrolytes have progressively addressed challenges like poor rechargeability and limited cycle life [46,47]. The essence of MABs lies in their innovative design, where electrical energy is generated through the reaction of atmospheric oxygen with a metallic electrode. The process underscores the pivotal role of air in facilitating the electrochemical reactions within these batteries, resulting in efficient energy conversion and storage. This dynamic offers a promising avenue for advancing energy storage technologies toward higher performance and sustainability through the unique interplay between air and metal electrodes. Thus, MABs are anticipated to have a significant impact in the markets of the automobile, consumer electronics, stationary storage, defense, marine, and aviation markets [48,49].

MABs have become prominent because of their efficient operation in outdoor settings, making them ideal for meeting the high energy requirements of EVs. Researchers anticipate that specific MABs will directly replace Li-ion batteries, while others will act as range extenders. This feature reduces range anxiety, enhances safety, and improves cost efficiency by decreasing the size of Li-ion batteries when they are not replaced directly. Figures 2 and 3 give a visual summary of essential parameters influencing the selection and utilization of various metals in MAB technologies. Figure 2 presents a comparative analysis of the energy parameters of different metals used in MABs, highlighting the theoretical volumetric energy density in Wh L^{-1}, specific energy in Wh kg^{-1}, and nominal cell voltage in V. Concurrently, Figure 3 illustrates the cost and abundance of these elements, offering insights into their economic feasibility and natural availability.

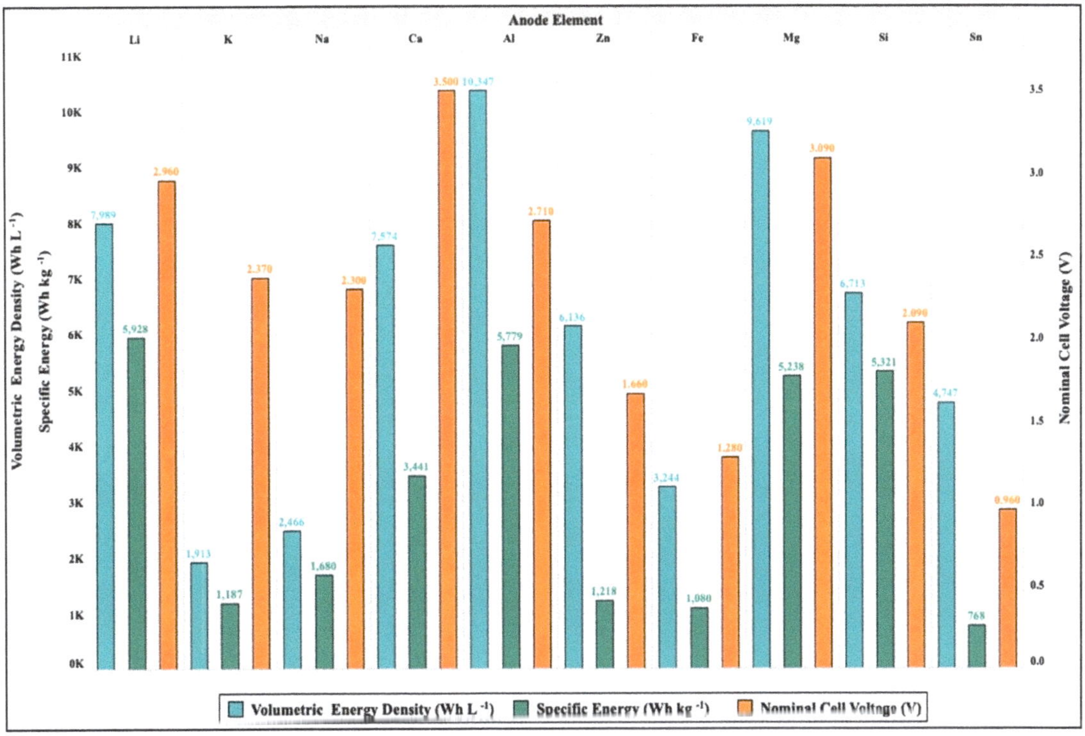

Figure 2. Comparison of the performance of several metal-air batteries, highlighting the theoretical specific energy, volumetric energy density, and nominal cell voltage.

Together, these figures present a holistic perspective on the essential parameters that must be considered when evaluating the practicality and efficiency of using these different metals. From the graph, lithium appears to be a very effective anode because of its superior theoretical specific energy of 5928 Wh kg^{-1} and an open circuit voltage (OCV) of 2.96 V. Nevertheless, the very low abundance (<0.0017 %) and high cost of lithium metal (46 USD kg^{-1}), combined with its intrinsic instability when it comes into contact with air and aqueous electrolytes, present significant challenges [50]. Subsequently, Al metal emerges as a promising candidate due to its exceptional theoretical specific energy of 5779 in Wh kg^{-1} and its energy density of 10,347 Wh L^{-1}. Al metal is also known for its high safety, cost-effectiveness, and abundance. However, its high corrosion rate in aqueous electrolytes, coupled with parasitic reactions, diminishes its overall performance [51]. The appeal of employing Mg metal in MABs lies in their ability to achieve uniform metal deposition and significant theoretical energy potentials (energy density of 9619 Wh L^{-1}, specific energy of 5238 Wh kg^{-1}, and OCV of 3.090 V). However, the ubiquitous usage of magnesium electrodes in aqueous electrolytes is impeded by significant corrosion challenges and high self-charge characteristics. The high corrosion rates compromise the electrode's longevity and performance, posing a substantial barrier to its practical application [52]. Additionally, the cost of magnesium, approximately 11.02 USD kg^{-1}, coupled with its relatively limited abundance of only 2.10%, further limits its viability as a sustainable material in large-scale energy storage solutions.

It should be noted that, as of this point, in the context of aqueous MABs, only MABs with Zn anodes can be electrically recharged, while Mg- and Al-employing air batteries require mechanical replacement of these metal anodes due to their inability to be directly converted from ions to metal [53]. Given this unique advantage, employing Zn as the metal anode presents promising opportunities for the development and enhancement of MAB technology. Zn is one of the most widely researched anodes for MABs with a high theoretical energy density of 6136 Wh L^{-1} and a specific energy of 1218 Wh kg^{-1}. However, limitations such as the formation of Zn dendrites and low cycle life need to be addressed to enhance its commercial viability [54,55]. While Na and K metals exhibit some chemical similarities to Li, their application as a viable battery alternative faces substantial hurdles. These hurdles include the lack of comprehensive understanding of the underlying electrochemistry, the development of stable electrolytes, and the design of high-performance electrode materials for these batteries [56,57]. Other metal anodes such as Fe, Si, Ca, and Sn, while less extensively researched, show promising potential due to their high crustal abundance and lower costs.

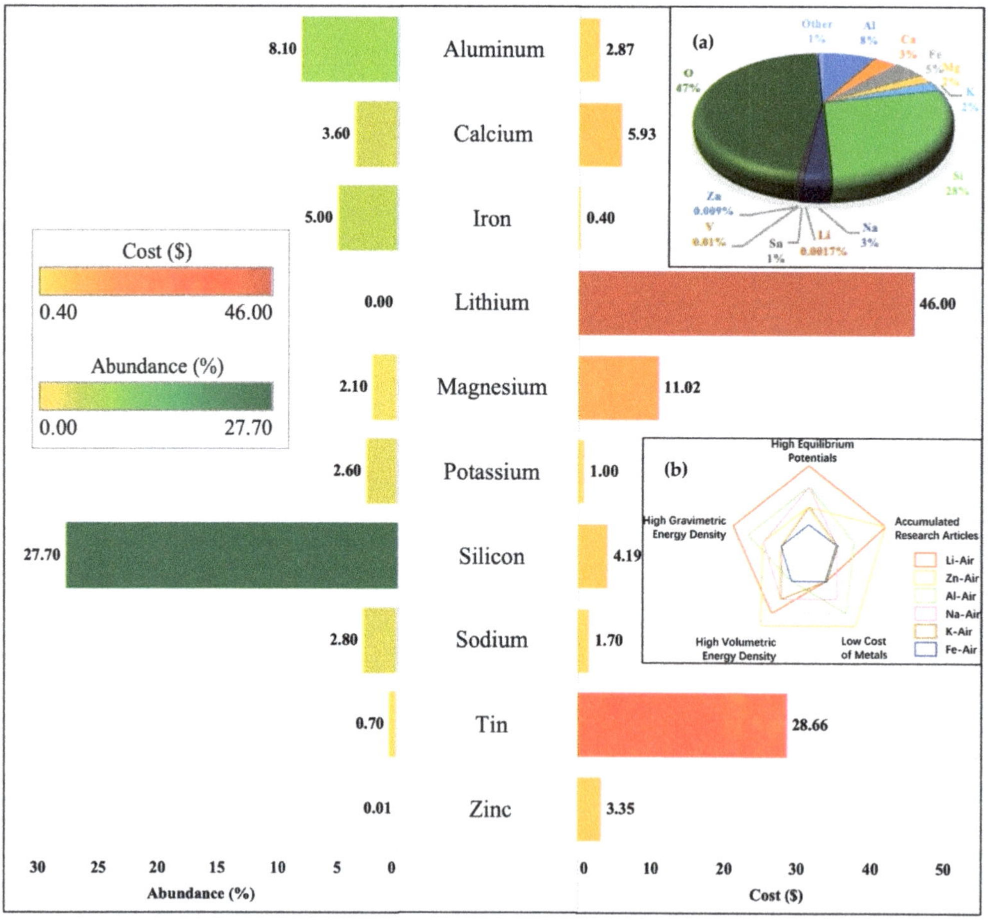

Figure 3. Abundance vs. cost of metal–air battery elements. Refer to the inset (a) for the earth's crustal distribution (based on data from [49,58,59]) and inset (b) for a radar chart illustrating the benefits and drawbacks of various types of metal–air batteries based on research articles collected from the Web of Science between 2000 and 2023 (adapted from [55]).

Currently, none of these battery chemistries have achieved the technological maturity required for widespread industrial adoption. As a result, their potential to either replace or supplement Li-ion batteries as the primary power source or range extenders for future EVs remains highly uncertain. Although some studies explore the potential of specific MABs like Li–air or Zn–air batteries as range extenders, they do not offer a comprehensive comparison of all the MABs currently being studied in the field. Thus, this comparative review aims to bridge this knowledge gap by offering a comprehensive analysis of different MABs as potential EV range extenders. The review will delve into their structural compositions, operational principles, performance metrics, and existing challenges. By critically evaluating the strengths and limitations of each battery type, the goal is to illuminate their suitability for integration into the rapidly evolving EV market. Section 2 introduces MABs as energy storage systems and delves into specific types with different anodes, describing in detail their types, working principles, advantages, and challenges. Section 3 provides a comparison of the practical properties of these MABs, focusing on factors such as driving range, battery pack volume, and total vehicle cost. Section 4 concludes with summarizing remarks, encapsulating the observations made throughout the review.

2. Metal–Air Batteries

The operational mechanisms of both aqueous and non-aqueous MABs and the various configurations are depicted in Figure 4 [60]. In essence, these batteries consist of three key parts: an anode (metal), an electrolyte solution positioned between the electrodes, and a permeable air-cathode. The categorization of MABs by Wang et al. [60] includes three distinct types: static battery setups with multiple cells, cell configurations with electrolyte flow, and innovative flexible battery designs.

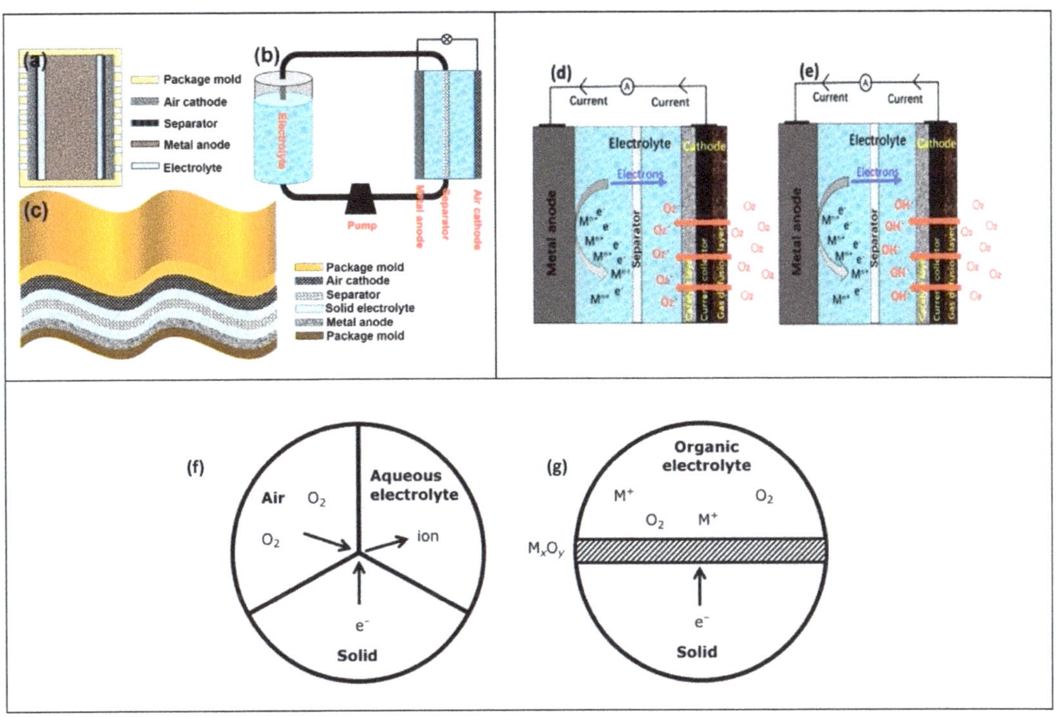

Figure 4. Illustration of different configurations of MABs, including a conventional static battery setup (**a**), a cell design with electrolyte flow (**b**), and a novel flexible cell design (**c**); (**d**,**e**) representations of non-aqueous and aqueous electrolyte MABs, respectively [60]; (**f**,**g**) a phase boundary diagram for both aqueous and non-aqueous electrolytes (modified from [48]).

In conventional static electrochemical setups, such as the one depicted in Figure 4a, four critical components are employed: a cathode, a separator, an electrolyte, and an anode. The anode reaction exhibits rapid kinetics, while the oxygen-reduction reaction (ORR) at the positive electrode is kinetically slow. This sluggishness is attributed to the reaction occurring at a three-phase boundary (solid catalyst, liquid electrolyte, and gaseous oxygen) that is crucial for the ORR. Furthermore, the oxygen-evolution reaction (OER) proceeds concurrently on the cathode at a two-phase boundary, consisting solely of the catalyst and electrolyte. Achieving optimal performance for both the OER and ORR necessitates the use of highly efficient dual-functional catalysts. However, a significant obstacle for conventional static designs is the accumulation of solid byproducts on the surfaces of both electrodes during the cycling of the cell. These deposits can block electrode pores, impeding air diffusion and ultimately reducing battery performance. Flow batteries (Figure 4b) include electrodes, an electrolyte, a separator, and an additional reservoir for storing the electrolyte, often accompanied by a pump to facilitate electrolyte flow. This flowing-electrolyte configuration mitigates some issues related to both positive and negative electrodes.

For example, in Zn–Ni battery types, a large volume of flowing electrolyte helps to prevent dendrite growth and uneven zinc geometry variations, thus mitigating passivation through enhanced current uniformity and reduced concentration disparities. Despite these advantages, the complex configuration of flowing electrolytes introduces drawbacks such as decreased energy efficiency and volumetric density, along with the necessity for auxiliary flow channels and pumps to enhance electrolyte circulation during cell operation. Han et al. [61] conducted an extensive review comparing static and flow batteries, providing a detailed analysis of their respective advantages and challenges. Flexible batteries (Figure 4c) have gained significant attention recently due to the rising demand for portable electronics. These miniaturized batteries prioritize high energy density within a lightweight form factor. They achieve this through a combination of electrodes, a separator, and a high-conductive (often solid-state) electrolyte. The small thickness of the metal anode further reduces the overall mass of the cell. Different nanomaterials like graphene, carbon nanotubes, and fibers are being investigated for optimal cathode performance. Flexible battery technology is transitioning beyond basic chemistries, with the emerging lithium-based and metal–air options. Notably, Zn and Al- based MABs are currently favored for the application of flexible electronics owing to their massive specific energy, safety profile, and cost-effectiveness [62–67].

Regarding the categorization of MABs depending on the electrolyte type, there are various kinds of electrolytes employed: aqueous (water-based), non-aqueous (organic solvent-based), solid-state (no liquid component), and hybrid (combinations of the above). Metal electrodes that are highly water-sensitive, such as those in lithium–, potassium–, and sodium–air batteries, cannot utilize aqueous electrolytes, as these alkali metals react vigorously with water, often leading to hazardous situations. In contrast, other types of MABs, including those with aluminum, zinc, iron, or magnesium electrodes, which are compatible with water, can use both aqueous and non-aqueous electrolytes, among others [62,68–70]. The distinctions between reactions in aqueous and non-aqueous MABs pose unique challenges for technological implementation, as elucidated in the review by Hardwick et al. [48]. In non-aqueous MABs, the ORR leads to the generation of metal oxides that precipitate on the surface of the permeable air-cathode itself. Conversely, in alkaline-based aqueous MABs, the anodic metallic ions do not arrive at the positive electrode; instead, the precipitation of these metal ions occurs by the reaction with hydroxide ions (OH^-), or they form a complex with a negative charge, such as the zincate anion $(Zn\left(OH\right)_4^{2-})$ near the anode, preventing its migration to the cathode. Aqueous MABs encounter the additional hurdle of low oxygen solubility in water (0.2 mM) under atmospheric pressure, necessitating the utilization of gaseous oxygen. Atmospheric oxygen permeates into the air-cathode propelled by the pressure gradient between the cell's interior and exterior. Catalysts enable the formation of hydroxyl (OH^-) ions in the basic media by oxygen reduction, with electrons produced at the negative electrode, potentially resulting in the production of metal hydroxides in solid form. This three-phase reaction involves a solid catalyst, a liquid electrolyte, and gaseous oxygen, as depicted in Figure 4f. Meanwhile, Figure 4g portrays the boundary diagram of non-aqueous MABs, facilitating a comparison between them [71–77]. Within aqueous MABs, the anode undergoes an oxidation reaction during discharge, releasing electrons and simultaneously forming metal hydroxide as the metal cations react with OH^- ions,

$$M + nOH^- \rightarrow M(OH)_n + ne^- \quad (1)$$

Here, M represents the metal, and n stands for the valence metal ions. The liberated electrons (e^-) travel via the closed circuit and arrive at the positive electrode to engage in the ORR. There, the oxygen molecules present in the ambient air, coupled with the e- and water, undergo reduction to produce OH^- ions.

$$\frac{n}{4}O_2 + ne^- + \frac{n}{2}H_2O \rightarrow nOH^- \quad (2)$$

The produced OH^- ions move from the cathode to the metal anode, closing the loop. By combining Equations (1) and (2), the overall reaction can be derived [44].

$$M + \frac{n}{4}O_2 + \frac{n}{2}H_2O \rightarrow M(OH)_n \quad (3)$$

Similarly, in non-aqueous MABs, the metal undergoes oxidation to form metal cations and electrons at the anode,

$$M \leftrightarrow M^+ + e^- \quad (4)$$

whereas, at the cathode, the metal cations react with O_2 and e^- to give metal oxides [45],

$$xM^+ + O_2 + xe^- \leftrightarrow M_xO_2 \; (x = 1 \; or \; 2) \quad (5)$$

To quantify the performance of MABs, parameters like cell voltage, specific capacity, and energy density are used. These are crucial parameters for evaluating MAB performance. To calculate these metrics and gain a deeper understanding of MAB behavior, various formulas based on the underlying electrochemical reactions are utilized. Table 1 summarizes the formulae employed for calculating these key performance indicators of MABs.

Table 1. Key performance indicators and corresponding formulae for metal–air batteries.

Parameter	Formula	Units	Explanation	Ref.
Specific capacity	$C = \frac{(I \times t)}{m}$	mAh g^{-1}	I is the discharge current, t is the discharge time, and m is the mass of the electrode	[78,79]
Capacity density	$Q = \frac{It}{\Delta m}$	mA cm^{-2}	I is the current in mA, t is the discharge time in hours, and Δm is the weight loss in g	[38]
Power density	$P = \frac{EI}{S}$	mW cm^{-2}	E is the average discharge voltage (V), I is the current in mA, and S is the active surface area in cm^2	[38]
Specific energy	$SE = \frac{V \times C}{M_B}$	Wh kg^{-1}	SE is the specific energy, V is the nominal battery voltage in V, C is the rated battery capacity in Ah, and M_B is the battery mass	[80]
Energy density	$W = \frac{EIt}{\Delta m}$	mWh g^{-1}	E is the average discharge voltage (V), I is the current in mA, t is the discharge time in hours, and Δm is the weight loss in g	[38]
Instantaneous specific energy (including oxygen)	$\frac{F \cdot n \cdot OCV}{3.6}\left[\frac{\ln[(M_{M+O}/M_{Metal})]}{M_{M+O} - M_{Metal}}\right]$	Wh kg^{-1}	F is Faraday's constant, n is the number of electrons transferred per metal ion, OCV is the nominal voltage, M_{Metal} is the molar mass of metal anode, and M_{M+O} is the combined molar mass of metal anode and stoichiometric amount of oxygen that enters the battery	[50]
Energy density (including oxygen)	$Volumetric\ Energy\ Density = (SE) \cdot (OAD)$	Wh L^{-1}	SE is the specific energy in $Whkg^{-1}$ and OAD is the oxidizing anode density in kgL^{-1}	[50]
Nernst equation	$E = E^{\circ} - \frac{RT}{nF} lnQ$	V	R is the gas constant, T is the temperature in K, F is Faraday's constant, n is the number of moles of electrons transferred, and Q is the reaction quotient	[81]
Gibbs free energy	$\Delta G^{\circ} = -nFE^{\circ} = \Delta H^{\circ} - T\Delta S^{\circ}$	J	T is the temperature in K, F is Faraday's constant, n is the number of moles of electrons transferred, S is the entropy, and H is the enthalpy	[81]
Current efficiency	$\eta_a = \frac{Q}{Q_0}$	%	Q is the actual capacity density in $mAcm^{-2}$ and Q_0 is the theoretical capacity density in $mAhg^{-1}$	[38]
Energy efficiency	$\eta_{fuel} = \frac{W}{W_0} \times 100$	%	W is the actual energy density in $mWHg^{-1}$ and W_0 is the theoretical energy density in $mWhg^{-1}$	[38]
Charging energy efficiency	$\eta_{charge} = \frac{\int_{SOC(0)}^{SOC(t)} U_{OCV} C_n d\ SOC}{\int_{SOC(0)}^{SOC(t)} U_{charge} C_n d\ SOC}$	%	Numerator = net energy; denominator = charge energy; SOC = state of charge; U = battery voltage; C_n = battery standard capacity	[81]

Table 1. *Cont.*

Parameter	Formula	Units	Explanation	Ref.
Discharging Energy efficiency	$\eta_{disch} = \frac{\int_{SOC(0)}^{SOC(t)} U_{disch} C_n d\ SOC}{\int_{SOC(0)}^{SOC(t)} U_{OCV} C_n d\ SOC}$	%	Numerator = discharge energy; denominator = net energy; SOC = state of charge; U = battery voltage; C_n = battery standard capacity	[81]
Charge-discharge Energy Efficiency	$\eta_{battery} = \frac{\int_{SOC(0)}^{SOC(t)} U_{disch} C_n d\ SOC}{\int_{SOC(0)}^{SOC(t)} U_{charge} C_n d\ SOC}$	%	Numerator = discharge energy; denominator = charge energy; SOC = state of charge; U = battery voltage; C_n = battery standard capacity	[81]
Vehicle Range	$R_V = \frac{E_B}{ECE_V\left(M_V + \frac{E_B\ k_{m,B}}{SE_{BC}}\right)}$	km	E_B is the battery pack energy in Wh, ECE_V is the energy consumption efficiency of the vehicle in Wh km^{-1} kg^{-1}, M_V is the vehicle mass without a battery pack in kg, $k_{m,B}$ is the battery mass overhead, and SE_{BC} is the specific energy of the battery cell in $Whkg^{-1}$	[20]
Total Vehicle Cost	$C_{V,T} = C_V + C_B E_B$	USD	C_V is the vehicle cost without battery pack in USD and C_B is the battery pack cost in USD kWh^{-1}	[19]
Battery Pack Volume	$Vol_B = \frac{E_B \times k_{vol,B}}{ED_{BC}}$	L	$k_{vol,B}$ is the battery volume overhead and ED_{BC} is the energy density of the battery cell in WhL^{-1}	[19]

MABs have a clear advantage in terms of energy storage capacity. When compared to a standard lithium/manganese dioxide (Li/MnO_2) battery with an assumed voltage of 3 V, MABs boast a much higher theoretical specific energy (Wh g^{-1}) and volumetric energy density (Wh cm^{-3}). This is illustrated in Figure 5a,b, where the values for MABs are significantly greater. It is important to note that for rechargeable MABs (either electrically or mechanically), a more precise comparison would involve using data from discharged states. This is because the discharged material tends to be heavier and larger than its charged counterparts. Figure 5c,d showcase the theoretical energy storage potential including the discharged materials. Here, this potential is reduced as the by-products dissolve in the electrolyte. For comparison, the theoretical values for a common Li-ion battery ($LiCoO_2$) are 2.9 Wh cm^{-3} and 0.78 Wh g^{-1}, which in turn highlights the significant theoretical advantage of MABs for energy storage capacity. Accompanying this discussion is a comprehensive table that outlines the key theoretical parameters for each anode material, serving as a reference point for understanding the performance expectations of these technologies. In Table 2, specific energy values were determined by numerically integrating the oxygen uptake between the completely charged and completely discharged states of the battery. Volumetric energy densities were evaluated based on the anode density in its completely discharged state. The equations used to calculate the specific energy values are provided in Table 1. Refer to Supplementary Table S1 for further calculations and references. However, a significant drawback lies in the discharge behavior of MABs. Unlike Li-ion batteries, where electrode volume changes are minimal, the volume of MABs can increase by several tens of percent during discharge. This can pose challenges for their reusability. Interestingly, some MAB configurations can be manufactured pre-discharged (for instance, an anode built of metal oxide) and then activated through charging. The voltage output of MABs is typically stable and flat, reflecting a straightforward two-phase reaction between the metal anode and its products. The non-conductive nature of these discharged materials, compared to the highly conductive metals in their initial state, leads to voltage drops during discharge cycles [82–85]. Also, the presence of CO_2 in the ambient air poses several challenges. CO_2 reacts with the metal ions in the battery's cathode to form metal carbonates, which can block the active sites on the cathode, reducing the battery's efficiency and capacity. Additionally, the interaction between CO_2 and the electrolyte can lead to the formation of unwanted by-products, degrading the electrolyte and affecting the overall performance and lifespan of the battery. The presence of CO_2 also complicates the recharge process, as MABs using a mixture of O_2 and CO_2 as the active material in the cathode require specific measures to prevent electrolyte decomposition during recharge. Furthermore, the environmental impact of CO_2 in MABs is a concern, as the reactivity of CO_2 with the electrolyte and metal anode influences the discharge mechanism and stability, affecting the battery's overall environmental footprint [86,87].

To learn more about the potential and limitations of MABs, it is essential to explore the specific characteristics of each type of battery chemistry. While the general principles of MABs provide a foundational overview, the unique properties and performance metrics of individual MAB types are critical in determining their suitability for various applications, particularly in EVs. This detailed examination will not only highlight the distinct advantages and challenges associated with each type but will also provide valuable insights into their theoretical performance parameters. To facilitate this analysis, the following sections will delve into the specific MAB chemistries, beginning with Li–air batteries, to examine their design, functionality, and practical implications.

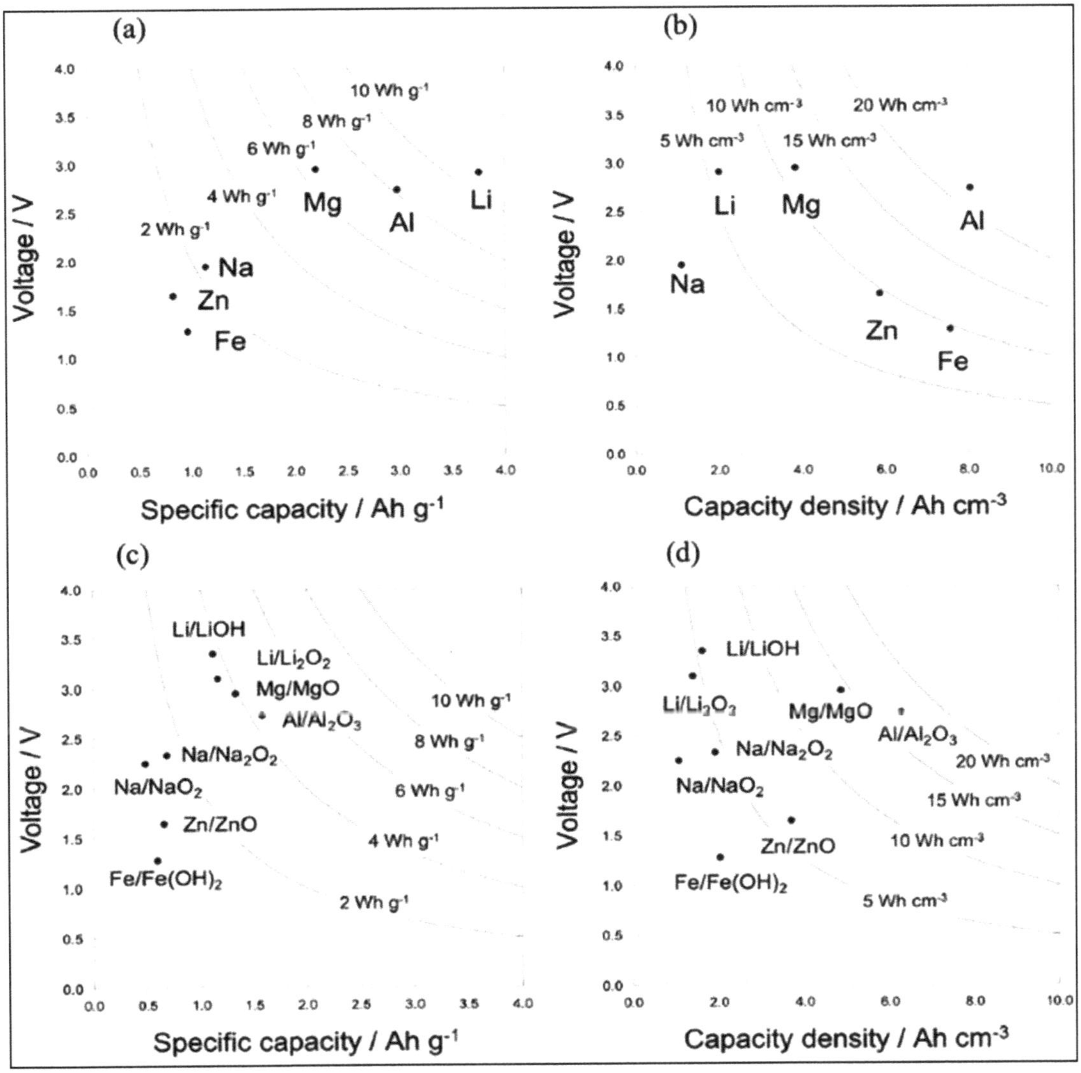

Figure 5. The metal's mass and volume define the theoretical (**a**) specific energy and (**b**) volumetric energy density of MABs, whereas the plots (**c**,**d**) include the discharged products [82].

Table 2. Theoretical parameter comparison across various metal–air batteries.

Parameter	Li–Air	K–Air	Na–Air	Mg–Air	Al–Air	Zn–Air	Ca–Air	Fe–Air	Sn–Air	Si–Air	Ge–Air
Specific Energy (Wh kg^{-1})	5928	187	680	5238	5779	1218	3441	1080	768	5321	2214
Specific Power (W kg^{-1})	-	-	12–300	20–550	15–130	10–200	-	-	-	-	-
Specific Capacity (Ah kg^{-1})	2003	501	730	1695	2132	733	983.1	844	800	2548	1223
Volumetric Capacity density (Ah L^{-1})	2699	807	1072	3112	3817	3694	2164	2537	4944	3212	2633
Volumetric Energy Density (Wh L^{-1})	7989	1913	2466	9619	10347	6136	7574	3244	4747	6713	4766
Theoretical Open Circuit Voltage (V $cell^{-1}$)	2.96	2.37	2.3	3.09	2.71	1.66	3.5	1.28	0.96	2.09	1.81
Energy Efficiency (%)	68–94	-	-	-	70	-	-	96	70–90	-	-
Operational Temperature Range (°C)	10–40	-	105–110	420–620	30–50	−20–70	-	-	-	-	-
Cost (USDkg^{-1})	46	13.02	1.7	11.02	2.866	3.351	5.93	0.4	28.66	4.19	1400

Refer to Supplementary Table S1 for the references and detailed calculations.

2.1. Lithium–Air Batteries

Lithium–air (Li–air) batteries were first conceptualized in 1976, though it was not until 1996 that Abraham et al. successfully demonstrated a practical Li–air battery using a non-aqueous electrolyte [88]. This technology has garnered significant attention due to its unique properties. Li–air batteries utilize lithium, the lightest element, and boast a remarkable theoretical specific energy. This value can reach 13,000 Wh kg^{-1} calculated solely on the mass of lithium metal. However, considering the weight of the entire oxidized Li anode, the specific energy is still high at 5928 Wh kg^{-1}. Additionally, Li–air batteries possess an exceptional specific capacity of 2003 Ah kg^{-1} and operate at a voltage of 2.91 V [50,89–91]. Several Li–air battery architectures have been researched, including liquid electrolytes (aqueous and non-aqueous), hybrid systems (combining liquid and solid-state approaches), and the solid-state electrolyte system.

Driven by their high theoretical energy capacity and comparatively simpler construction, Li–air batteries with non-aqueous electrolytes have accumulated immense research effort [92–94]. These batteries (Figure 6) rely on Li metal for the anode and an air electrode that reacts with oxygen from the atmosphere. The components are isolated by a thin membrane, while an organic electrolyte enables the mobility of Li ions during operation. The reaction produces solid, insoluble Li_2O_2 as the discharge product and the corresponding reactions are shown in Table 3 [95,96]. Though these batteries possess substantial theoretical energy densities (3505 Wh kg^{-1} and 3436 Wh L^{-1}), they exhibit a high decomposition overpotential during the charging cycle. Additionally, during the discharge cycle, pore clogs of lithium oxide (Li_2O_2) within the cathode structure reduce the overall performance, and there is a necessity for pure oxygen as the fuel limiting the design configurations. Moreover, unfavorable side reactions are observed, stemming from reactions with water and carbon dioxide at both electrodes.

The aqueous Li–air battery (Figure 6) offers an alternative approach to the non-aqueous design. The electrolyte is placed in the cathode compartment to address safety concerns and prevent lithium from reacting with water. This battery also employs a special Li ion conducting separator between the anode and the electrolyte [97–99]. The reactions at the cathode differentiate between aqueous and non-aqueous Li–air batteries. This complexity arises from the dependence of these cathodic reactions on the electrolyte pH. At low pH (acidic environment), oxygen consumes H^+ ions to form water, while an alkaline environment promotes a reaction with H_2O to produce OH^- ions. This contrasting behavior significantly impacts the standard reaction potential (refer to Table 3 for the relevant reactions).

While the theoretical energy density of aqueous Li–air batteries (1910 Wh kg^{-1}) falls short of their non-aqueous counterparts (3505 Wh kg^{-1}), it still surpasses the energy potential of gasoline. However, the key strength of aqueous Li–air batteries lies in their inherent compatibility with atmospheric air. Unlike non-aqueous Li–air systems, where severe water contamination from the environment can occur, aqueous designs are less susceptible to this issue. Furthermore, the discharge product, LiOH, exhibits solubility within the aqueous electrolyte. This characteristic offers the potential for significantly reduced overpotential during the charging cycle and improved areal discharge capacity [100]. However, the lithium anode struggles with long-term stability, forcing efforts to maintain a neutral pH and minimize corrosion of the separator. Additionally, the battery's discharge capacity is reduced because of the limited conductivity of the separator at room temperature. While high temperatures can improve conductivity, they introduce other problems like electrolyte evaporation and higher corrosion rates. Furthermore, LiOH precipitation on the cathode during discharge hinders performance, requiring novel cathode designs with improved air channels [101–109]. By examining and comparing the pros and cons of both aqueous and non-aqueous Li–air battery technologies, it becomes evident that despite their promising theoretical and practical aspects, they encounter substantial challenges that necessitate significant advancements before achieving widespread commercialization.

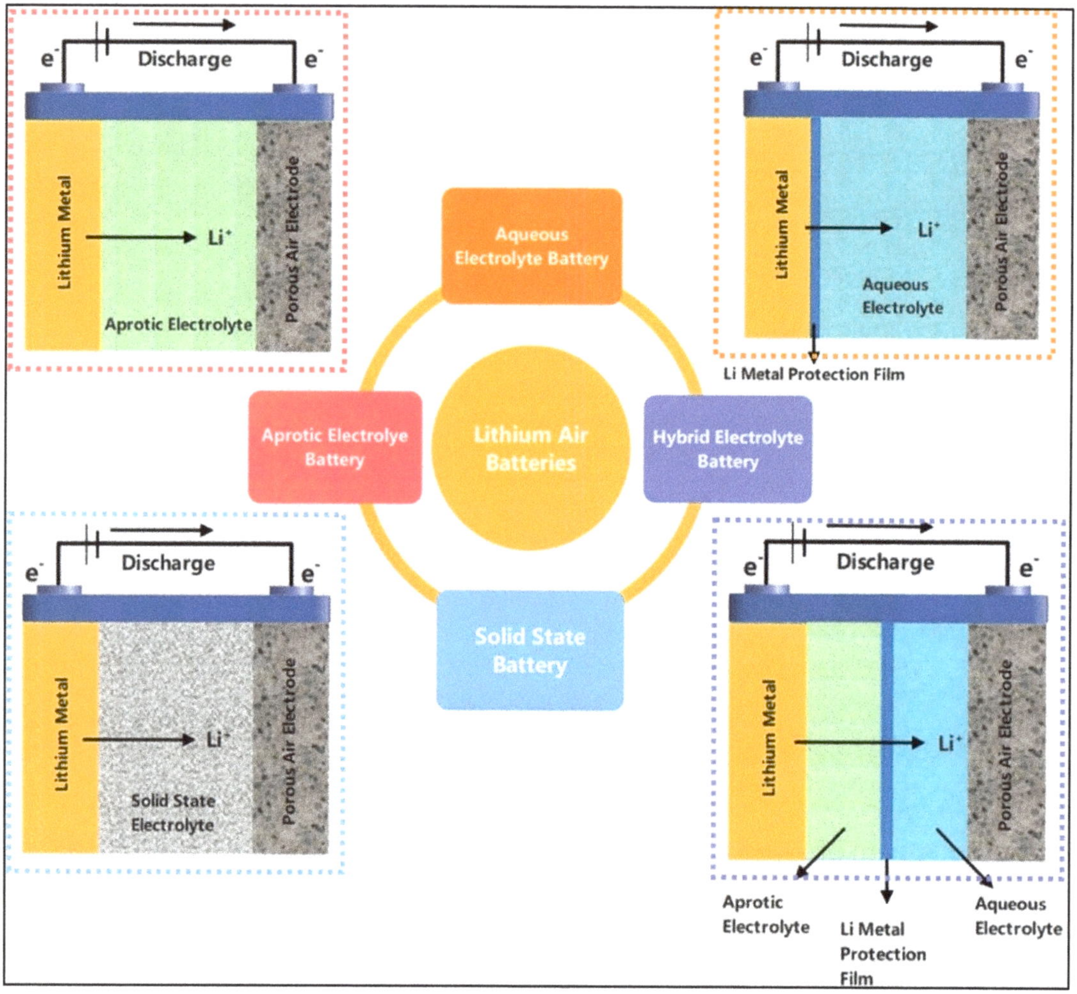

Figure 6. Classification of Li-air battery technologies with schematic representations. Modified from [91].

To address some of the above-mentioned limitations, a hybrid electrolyte system was put forward in 2010 [102,104]. This design combines a Li negative electrode in a non-aqueous electrolyte (organic) with an oxygen-reduction cathode in an aqueous electrolyte, separated by a LISICON (lithium superionic conductor) plate. This innovative strategy offers the potential to overcome some of the significant challenges plaguing conventional non-aqueous Li–air batteries. These challenges include lithium metal corrosion, Li_2O_2 product clogging of cathode pores, and the comparatively lower voltage output. By strategically separating the Li anode and the oxygen-reduction cathode using a LISICON plate, the hybrid system appears to circumvent these limitations [102,104]. However, a critical hurdle lies in the mechanical and chemical stability of the LISICON separator. For instance, at higher pH levels, the corrosion rate of the LISICON plate accelerates, significantly compromising its stability and long-term performance within the battery. The Li dendrites developed during the charging of the battery negatively react with LISICON. Thus, the plate's durability decreases and has a short-term functionality within the battery system. Furthermore, developing low-cost, highly efficient carbon-based composite catalysts for these batteries remains a challenge [110].

Solid-state electrolytes present a paradigm shift in Li–air battery development, offering a confluence of higher energy density and safety. These solid-state Li–air batteries typically comprise three key components: an air cathode, an electrolyte, and a Li anode (as illustrated in Figure 6). Their exceptional mechanical strength and electrochemical stability enable the strategic integration of high-energy-density Li metal anodes and higher-potential cathode materials. This synergy unlocks the potential for significant advancements in energy density without compromising safety. Additionally, solid-state electrolytes simplify the manufacturing process due to their inherent ease of preparation, shaping, and large-scale production. This translates to streamlined battery management systems and potentially reduced production costs. Finally, the thermal stability and safety of solid electrolytes eliminate the need for intricate encapsulated cooling systems, further contributing to cost reductions and overall system optimization [111–113]. While solid-state electrolytes present a favorable avenue for futuristic Li–air batteries, their widespread adoption hinges on overcoming several critical limitations. A significant challenge lies in the inherently low electrolytic conductivity of many solid-state electrolytes, particularly at lower temperatures. This limitation directly impacts the battery's ability to charge and discharge efficiently. Furthermore, high interfacial resistance at the solid-electrolyte interface hinders the flow of ions, further compromising overall battery performance. Additionally, compatibility issues can arise between certain solid-state electrolytes (SSE) and Li metal cathodes, leading to degradation and a decline in performance over time. Finally, the weak physical stability observed in some electrodes can induce significant interfacial stress changes, ultimately contributing to mechanical failure and a shortened battery lifespan [114–116]. In recognition of the crucial role that solid-state electrolytes play in enabling efficient ionic transport between the anode and cathode, researchers [117–125] are diligently pursuing solutions to address the limitations. This multi-pronged approach encompasses advancements in SSE materials, cathode materials, and anode materials. Notably, optimizing the properties of SSEs remains a key factor in unlocking the utmost capability of solid-state Li–air batteries [126].

Table 3. Various metal anodes with electrolytes used in MABs: cell descriptions, electrode reactions, redox reactions, and electrochemical potentials.

Metal	Electrolyte	Cell Description	Reaction	Potential	Cell Reaction (Overall)	Refs.
Aluminum (Al)	Aqueous—Neutral	Anode: Aluminum Cathode: Air-breathing	Anode: $Al + 3H_2O \rightarrow Al(OH)_3 + 3H^+ + 3e^-$ Cathode: $O_2 + 4e^- + 2H_2O \rightarrow 4OH^-$	Anode: −1.47 V Cathode: 0.81 V	$Al + \frac{3}{4} O_2 + \frac{3}{2} H_2O \rightarrow Al(OH)_3$ (2.28 V)	[2,127,128]
	Aqueous—Alkaline	Anode: Aluminum Cathode: Air-breathing	Anode: $Al + 4OH^- \rightarrow Al(OH)_4^- + 3e^-$ Cathode: $O_2 + 4e^- + 2H_2O \rightarrow 4OH^-$ Side: $Al + 3H_2O + OH^- \rightarrow Al(OH)_4^- + \frac{3}{2}H_2$	Anode: −2.35 V Cathode: 0.39 V	$Al + \frac{3}{4} O_2 + \frac{3}{2} H_2O + OH^- \rightarrow Al(OH)_4^-$ (2.75 V)	[2,51,129]
	Aqueous—Acidic	Anode: Aluminum Cathode: Air-breathing	Anode: $Al \rightarrow Al^{3+} + 3e^-$ Cathode: $O_2 + 4e^- + 4H^+ \rightarrow 2H_2O$ Side: $Al + 3H^+ \rightarrow Al^{3+} + \frac{3}{2}H_2$	Anode: −1.66 V Cathode: 1.23 V	$Al + \frac{3}{4} O_2 + 3H^+ \rightarrow Al^{3+} + \frac{3}{2} H_2O$ (2.89 V)	[2]
	Organic Solvent Based: Solid-State Gel Polymer Electrolyte	Anode: Aluminum Cathode: Air-breathing (carbon and polymer-based) Electrolyte: PVA/LiCl polymer	Anode: $Al + 3OH^- \rightarrow Al(OH)_3 + 3e^-$ Cathode: $O_2 + 4e^- + 2H_2O \rightarrow 4OH^-$	Anode: −2.3 V Cathode: 0.81 V	$Al + 3H_2O \rightarrow Al(OH)_3 + 3H^+$ (3.31 V)	[2,130,131]
	Room Temperature Ionic Liquid (RTIL)-Based	Anode: Aluminum Cathode: Air-breathing Electrolyte: Aprotic room temperature ionic liquids	Anode: $Al + 7AlCl_4^- \rightarrow 4Al_2Cl_7^- + 3e^-$ Cathode: $H_2O_2 + 2H^+ + 2e^- \rightarrow 2H_2O$	Anode: −1.3 V Cathode: 0.695 V	$Al + 7AlCl_4^- + \frac{3}{2}H_2O_2 + 3H^+$ $\rightarrow 4Al_2Cl_7^- + 2H_2O$ (1.995 V)	[129,131–134]
Germanium (Ge)	Aqueous—Alkaline: 1–6 M KOH	Anode: Germanium wafers Cathode: Air-breathing Electrolyte: Aqueous KOH	Anode: $Ge + 4OH^- \rightarrow Ge(OH)_4 + 4e^-$ Cathode: $O_2 + 4e^- + 2H_2O \rightarrow 4OH^-$ Passivation: $Ge(OH)_4 \rightarrow GeO_2 + 2H_2O$	Anode: 1.0 V Cathode: 0.81 V	$Ge + O_2 + 2H_2O \rightarrow Ge(OH)_4$ (1.81 V)	[135]
Calcium (Ca)	Non-Aqueous—(Aprotic) Organic Electrolyte	Anode: Calcium Cathode: Air-breathing Electrolyte: Aprotic liquid	Anode: $Ca \rightarrow Ca^{2+} + 2e^-$ Cathode 1: $Ca^{2+} + 2O_2 + 2e^- \rightarrow Ca(O_2)_2$ Cathode 2: $Ca^{2+} + O_2 + 2e^- \rightarrow CaO_2$ Cathode 3: $Ca^{2+} + \frac{1}{2}O_2 + 2e^- \rightarrow CaO$	Anode: −2.368 V Cathode: varies	$Ca + 2O_2 \rightarrow Ca(O_2)_2$ (3.25 V) $Ca + O_2 \rightarrow CaO_2$ (3.28 V) $Ca + \frac{1}{2}O_2 \rightarrow CaO$ (3.13 V)	[136–138]
	Non-Aqueous— Ionic Liquid	Anode: Calcium Cathode: Carbon nanotube sheet Electrolyte: Ca salt in 1:1 EMIM-BF_4/DMSO solution	Anode: $Ca \rightarrow Ca^{2+} + 2e^-$ Cathode 1: $O_2 + 2Ca^{2+} + 4e^- \rightarrow 2CaO$ Cathode 2: $2O_2 + 2Ca^{2+} + 4e^- \rightarrow 2CaO_2$	Anode: −2.868 V Cathode: 0.715 V Cathode: 0.265 V	$O_2 + 2Ca \rightarrow 2CaO$ (3.583 V) $2O_2 + 2Ca \rightarrow 2CaO_2$ (3.133 V)	[131,139]
Iron (Fe)	Aqueous—Alkaline	Anode: Iron Cathode: Air-breathing Electrolyte: Basic media	Anode: $3Fe + 8OH^- \rightarrow Fe_3O_4 + 4H_2O + 8e^-$ Cathode: $O_2 + 4e^- + 2H_2O \rightarrow 4OH^-$ Side: $2H_2O + 2e^- \rightarrow H_2 + 2OH^-$	Anode: −0.88 V Cathode: 0.401 V	$3Fe + 2O_2 \rightarrow Fe_3O_4$ (1.28 V)	[140,141]
	Solid Oxide	Anode: Ni-Fe alloy Cathode: $Ba_{0.6}La_{0.4}CoO_3$ Electrolyte: LSGM Fuel: Iron powder	Anode: $H_2 + O^{2-} \rightarrow H_2O + 2e^-$ Cathode: $\frac{1}{2}O_2 + 2e^- \rightarrow O^{2-}$ Iron Powder: $3Fe + 4H_2O \rightarrow Fe_3O_4 + 4H_2$	Anode: - Cathode: -	$3Fe + 2O_2 \rightarrow Fe_3O_4$ (0.97 V)	[142–146]

Table 3. *Cont.*

Metal	Electrolyte	Cell Description	Reaction	Potential	Cell Reaction (Overall)	Refs.
Lithium (Li)	Aprotic Solvent	Anode: Lithium Cathode: Porous carbon Electrolyte: Aprotic liquid	Anode: $Li \rightarrow Li^+ + e^-$ Cathode: $O_2 + Li^+ + e^- \rightarrow LiO_2$ Cathode: $LiO_2 + Li^+ + e^- \rightarrow Li_2O_2$ Cathode: $O_2 + 2Li^+ + 2e^- \rightarrow Li_2O_2$ Cathode: $Li_2O_2 + 2Li^+ + 2e^- \rightarrow 2Li_2O$ Cathode: $O_2 + 4Li^+ + 4e^- \rightarrow 2Li_2O$	Anode: −3.0401 V Cathode: varies	$O_2 + Li \rightarrow LiO_2$ (3.0 V) $LiO_2 + Li \rightarrow Li_2O_2$ (2.96 V) $O_2 + 2Li \rightarrow Li_2O_2$ (3.10 V) $Li_2O_2 + 2Li \rightarrow 2Li_2O$ (2.91 V) $O_2 + 4Li \rightarrow 2Li_2O$ (2.91 V)	[147–150]
	Aqueous—Acidic	Anode: Lithium Cathode: Air-breathing Electrolyte: Acidic media	Anode: $Li \rightarrow Li^+ + e^-$ Cathode: $\frac{1}{2}O_2 + 2e^- + 2H^+ \rightarrow H_2$	Anode: −3.0401 V Cathode: 1.229 V	$2Li + \frac{1}{2}O_2 + 2H^+ \rightarrow H_2O + 2Li^+$ (4.274 V)	[131,147,151]
	Aqueous—Alkaline	Anode: Lithium Cathode: Air-breathing Electrolyte: Basic media	Anode: $Li \rightarrow Li^+ + e^-$ Cathode: $\frac{1}{2}O_2 + H_2O + 2e^- \rightarrow 2OH^-$	Anode: −3.0401 V Cathode: 0.401 V	$2Li + \frac{1}{2}O_2 + H_2O \rightarrow 2LiOH$ (3.44 V)	[131,151]
Magnesium (Mg)	Aqueous—Neutral	Anode: Magnesium (or alloy) Cathode: Air-breathing Electrolyte: Saline	Anode: $Mg \rightarrow Mg^{2+} + 2e^-$ Cathode: $O_2 + 4e^- + 2H_2O \rightarrow 4OH^-$ Side: $2H_2O + 2e^- \rightarrow H_2 + 2OH^-$	Anode: −2.37 V Cathode: 0.81 V	$2Mg + O_2 + 2H_2O \rightarrow 2Mg(OH)_2$ (3.1 V)	[53,131]
Potassium (K)	K Salt Ether Solvent, Potassium Superoxide	Anode: Potassium Cathode: Air-breathing Electrolyte: Organic	Anode: $K \rightarrow K^+ + e^-$ Cathode: $O_2 + e^- \rightarrow O_2^-$	Anode: −2.931 V Cathode: −0.451 V	$K + O_2 \rightarrow KO_2$ (2.48 V)	[131,152]
Silicon (Si)	Non-Aqueous—RTIL: EMIm (HF) 2.3 F	Anode: Silicon Cathode: Air-breathing Electrolyte: Room temperature ionic liquids	Anode: $Si + 12(HF)_2F^- \rightarrow$ $SiF_4 + 8(HF)_3F^- + 4e^-$ Cathode: $O_2 + 12(HF)_3F^- + 4e^- \rightarrow$ $2H_2O + 16(HF)_2F^-$ SiO_2 formation: $SiF_4 + 2H_2O + 4(HF)_2F^-$ $\rightarrow SiO_2 + 4(HF)_3F^-$	Anode: −2.71 V Cathode: 0.40 V	$Si + O_2 \rightarrow SiO_2$ (2.21 V)	[153]
	Aqueous—Alkaline: KOH	Anode: Silicon Cathode: Air-breathing	Anode: $Si + 4OH^- \rightarrow Si(OH)_4 + 4e^-$ Cathode: $O_2 + 4e^- + 2H_2O \rightarrow 4OH^-$	Anode: −2.71 V Cathode: −0.44 V Alternate: −0.38 V	$Si + O_2 + 2H_2O \rightarrow Si(OH)_4$ (2.09 V)	[153]
Sodium (Na)	Aqueous—Neutral	Anode: Sodium Cathode: Air-breathing Electrolyte: Saline	Anode: $Na \rightarrow Na^+ + e^-$ Cathode: $O_2 + 4e^- + 2H_2O \rightarrow 4OH^-$	Anode: −0.96 V Cathode: 0.33 V	$4Na + O_2 + 2H_2O \rightarrow 4NaOH$ (3.11 V)	[154]
	Aprotic Solvent	Anode: Sodium Cathode: Air-breathing	Anode: $Na \rightarrow Na^+ + e^-$ Cathode: $O_2 + e^- \rightarrow O_2^-$ Alternate Cathode: $O_2 + 2e^- \rightarrow O_2^{2-}$	Anode: −2.71 V Cathode: −0.44 V Alternate: −0.38 V	$Na + O_2 \rightarrow NaO_2$ (2.27 V) $2Na + O_2 \rightarrow Na_2O_2$ (2.33 V)	[154,155]
Tin (Sn)	Solid Oxide	Anode: Tin alloy Cathode: LSM/LSM-GDC Electrolyte: Ethylene glycol with tin powder (high temperature)	Anode: $Sn + 2O_2^{2-} \rightarrow SnO_2 + 4e^-$ Cathode: $O_2 + 2e^- \rightarrow O_2^{2-}$	Anode: −0.96 V Cathode: 0.33 V	$Sn + 2O_2 \rightarrow SnO_2$ (1.29 V)	[156,157]

Table 3. *Cont.*

Metal	Electrolyte	Cell Description	Reaction	Potential	Cell Reaction (Overall)	Refs.
Vanadium (Va)	Aqueous—Acidic	Anode: Zinc Cathode: Air-breathing Electrolyte: Acidic media	Anode: $V^{2+} \rightarrow V^{3+} + e^{-}$ Cathode: $VO_2{}^{+} + 2H^{+} + 2e^{-} VO^{2+} + H_2O$	Anode: −0.26 V Cathode: 1.0 V	$V^{2+} + VO_2{}^{+} + 2H^{+} \rightarrow V^{3+} + VO^{2+} + H_2O$ (1.26 V)	[158,159]
	Vanadium–Oxygen Fuel Cell	Anode: Vanadium Cathode: Oxygen diffusion electrode	Anode: $V^{2+} \rightarrow V^{3+} + e^{-}$ Cathode: $O_2 + 4e^{-} + 4H^{+} \rightarrow 2H_2O$	Anode: −0.26 V Cathode: 1.23 V	$4V^{2+} + O_2 + 4H^{+} \rightarrow 4V^{3+} + 2H_2O$ (1.49 V)	[143]
Zinc (Zn)	Aqueous—Neutral	Anode: Zinc Cathode: Air-breathing Electrolyte: Any neutral media	Anode: $Zn \rightarrow Zn^{2+} + 2e^{-}$ Cathode: $O_2 + 4e^{-} + 2H_2O \rightarrow 4OH^{-}$	Anode: −0.762 V Cathode: 0.817 V	$2Zn + O_2 + 2H_2O \rightarrow 2Zn(OH)_2$ (1.579 V)	[160]
	Aqueous—Alkaline	Anode: Zinc Cathode: Air-breathing Electrolyte: Any basic media	Anode: $Zn + 2OH^{-} \rightarrow ZnO + H_2O + 2e^{-}$ Cathode: $O_2 + 4e^{-} + 2H_2O \rightarrow 4OH^{-}$	Anode: −1.26 V Cathode: 0.40 V	$2Zn + O_2 \rightarrow 2ZnO$ (1.66 V)	[50,161]
	Non-Aqueous: Protic Room Temperature Ionic Liquids	Anode: Zinc Cathode: Air-breathing Electrolyte: Protic room temperature ionic liquid	Anode: $Zn \rightarrow Zn^{2+} + 2e^{-}$ Cathode: $O_2 + 2H^{+} + 2e^{-} \rightarrow H_2O_2$	Anode: −0.7618 V Cathode: 0.695 V	$Zn + O_2 + 2H^{+} \rightarrow Zn^{2+} + H_2O_2$ (1.457 V)	[131,162]

2.2. Zinc–Air Batteries

Zinc–air (Zn–air) batteries currently hold the distinction of being the sole commercially available metal–air battery technology worldwide. Their extensive history underscores their efficacy in low-current, long-duration applications such as hearing aids. Early prototypes of Zn–air batteries date back to the 1870s. Advancements in electrode design soon followed, paving the way for the primary Zn–air batteries onto the market in the 1930s, ultimately leading to their widespread adoption in hearing aids by the 1970s [163,164]. While zinc exhibits a lower specific capacity (733 Ah kg^{-1}) compared to certain alternative metals investigated for MABs (e.g., lithium and aluminum), it continues to demonstrate a notable performance. This is evidenced by its impressive theoretical specific energy and energy density as mentioned earlier. Furthermore, Zn–air batteries boast high anode utilization and compatibility with alkaline electrolytes. This compatibility offers a distinct advantage, as the comparatively slower reaction rate between the alkaline electrolyte and zinc minimizes hydrogen evolution, a significant challenge encountered in other metal–air chemistries (like Al- and Mg-based MABs). These properties position them as strong contenders in the development of next-generation portable, flexible electronics, and in the electrification of automobiles [163,165–168]. Zn–air batteries can be broadly classified into primary, mechanically rechargeable, and electrically rechargeable types. A typical primary Zn–air cell comprises several key components such as a zinc anode, a separator, an electrolyte, and an air cathode [169]. Mechanically rechargeable Zn–air batteries operate as functionally similar to primary batteries, relying on replacing the Zn anode and electrolyte during recharge. This distinct characteristic necessitates the use of only ORR electrocatalysts at the cathode [170]. Commercially proven primary Zn–air batteries, known for their high energy density, are used in power hearing aids (in the form of button cells) and even telecommunications and satellite systems in prismatic pouch formats. One key challenge associated with primary or replenishable primary Zn–air batteries is the potential for electrolyte evaporation. This evaporation can disrupt ionic conductivity within the battery, ultimately leading to performance degradation [171–174].

Rechargeable Zn–air batteries come in various arrangements, differing in their electrolyte types and design configurations. The most common design, depicted in Figure 7a, utilizes a simple "sandwich" framework with four key components. The first is a zinc anode. The second component, the air electrode, incorporates bifunctional oxygen electrocatalysts, a gas diffusion layer (GDL), and a current tab. Finally, the design is completed by a liquid electrolyte and a separator. This configuration allows for easy assembly in research labs using readily available materials like rubber gaskets, steel or plastic plates, and chambers.

Although planar Zn–air batteries can be arranged either horizontally or vertically, the vertical configuration is preferred to avoid electrolyte evaporation issues and ensure better ionic connectivity. While commercially available rechargeable Zn–air batteries are not yet widespread, their simple design offers potential advantages for EVs and other applications, owing to their high energy density and lightweight properties. In contrast, the planar design can lead to challenges, such as electrolyte loss in horizontal setups, which hampers the commercial viability of these batteries [172,174,175].

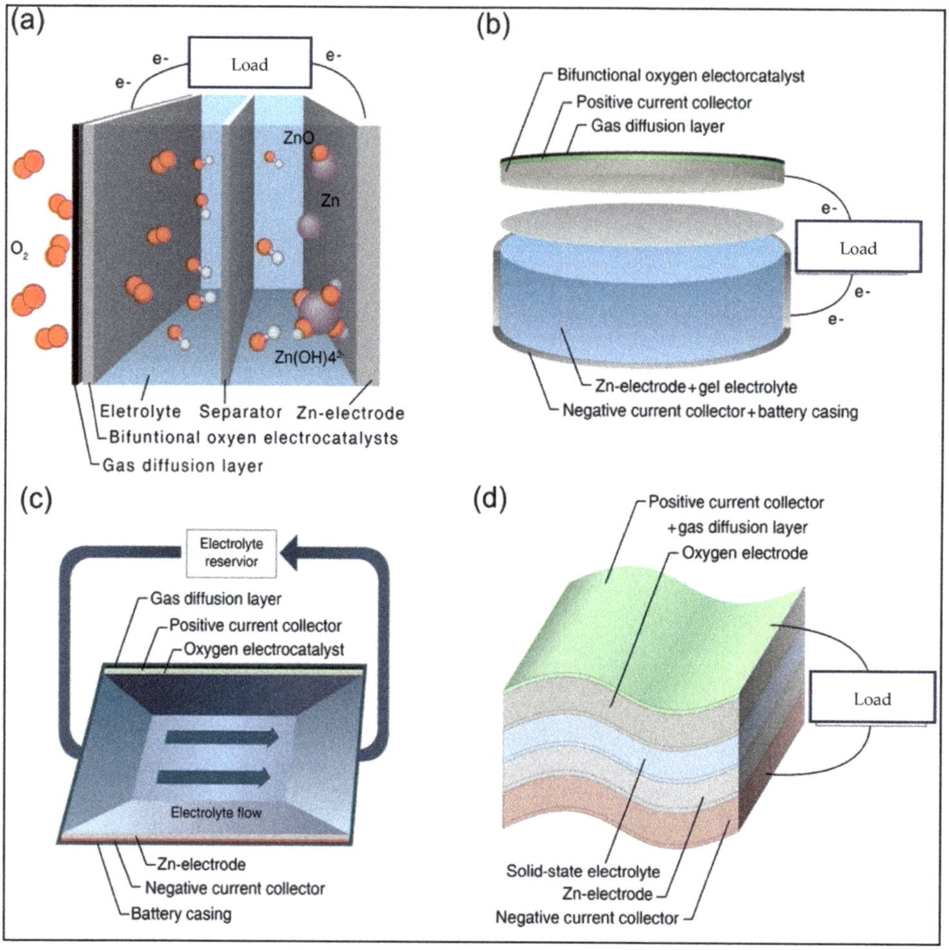

Figure 7. Rechargeable Zn-air battery designs: (**a**) Planar architecture with aqueous electrolyte. (**b**) Planar architecture with gel electrolyte. (**c**) Electrolyte flow-through design. (**d**) Flexible configuration (adapted from [176]).

Building on the foundational principles of conventional planar Zn–air batteries, the planar gel electrolyte architecture shown in Figure 7b represents an advanced iteration that addresses electrolyte evaporation loss and enhances efficiency using gel polymer electrolytes (GPEs). GPEs possess properties intermediate between liquid and solid states. Notably, they have high ionic conductivity (10^{-4} to 10^{-3} S cm^{-1} at ambient temperature), flexibility, and efficient electrolyte retention. As ionic conductors and separators, GPEs facilitate ion transport through a mechanism similar to liquid electrolytes. Despite significant advancements in GPEs, they still face challenges in making Zn–air batteries suitable for consumable electronics and EVs. Alkaline GPEs tend to generate carbonates which diminishes battery reversibility. Given the ongoing research into electrolytes, further advancements are essential to address these limitations [177,178].

A Zn–air flow battery (Figure 7c) utilizes a circulating electrolyte design to enhance performance and mitigate degradation issues associated with both the anode and cathode. On the metal anode side, the larger volume flowing electrolyte improves the current distribution and reduces concentration gradients, which helps to prevent shape change, dendrite formation, and passivation. The precipitated carbonate layer on the cathode can be washed away by the flowing electrolyte and removed using an external filter. This design is comparable to the approach utilized in alkaline fuel cells, where circulating electrolytes are employed to prevent carbonate precipitation on the cathode surface. Consequently, rechargeable Zn–air flow batteries are expected to offer a longer cycle life compared to traditional static electrolyte configurations. However, Zn–air flow batteries also come with certain disadvantages. The increased complexity of the system, which includes the need for pumps, tubing, and circulating the electrolyte, results in decreased energy efficiency. Additionally, the excess volume of electrolytes and the associated components lead to lower specific and volumetric energy density [179–182].

A Zn–air flexible battery (Figure 7d) is an innovative power source designed to be flexible, making it suitable for various applications such as wearable technology. They are preferred due to their cost-effectiveness, high energy density, and intrinsic safety, which obviates the need for a rigid protective casing. Many studies are exploring methods and materials to produce flexible Zn–air batteries, including thin-film "wearable" designs, cylindrical cable-type batteries, and biocompatible flexible batteries. The key challenge in developing flexible Zn–air batteries is their exposure to the air, making the use of liquid electrolytes undesirable due to risks of evaporation or leakage onto sensitive electronic equipment. Consequently, current research efforts are centered on developing a solid-state electrolyte that is mechanically flexible, durable, and exhibits adequate ionic conductivity. Additionally, the electrodes and other cell components must be capable of withstanding and maintaining operation under significant bending. Despite these challenges, the advantages of flexibility, safety, and high energy density make Zn–air flexible batteries a promising choice for future energy storage solutions [183–186].

2.3. Aluminum–Air Batteries

Aluminum–air batteries (Al–air) have been the subject of extensive research since the 1960s. The concept of utilizing an Al anode was initially introduced by Zaromb in 1962 within the context of an Al/O_2 system renowned for its high energy density. Subsequent research has explored various applications, such as EVs, military communications, and unmanned aerial and underwater vehicles. A significant advancement occurred in 2016, creating a 100 kg Al–air stack battery capable of extending an EV's range beyond 3000 km. This achievement underscores the substantial potential of Al–air technology for the transportation sector [187]. Al–air batteries offer a cutting-edge energy storage approach due to aluminum's high volumetric capacity, abundance, and low cost. Typically, these primary batteries consist of an Al metal anode, an air cathode, and an aqueous electrolyte, which is often sodium chloride (NaCl), potassium hydroxide (KOH), or sodium hydroxide (NaOH). Additionally, AABs include components such as external loads, pumps, anolyte and catholyte chambers, current collectors, separators, membranes, and electronic systems, with configurations varying by design. Aluminum is oxidized at the negative electrode by hydroxide ions during discharge, producing aluminate ions and electrons. Simultaneously, O_2 from the air is reduced at the air cathode by these electrons, and hydroxide ions migrate through the electrolyte to maintain charge balance [188–191]. The corresponding reactions with various electrolyte types are summarized in detail in Table 3.

Al–air batteries exhibit diverse configurations, varying in both electrolyte composition and design structure. As depicted in Figure 8a, a planar structure refers to a flat, layered design where the components of the Al–air battery are arranged in flat sheets stacked either vertically or horizontally. Three types of aqueous electrolytes are used based on their pH: neutral, acidic, and alkaline electrolytes. Neutral Al–air batteries are eco-friendly, utilizing non-toxic raw materials and generating non-hazardous by-products. The generated OH^-

during the operation quickly combines with aluminum ions to form aluminum hydroxide, keeping the electrolyte neutral. This neutrality offers a high level of safety to Al–air batteries, making them suitable for marine and consumer applications. Since the 1970s, large neutral Al–air battery stacks (Figure 8b) have been developed for marine use, leveraging seawater as a natural electrolyte. This allows offshore Al–air batteries to operate with only aluminum plates and air cathodes, as seawater can be utilized directly as an electrolyte. Neutral Al–air batteries have seen applications in consumer products, including wearable devices, small electronics, and other household appliances. However, these batteries are hindered by significant challenges, such as the intense polarization of the Al anode due to the passivation of the aluminum surface. This passivation layer impedes the migration of aluminum ions and electron transfer, reducing the battery's discharge performance. The presence of chloride ions can help to reduce polarization by assisting in the dissolution of the oxide film [192].

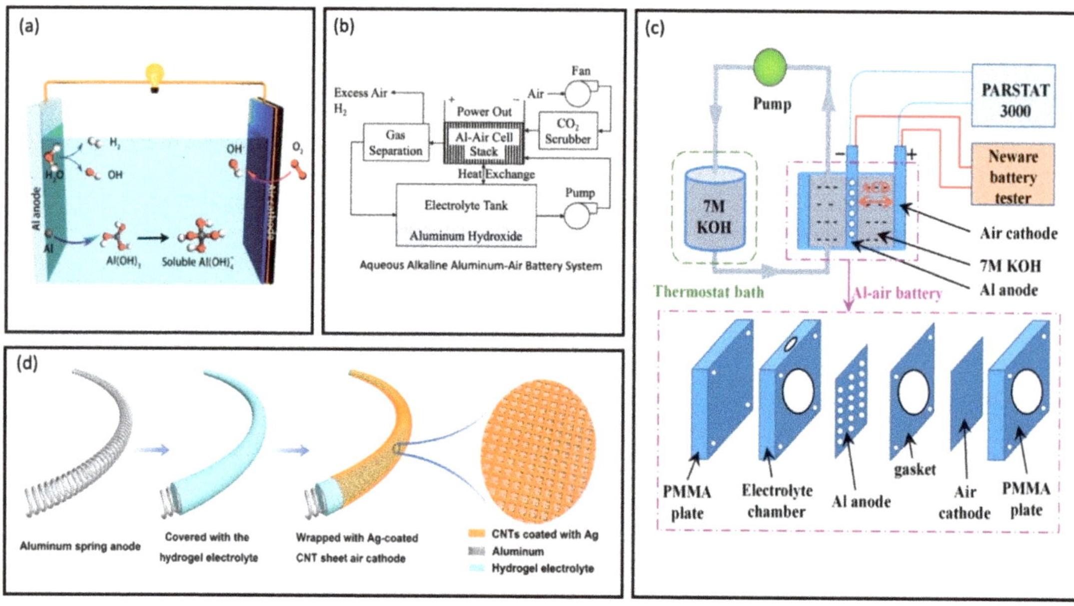

Figure 8. Al–air battery configurations: (**a**) Planar architecture with aqueous electrolyte [192]. (**b**) Al–air battery stack architecture [51]. (**c**) Electrolyte flow-through design [38]. (**d**) Flexible configuration [193].

To combat the disadvantages of anode passivation in neutral electrolytes, acidic electrolytes were proposed. Hydrochloric acid (HCl) and sulphuric acid (H_2SO_4) have been studied as possible options, offering the potential for higher voltages due to the high theoretical positive potential of the air cathode in low-pH environments. However, the aluminum anode's thermodynamic properties in acidic electrolytes are unsatisfactory, with severe H_2 evolution side reactions occurring alongside the main reaction. The time needed for the disintegration of the oxide film in lower concentrations of acidic electrolytes is higher than in alkaline electrolytes, leading to longer activation processes and increased polarization. Moreover, acidic electrolytes can cause pitting corrosion, which is detrimental to smooth discharge. These practical limitations have hindered the usage of acidic electrolytes in Al–air batteries, though ongoing research aims to address these issues [194–196].

The introduction of alkaline electrolytes marked a breakthrough for Al–air batteries. KOH or NaOH-based electrolytes, quickly disintegrate the passivation layer on the aluminum, greatly increasing the kinetics of the charge transfer reactions. Insoluble aluminum hydroxide is converted into soluble tetrahydroxy aluminate ions, reducing polarization, and maintaining high ionic conductivity. Alkaline Al–air batteries can maintain high voltages under high current densities due to the intense destructive capacity of hydroxide ions on the oxide film, facilitating smoother ion mobility at the solid-electrolyte interface at the anode. However, alkaline electrolytes also have inherent flaws. The reactive aluminum is exposed to the electrolyte as the passivation layer dissolves during discharge, leading to hydrogen evolution, which poses safety risks and reduces the energy density and specific capacity of the battery. Although high-concentration electrolytes and additive compounds are used to mitigate these issues, the actual anode potential in alkaline solutions remains lower than the theoretical value, indicating significant anode polarization [38,197]. Despite these challenges, alkaline Al–air batteries show promise for vehicle power supplies. Phinergy developed an electric car combining Al–air batteries and Li-ion batteries, extending the range of the Li-ion batteries from 60 km to 350 km [198]. Recently, a short-range aircraft powered by Al–air batteries was designed, showcasing significant advantages in weight and cost with a theoretical 950 nautical miles of mileage [199]. The primary barrier to commercialization lies in high anodic corrosion and complicated stack design. Many effective options to inhibit self-corrosion are studied, but achieving easy mechanical charging, simple, and safe replacement of aluminum anodes and electrolytes remains crucial for industrialization.

Al–air flow batteries (Figure 8c) employ an electrolyte flow system to remove reaction by-products, thereby improving energy efficiency. This design leads to exceptional electrochemical performance, including excellent discharge capabilities and extended cycle life. There are two primary methods for facilitating electrolyte flow in Al–air flow batteries: external mechanical pump-assisted driving and paper-based capillary action driving. The mechanical pump-assisted method, utilizing devices like peristaltic or syringe pumps, ensures a fast and stable electrolyte flow, significantly enhancing the performance of Al–air flow batteries. However, the bulky and complex structure of this system makes it suitable only for large-scale applications. To reduce size and complexity, researchers have developed a paper-based microfluidic Al–air battery. In this system, the solution flow is driven by capillary action within the microchannels of cellulose paper. This approach offers a compact and simplified design but comes with limitations such as reduced energy storage capacity and shorter battery shelf life, resulting in weaker discharge behavior compared to their larger counterparts [67,200]. Wen et al. [38] reported significant results using an Al–air flow battery with a 7 M KOH electrolyte flowing externally. They reached a peak power density of 545 mW cm^{-2} in a pure O_2 environment, with optimal performance parameters including a 0.5 mm anode-cathode distance (physical distance between the two electrodes to prevent short-circuit), an operating temperature of 60 °C, and an electrolyte flow rate of 15 mL min^{-1}. The battery demonstrated an anode efficiency of 96.2% and energy efficiencies of 44.4% in air and 42.6% in pure O_2. Furthermore, their research indicated that the technology can be scaled to kilowatt-level packs, with energy densities of 700 Wh L^{-1} and 900 Wh kg^{-1}, making it promising for applications in EVs and large-scale energy storage systems. Huang et al. developed a novel acoustofluidic saltwater Al–air flow battery that uses ultrasonic vibrations to drive the electrolyte flow and create an ultrasonic field within the reaction chamber. This innovative approach led to a peak power density of 43.88 mW cm^{-2}, which is up to 7.5 times higher than that with static electrolytes and comparable to mechanical pumping systems. The optimization of reaction vessel thickness, ultrasonic vibration velocity, and electrolyte concentration contributed to this performance enhancement. Additionally, the acoustofluidic system significantly decreased the energy usage, mass, and size of the electrolyte-driven component compared to traditional methods. This research highlights the potential of ultrasonic pump-based Al–air flow batteries for energy storage applications requiring eco-friendliness, cost efficiency, safety, and stable discharge performance [201].

A flexible Al–air battery (Figure 8d) is an advanced type of battery designed to offer both high energy density and mechanical flexibility. Unlike traditional rigid batteries, flexible Al–air batteries are constructed to be pliable and stretchable, allowing them to be integrated into various applications such as smart textiles and wearable electronics. Xu et al. [193] presented a fiber-shaped flexible Al–air cell with notable advantages and performance. This design achieved an impressive energy density of 1168 Wh kg^{-1}, over 12 times greater than fibrous Li-ion batteries, and a specific capacity of 935 mAh g^{-1}, about 6 times higher than fiber-type zinc-carbon batteries. It maintained an output voltage above 1 V even when stretched by up to 30%, demonstrating excellent mechanical stability. The fiber design also allows for integration into textiles, making it suitable for flexible and wearable electronics. Key advantages include its environmental friendliness and the usage of a porous design with aligned carbon nanotubes and silver nanoparticles to enhance performance. However, practical energy density remains lower than theoretical values due to limitations in electrode materials, and there are concerns about the aluminum anode's self-corrosion and the complexity of the manufacturing process. Despite these challenges, the battery offers promising potential for innovative applications in smart textiles and wearable technology.

2.4. Magnesium–Air Batteries

In 1966, Carson and Kent conducted conceptual and practical investigations into high-performance magnesium–air (Mg–air) primary cells, termed the "Magair" cell. This system employed an air cathode, a Mg alloy anode, and a neutral salt electrolyte, achieving a nominal voltage of 1.60 V. During discharge, the cell exhibited voltages of 1.30 V and 1.00 V at current densities of 10 and 60 mA cm^{-2}, respectively. The authors also demonstrated the concept of mechanical rechargeability through periodic replenishment of the anode and electrolyte salt [202]. While Mg–air batteries have received comparatively less attention than other MAB technologies, their potential as energy storage systems is compelling due to magnesium's abundance and the battery's high theoretical energy density. Moreover, when considering practical performance metrics, Mg–air batteries exhibit competitive advantages over Zn–air and Al–air technologies. Since their inception in the 1960s, studies on Mg–air cells have been mostly centered around the primary, or reserve-type, configurations. These batteries require activation before use and offer continuous discharge until anode depletion. Conversely, developing rechargeable Mg–air batteries has proven significantly more challenging [203].

In general, Mg–air batteries face several hurdles, including low working voltage and practical specific energy, substantially below theoretical values. These limitations stem largely from sluggish oxygen-reduction kinetics and anode corrosion, common challenges in MAB systems. Magnesium's unique corrosion behavior, exacerbated by impurities, further complicates the issue. Unlike most metals, magnesium exhibits a negative difference effect, where increasing cathodic current accelerates hydrogen evolution. Additionally, the poor conductivity and electrochemical activity of discharge products (MgO and MgO_2) hinder rechargeability [204]. Figure 9 illustrates various configurations of Mg–air batteries, categorized by cell architecture and electrolyte types.

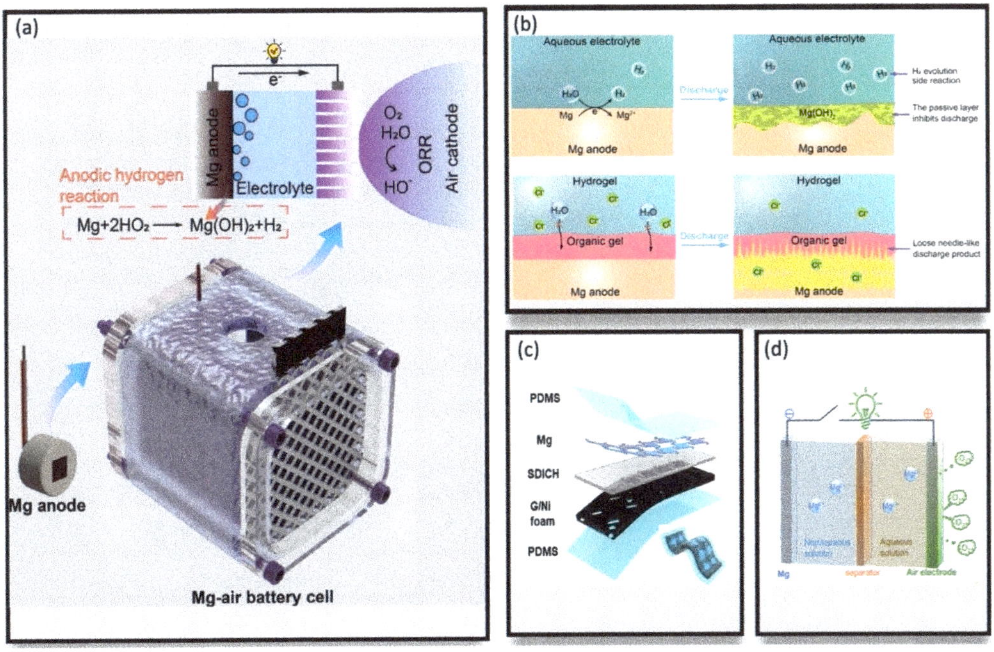

Figure 9. Mg-air battery configurations: (**a**) Planar architecture with aqueous electrolyte [205]. (**b**) Two-layer gel electrolyte structure [206]. (**c**) Flexible configuration [207]. (**d**) Hybrid electrolyte design [208].

Chen et al. [205] devised a planar structure utilizing ultra-high-purity (UHP) Mg-Ge anodes for Mg–air cells, enhancing their performance significantly as shown in Figure 9a. The inclusion of Ge in the anode composition improves anodic reaction activity while effectively suppressing the anode hydrogen evolution reaction (HER). The electrolyte utilized in this system is a 3.5 wt. % NaCl solution, which facilitates ion transport during the electrochemical processes. The UHP Mg-0.5Ge anode possessed an impressive cell voltage of 1.69 V at a current density of 0.5 mA cm^{-2} during discharge and achieved a high energy density of 2272 mWh g^{-1} at 20 mA cm^{-2}. These performance metrics surpass those of all previously researched Mg-based anodes. Moreover, the low content of impurity and the formation of a Ge-bearing surface layer results in a remarkable 95% discharge efficiency. This is 3.5 times greater than that of the frequently used AZ31 anode. Thus, the advantages include enhanced discharge characteristics and improved battery efficiency due to the low impurity content and a Ge-bearing surface film as mentioned. However, potential disadvantages may arise from the corrosive nature of the NaCl electrolyte, which might impact the longevity of the cell components, alongside the complexities associated with manufacturing high-purity materials, raising concerns about cost and scalability. Overall, the exceptional properties of the UHP Mg-Ge anode pave the way for significant advancements in Mg–air batteries.

Li et al. [206] employ a two-layer gel electrolyte structure that addresses significant performance limitations associated with Mg anodes. The architecture consists of a Mg anode, an air-breathing cathode, and a two-layer gel electrolyte as described in Figure 9b. The two layers built of an organic gel and hydrogel were evenly coated on the surface of the anode. The two-layer serves two primary functions: the first layer prevents corrosion of the Mg anode, while the second layer inhibits the development of a thick, insoluble passive layer that forms on the Mg's surface during discharge and corrosion processes. This design enables effective interaction between the Mg anode and the electrolyte, facilitating the battery's operation. The working mechanism involves the reaction of Mg with water in the electrolyte, producing magnesium hydroxide and releasing H_2 gas, with electrons being transferred directly to the electrolyte. Key results from this study indicate that these Mg–air batteries achieved an impressive specific capacity of 2190 mAh g^{-1} based on the anode alone and an anode efficiency of 99.3%, alongside an energy density of 2282 Wh kg^{-1} based on the Mg anode and air electrode. The dual-layer architecture presents several advantages, including enhanced battery performance, improved efficiency, and high utilization of the Mg anode. However, the design also has drawbacks, such as increased complexity in the cell's design and potential concerns regarding the long-range stability of the gel electrolyte in various environmental conditions.

The architecture of the cell is based on a completely flexible Mg–air battery that employs dual-ion-conducting hydrogels (SDICH) as illustrated in Figure 9c. This design allows for significant mechanical flexibility and stretchability, making it suitable for integration into biomedical devices. The materials used include a Mg anode, an air-breathing cathode, and an SDICH electrolyte. This combination produces electricity while maintaining the necessary flexibility and stretchability for biomedical applications. The battery successfully powered 120 light-emitting diodes (LEDs) for over 5 h, demonstrating its capacity and efficiency. It exhibited stable electrical output even under mechanical stretching, indicating resilience to deformation without performance degradation. The biocompatibility of the battery was validated through in vitro and in vivo studies, showcasing its ability to be used in medical applications. The stretchable design allows the battery to be integrated into wearable and flexible devices, making it safe for use in biomedical applications, particularly in systems that are in contact with human tissue, such as wound dressings. However, the long-term stability of the dual-ion-conducting hydrogel under various environmental conditions may require further investigation. Additionally, the production of such advanced materials and architectures may be more complex and costly compared to simpler battery designs [207].

Hybrid electrolyte Mg–air batteries (Figure 9d) use both aqueous and non-aqueous electrolytes to improve the electrochemical reactions within the battery. This approach improves the characteristics of traditional Mg–air batteries by enhancing charge transport reactions, reducing polarization effects, and enabling recharging capabilities. Hybrid electrolyte Mg–air batteries offer several advantages. They can achieve up to 90% round-trip efficiencies, making them an attractive option for energy storage solutions. Their enhanced power density results from the integration of materials that facilitate fast charge transport and high current density. Furthermore, the design of hybrid electrolytes contributes to a longer battery lifespan by addressing issues such as self-corrosion and passivation of the Mg anode. However, there are notable disadvantages such as the Mg's self-corrosion which can produce insulating products like MgO and impede battery performance by increasing overpotential and polarization. Additionally, parasitic reactions that produce hydrogen gas can disrupt the pressure balance within sealed cells, further increasing cell polarization. Thus, hybrid electrolytes carry inherent disadvantages, along with the complexity of managing both aqueous and non-aqueous components and the need for advanced electrocatalysts for effective oxygen reduction. These factors can reduce the operational stability of the batteries, impacting their overall efficiency and longevity [204].

2.5. Sodium–Air and Potassium–Air Batteries

Sodium–air (Na–air) and potassium–air (K–air) batteries demonstrate substantial energy densities in contrast to other battery technologies. Na–air batteries offer a specific energy of 1680 Wh kg^{-1} and a volumetric energy density of 2466 Wh L^{-1}, while K–air batteries exhibit a specific energy of 1187 Wh kg^{-1} and a volumetric energy density of 1913 Wh L^{-1}. These impressive figures underscore the significant potential of sodium- and potassium-based systems for future energy storage applications. However, despite concerted research efforts during the last few decades, advancements in the research area of alkali MABs remain in their early stages. Peled et al. [209] introduced the theories of Na–air batteries in 2011. This initial cell configuration employed a liquid sodium anode, a polymer electrolyte, and a commercial air cathode. To prevent sodium metal buildup on the air electrode, the system was run at a high temperature of 105 °C to 110 °C, exceeding sodium's melting point. Ren et al. [70] first illustrated the idea of a K-O_2 battery in 2013, showcasing decreased overpotentials by leveraging the reversibility of the O_2/O_2^- redox couple. For the first time, they reported a charge/discharge potential gap smaller than 50 mV at a current density of 0.16 mA cm^{-2}. Raman analyses and XRD verified the generation and decomposition of KO_2 during the battery cycle test. In the literature, K–air and K-O_2 batteries are often used interchangeably, but most research has been conducted with oxygen, with few works reported on dry ambient air [210].

Among the various types of Na–air cells, hybrid electrolyte and organic electrolyte Na–air cells have been the most extensively studied. These two types differ primarily not only by the electrolyte type but also by their structural designs. Organic Na–air batteries constitute one primary research focus within the broader Na–air battery field. Their fundamental architecture, as depicted in Figure 10a, centers around a Na anode, a porous bifunctional catalyst cathode, and an organic electrolyte-saturated separator.

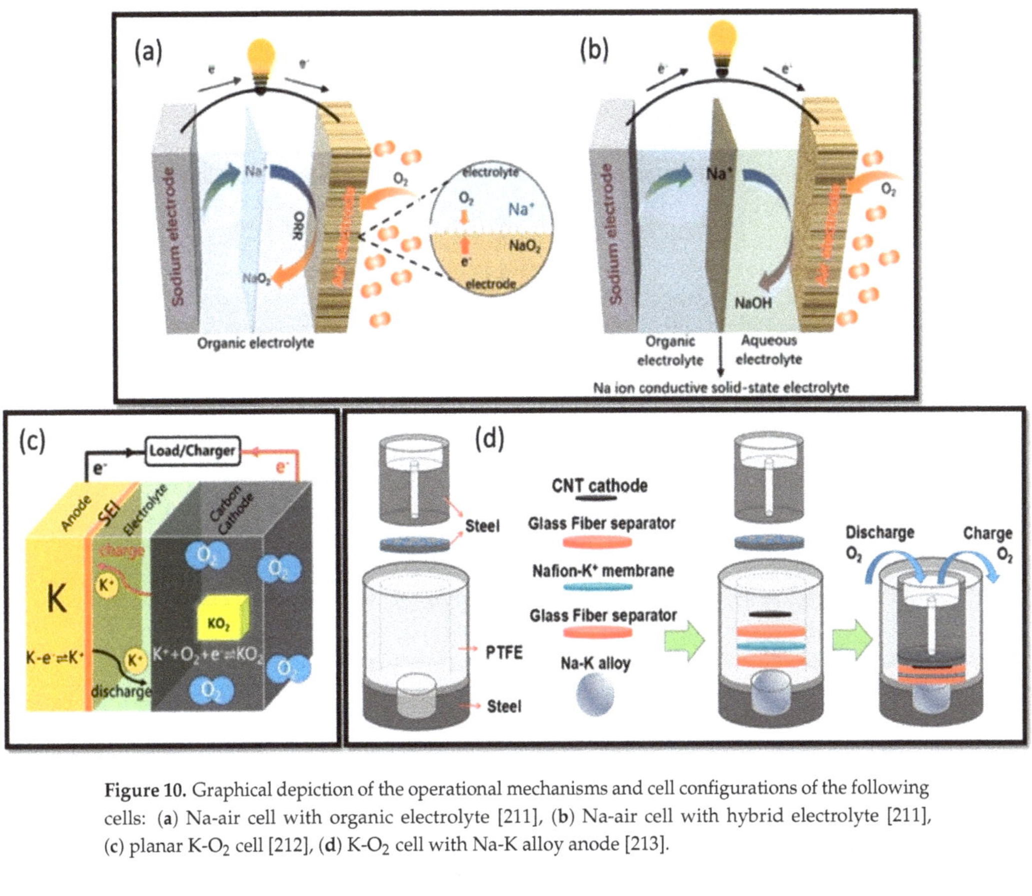

Figure 10. Graphical depiction of the operational mechanisms and cell configurations of the following cells: (**a**) Na-air cell with organic electrolyte [211], (**b**) Na-air cell with hybrid electrolyte [211], (**c**) planar K-O_2 cell [212], (**d**) K-O_2 cell with Na-K alloy anode [213].

A range of organic solvents like dimethyl ether (DME) and diethylene glycol dimethyl ether (DEGDME), along with salts such as sodium hexafluorophosphate ($NaPF_6$) and sodium bis(trifluoromethanesulfonyl)imide (NaTFSI), are used to modulate battery performance. The specific salt and its concentration significantly impact the battery's overall efficacy. Functionally akin to Li-O_2 due to the chemical parallels between sodium and lithium, organic Na–air batteries excel in lower overpotential and have a superior cycling stability [214–216].

In contrast, hybrid Na–air batteries come in a wide variety of designs, going beyond conventional structures. As shown in Figure 10b, they incorporate a dual-electrolyte system: an aqueous electrolyte at the anode and an organic electrolyte at the cathode, separated by a sodium ion-conductive solid electrolyte membrane, often NASICON ($Na_3Zr_2Si_2PO_{12}$). This configuration shields the anode from oxidative reactions while facilitating unidirectional sodium ion migration. The cathode electrolyte mirrors that of organic Na–air batteries, while the anode typically employs an aqueous NaOH solution. The energy storage mechanism in hybrid Na–air batteries is simpler, involving oxygen uptake and release at the positive electrode during the discharge and charge operations. This process corresponds to the ORR and OER, respectively. A key advantage of this design is the generation of a single discharge product, NaOH, which minimizes side reactions and enhances cycle life. Moreover, the solubility of NaOH in the electrolyte prevents accumulation on the air cathode, further improving efficiency. Unlike their organic counterparts, hybrid Na–air cells can function with atmospheric air and exhibit higher theoretical voltages. Both organic and hybrid Na–air batteries hold promise for diverse applications. However, they present distinct advantages and drawbacks. Organic Na–air batteries prioritize safety due to sodium's reactivity with water, but they suffer from higher costs, environmental concerns associated with organic electrolytes, and performance limitations caused by discharge product deposition. Additionally, they necessitate dry oxygen and gas purification systems, increasing operational expenses. In comparison, hybrid Na–air battery systems exhibit lower overpotentials, higher energy densities, and extended cycle life. Their ability to utilize ambient air and dissolve discharge products in the cathode electrolyte renders them more cost-effective and efficient. Consequently, hybrid Na–air batteries emerge as a compelling alternative to their organic counterparts [154,217–220].

Yu et al. [213] successfully showcased a novel K-O_2 battery that is dendrite-free operating at room temperature, utilizing a liquid Na-K alloy as the negative electrode, a Nafion-K^+ membrane separator, and a 0.5 M potassium hexafluorophosphate (KPF_6) in DEGDME electrolyte sandwiched between glass fibers and carbon nanotubes (CNTs) as the cathode (Figure 10b). The cathode is made of porous carbon paper coated with carbon nanotubes. The primary discharge product is KO_2, with some KOH also formed, demonstrating the efficacy of the Na-K alloy anode. Without the Nafion-Na^+ membrane, the battery possesses a discharge capacity of around 2.0 mAh and an initial voltage of approximately 2.45 V. However, the Nafion membrane's ion selectivity restricts K^+ diffusion, hindering discharge capacity in Na-O_2 cells. This configuration offers advantages like dendrite-free operation, effective anode performance, and reduced side reactions due to the Nafion membrane. Nevertheless, it suffers from limited discharge capacity due to ion selectivity, solvent decomposition, and sustainability concerns as the battery transitions from Na-O_2 to K-O_2 during discharge.

2.6. Other Metal–Air Batteries

In the realm of MABs, the utilization of semiconductors as anodes has opened new avenues for energy storage. Among these, silicon–air (Si–air) batteries are particularly noteworthy with a theoretical volumetric energy density of 6713 Wh L^{-1} and a specific energy of 5321 Wh kg^{-1}. The pioneering Si–air battery, developed by Cohn et al. [221] in 2009, maintained a consistent voltage between 1.0 and 1.2 V. This innovative design featured doped silicon wafers as the fuel, an air-breathing cathode, and a precisely engineered ionic liquid, 1-ethyl-3-methylimidazoliumoligofluorohydrogenate (EMI·$(HF)_2$·3F), as the electrolyte. Recent research has concentrated on aqueous Si–air batteries utilizing conventional KOH electrolytes. Initial research emphasized the importance of micro- or nano-porous Si surfaces to maintain extended discharge times, achieving up to 30 h at 0.05 mA cm^{-2} [222,223]. Without these surface modifications, the cell voltage drops sharply due to the formation of the passivation layer on the Si surface rapidly. The following studies demonstrated longer discharge times of up to 270 h at the same current density using a refill-type cell setup with an unmodified Si anode. In cases where a thicker Si wafer (3 mm) was employed as the anode, the alkaline Si–air battery, utilizing a 5 M KOH electrolyte, sustained operation for up to 1100 h at voltages above 1.1 V, delivering specific capacities of 115 mAh g^{-1} and anode mass conversion efficiencies around 3% [224]. More recent research has shown that reducing the KOH concentration to 0.5 M enhances Si anode utilization during discharge, with conversion efficiencies increasing to as much as 9% [225]. In 2024, Guerrero et al. [226] pioneered a study using both alkaline and non-aqueous Si–air batteries to power low-power transient electronics designed for partial self-destruction. The Si–air batteries demonstrate practical specific capacities and energies of 117 mAh g^{-1} and 123 Wh kg^{-1} for the alkaline system, and 1072 mAh g^{-1} and 841 Wh kg^{-1} for the non-aqueous system, until the active area of the Si anodes is fully consumed (Figure 11a). Nevertheless, Si–air batteries have several drawbacks. The silicon surface can quickly become passivated, causing a drop in voltage unless modified with micro- or nano-porous structures. The batteries also face mechanical stress due to silicon's expansion and contraction during use, which can shorten their lifespan. Additionally, the electrolyte can be consumed during reactions, leading to corrosion. Production costs are high because of the need for pure silicon and specialized processes. Si–air batteries also have a shorter cycle life and can be less energy-efficient [227].

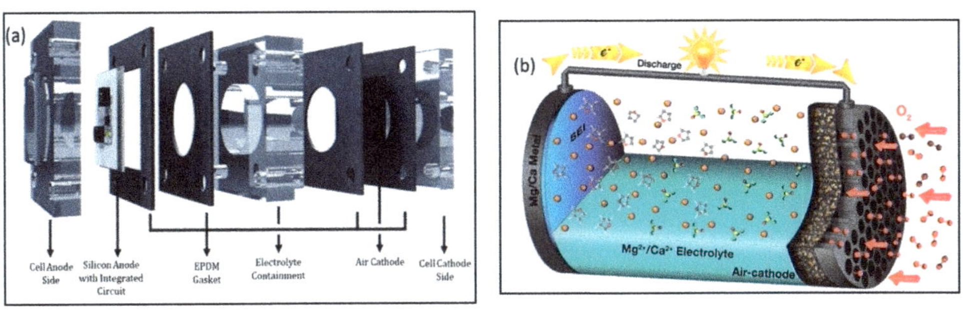

Figure 11. (**a**) Cell structure of a Si-air battery [226]. (**b**) Working illustration of a Mg/Ca-O_2 battery [228].

Germanium–air (Ge–air) batteries emerged shortly after Si–air batteries in 2013, described by Ocon et al. [229], leveraging germanium's widespread use in Li-ion batteries. The cell featured a nanoporous Ge anode, a KOH electrolyte, and an air-breathing cathode. It operated at a sustained discharge and maintained a stable cell voltage of 0.42 V, with a power density of 0.21 mW cm^{-2} and a current density of 0.5 mA cm^{-2}. Remarkably, there were no signs of polarization, even at elevated current densities, as evidenced by the slight increase in potential after 200 h of operation. However, since their introduction, there have been very few advancements in research on Ge–air batteries, and they remain a relatively unexplored area in the field.

The calcium–oxygen ($Ca\text{-}O_2$) battery presents a favorable avenue for rechargeable energy storage. These batteries are attractive due to their impressive theoretical volumetric energy density of 7574 Wh L^{-1}, specific energy of 3441 Wh kg^{-1}, and incredibly high nominal voltage of 3.5 V. This electrochemical system employs a calcium anode and an air-breathing cathode, where O_2 is reversibly converted to calcium peroxide. An ionic liquid-based electrolyte, consisting of EMIM-BF_4 and dimethyl sulfoxide, provides a stable operating environment. Recent studies have highlighted the exceptional performance of $Ca\text{-}O_2$ batteries, including their longevity (700 recharge cycles), stability in ambient air, and adaptability to flexible textile formats for wearable applications [230]. Lu et al. [228] introduced the concept of $Mg/Ca\text{-}O_2$ where a Mg/Ca metal is used as the anode. They conducted a comprehensive analysis of suitable electrolytes and identified key technological advancements required to address the challenges associated with this type of battery (Figure 11b). Challenges such as anode passivation and calcium ion transport limitations are yet to be addressed. $Ca\text{-}O_2$ batteries are in their initial research phase, and it will take considerable time and effort to realize their full potential.

The pioneering work on iron–air (Fe–air) batteries was first conducted by NASA in 1968. This concept was later adopted by the Swedish National Development Corporation in the 1970s to produce a 30 kWh battery for traction applications, achieving an energy density of 80 Wh/kg at a 5 h discharge rate [141]. Fe–air batteries boast several advantages over traditional batteries. Theoretically, they possess a volumetric energy density of 3244 Wh L^{-1}, a specific energy of 1080 Wh kg^{-1}, and an OCV of 1.28 V. Notably, in Fe–air batteries, the risk of iron dendrite formation during charging is eliminated, ensuring safety and longevity. Furthermore, the iron negative electrode offers a remarkably high theoretical specific capacity of 960 mAh g^{-1}, promising exceptional energy density [231]. Finally, their composition and operational characteristics make it a more environmentally friendly and sustainable energy storage solution with many recent works to solve the issues associated with this battery type.

In conclusion, while various MABs offer unique advantages such as high energy density and potential cost savings, their inherent complexity and current technological limitations make them unsuitable as stand-alone power sources in the context of the EV industry. Instead, their most promising application lies in being used as range extenders in EVs, where they can supplement existing battery systems and enhance driving range without the need for complete reliance on their performance. This hybrid approach maximizes their benefits while mitigating challenges related to efficiency, safety, and cost. To fully understand their potential, however, further study of practical experimental parameters is necessary to assess their viability in real-world applications.

3. Comparison of Performance Metrics

According to the International Energy Agency's (IEA) global EV outlook released in 2024, electric car sales continue to rise and could reach approximately 17 million in 2024, making up more than one in five cars sold globally and the pricing strategies of car manufacturers will perform a key role in increasing affordability, along with the rate at which EV battery prices decrease as evidenced by the trends shown in Figure 12. Under current projections, EV energy demand is expected to increase significantly by 2035. By 2030, demand could grow up to 4.5 times compared to 2023, and nearly 7 times by 2035. In more ambitious scenarios, demand may rise 5 to 7 times by 2030, and 9 to 12 times by 2035. For context, in 2035, the weekly EV battery demand could match the total annual demand recorded in 2019 [232]. These demands emphasize the importance of integrating MABs that offer higher specific energy, lower costs, and improved safety in the EV industry.

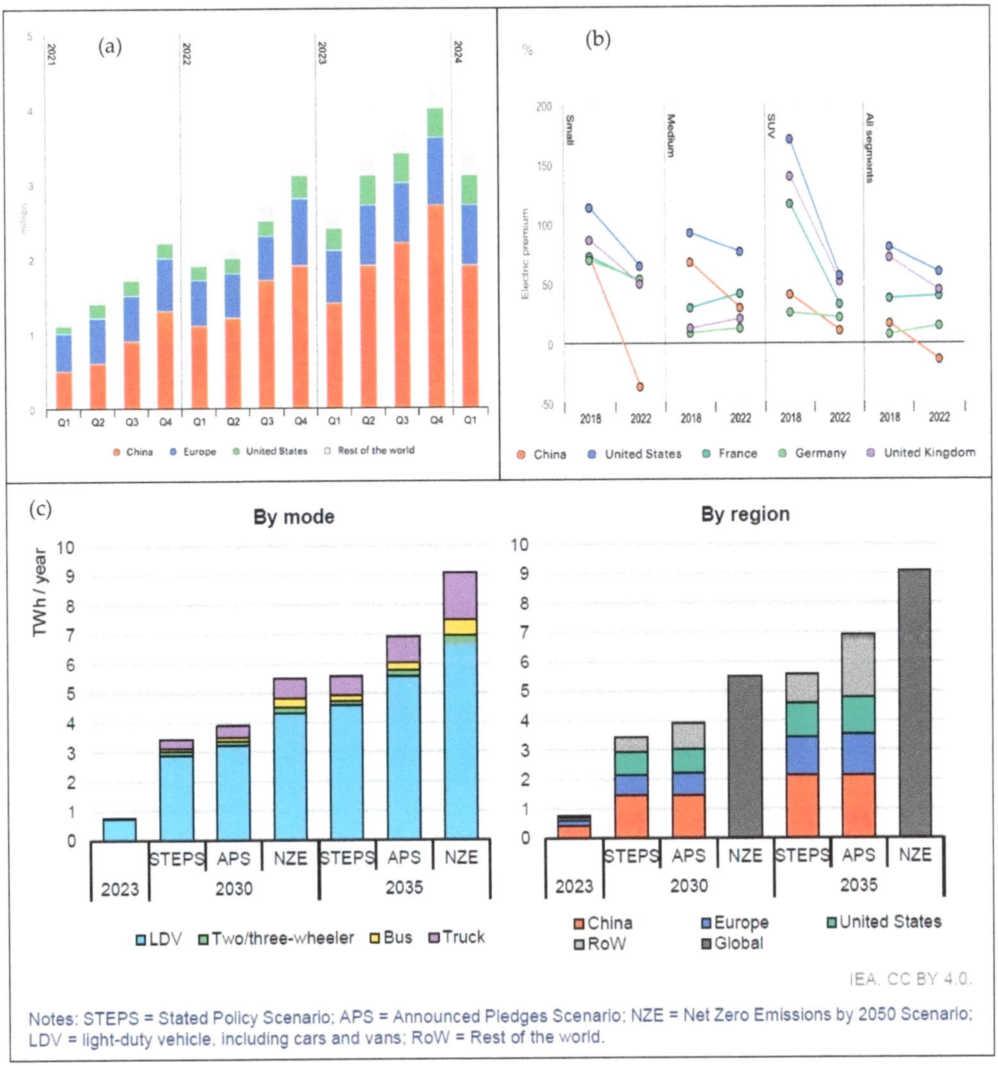

Figure 12. (**a**) Regional quarterly electric car sales (2021–2024), (**b**) the difference in sales-weighted average prices between traditional and electric vehicles in specific countries, before subsidies, by vehicle size (2018–2022), and (**c**) energy demand for electric vehicles, categorized by mode and region (2023–2035) [232].

As mentioned before, studying practical performance metrics is crucial as they provide real-world insights into the efficiency and limitations of MABs, beyond what theoretical parameters can predict. Thus, Figure 13 compares the practical characteristics of various MABs to better understand the potential of these candidates for next-generation EVs, alongside those of Li-ion batteries. The data are compiled from the latest literature showcasing the practical value ranges at the cell level, whereas the energy storage cost is shown as the price of a battery pack or system level. Safety ratings for various battery types are determined qualitatively on a scale of 1 to 10, with 1 being the least safe and 10 being the safest, based on factors like the risk of toxic material release, overheating, and electrolyte flammability. Li-ion, Li–air, and Zn–air batteries are rated directly based on these criteria, while other MAB types are evaluated by comparing their properties to those of the known Li-ion, Li–air, and Zn–air batteries. Adjustments are made to account for differences in metal reactivity, electrochemical properties, and cell components.

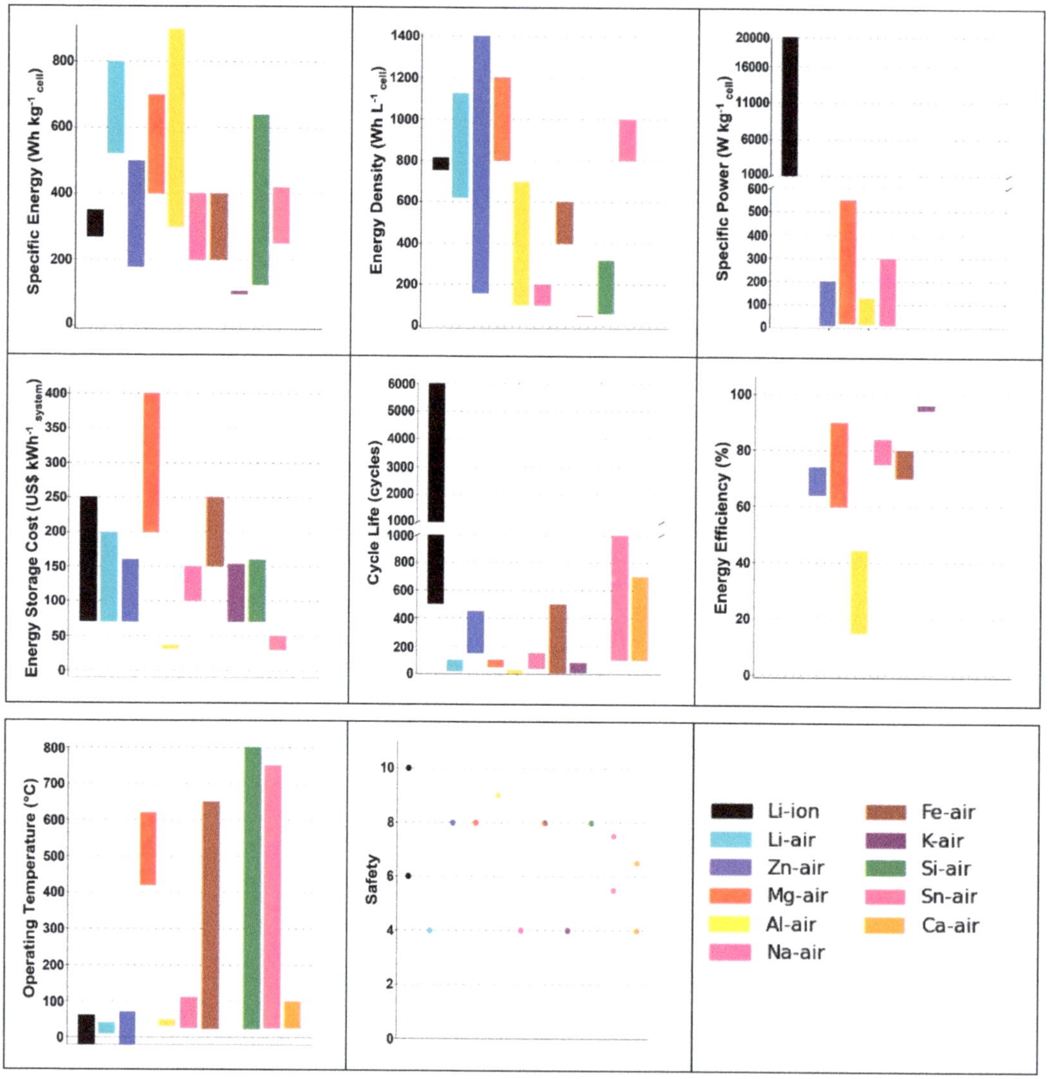

Figure 13. Comparative characteristics of various MABs and Li-ion cells. The data represent the practical value ranges at the cell level, derived from the existing literature. (For detailed references, see Supplementary Table S3).

This approach provides a logical framework for assigning relative safety values. Li-ion batteries have a variable safety profile, with ratings ranging from 6 to 10, indicating that their safety can vary significantly depending on specific conditions or configurations. Certain MABs show higher safety ratings, with Al–air batteries (aqueous electrolytes) rated at 9 and other types (Zn–air, Mg–air, Fe–air, and Si–air) rated at 8. Practical evaluation of certain battery metrics, such as energy storage cost, energy density, and specific energy, as shown in Figure 14, can be achieved by approximating their influence on the battery pack volume (Vol_B), total vehicle cost ($C_{V,T}$), and vehicle range (R_V). These metrics are directly correlated with the battery pack's energy (E_B) and can be calculated using the equations adapted from [20] and shown in Table 1. The overhead factors in Supplementary Table S3 quantify the necessary safety and air management equipment for each battery type, especially for MABs. This estimation approach for battery system mass overhead and battery system volume overhead uses comparative analysis based on the known properties of Li–air and Zn–air batteries and adjusts them according to the specific characteristics of other MABS such as the density of the metal used, packing efficiency, electrochemical properties, and cell components.

3.1. Cycle Life, Efficiency, and Safety Evaluation

Li-ion batteries, being a well-established and widely used technology, offer a highly balanced profile with their excellent cycle life (1000–6000) and consistently high energy efficiency (95%). Analyzing the capabilities of MABs and their effectiveness as range extenders is significantly shaped by their cycle life, energy efficiency, and safety profile. Zn–air batteries offer a balanced compromise with a cycle life ranging from 150 to 450 cycles and efficiency between 64% and 74%. This makes them a viable option as a range extender, given their moderate performance in both longevity and energy efficiency. In comparison, Fe–air batteries achieve a slightly higher maximum cycle life of up to 500 cycles and an efficiency of 80%. However, their relatively lower energy density and specific energy results in an unfavorable trade-off for a range extender. Among relatively high-efficiency, mechanically rechargeable batteries with a moderate cycle life, Li–air and Al–air batteries stand out. Currently, Li–air batteries offer up to 100 cycles (electrically rechargeable), while Al–air batteries exhibit efficiencies as high as 45%. Both types show impressive energy metrics and driving range, demonstrating significant potential. However, further research and development are needed to standardize these battery designs and address their current limitations. For enhanced safety in hybrid vehicles, Al–air and Zn–air batteries are preferable to Li–air, as seen from the safety profile. K–air batteries, while boasting a high efficiency of 94.2–96.1% but a poor cycle life of 8–80 cycles, also fall short in other key energy metrics, making them unsuitable for range-extending applications in the near term. Similarly, Na–air batteries, with their moderate cycle life and efficiency, also present less favorable trade-offs. Meanwhile, Si–air, Sn–air, and Ca–air batteries lack sufficient data for immediate application but could offer unique advantages or potential for future use as research and technology progress. (See Supplementary Table S3 for references to the numeric data discussed here.)

3.2. Cost Considerations

Figure 14 illustrates results for three vehicle categories: a mini vehicle (Figure 14a,b), typically found in markets that prioritize affordability; a mid-size vehicle (Figure 14c,d), prevalent in markets emphasizing range; and a semi-trailer truck (Figure 14e,f), representing heavy-duty applications. Figure 14a,c,e show the vehicle cost as it relates to the driving range, plotted until the battery volume surpasses the designated space limit within each vehicle. For the mini-vehicle, the vehicle cost is USD 10,000, with a mass of 650 kg and an energy consumption efficiency of 0.08 $Whkm^{-1}kg^{-1}$. For the mid-size vehicle, the vehicle cost is USD 25,000, with a mass of 1500 kg and an energy consumption efficiency of 0.06 $Whkm^{-1}kg^{-1}$; and the semi-trailer truck has a vehicle cost of USD 100,000, a mass of 24,000 kg, and an energy consumption efficiency of 0.0334 $Whkm^{-1}kg^{-1}$. These curves use midpoint values for battery system overhead ranges, energy storage cost, volumetric energy density, and specific energy (Supplementary Table S3). In Figure 14b,d,f, the cost of adding range per km is plotted against the vehicle's range. The curves are generated by calculating the tangents of the respective curves from Figure 14a,c,e. While these plots illustrate the inherent capability of each battery type to reduce costs and extend the driving range, the analysis does not consider factors such as power metrics, electric grid compatibility, self-discharge rate, cycle life, efficiency, and safety (as discussed in Section 3.2), which are essential for reliable operation in an electric vehicle. Na–air, K–air, and Si–air batteries have significant drawbacks due to their low energy densities, resulting in limited driving range extensions. Their cost per additional range is still greater than that of Li–air, Zn–air, and Al–air batteries, which offer comparable driving ranges at a significantly lower cost than Li-ion batteries across all three vehicle types. With the equipment to safeguard against atmospheric carbon dioxide and moisture, the utmost volumetric energy density of Li–air batteries at the vehicle level is projected to be only 384 WhL^{-1}. This volumetric limitation hinders their ability to achieve significantly longer driving ranges compared to Li-ion batteries [233]. On the flip side, Zn–air batteries, offering lower costs and greater range than Li–air batteries, are favored by researchers for forthcoming EVs because of their superior practical energy density [45,50]. Despite their potential, substantial improvements in energy efficiency, specific power, and cycling stability are necessary for the successful integration of Zn–air batteries, into electric vehicles in a dual-battery configuration. While Al–air batteries offer the lowest cost per additional kilometer, their low energy efficiency and complex mechanical recharge design may hinder their adoption. However, the combination of attributes such as low cost and high safety could indirectly boost range by powering auxiliary units, reducing the load on the main battery. The Fe–air battery has the second-highest cost for additional range, only surpassed by Mg–air. Although Mg–air batteries provide a driving range comparable to Li-ion batteries, their cost is significantly higher, exceeding what the average EV customer might expect to pay. Unless the estimated cost for both Fe–air and Mg–air is substantially reduced, these batteries may not be suitable for high-usage EVs.

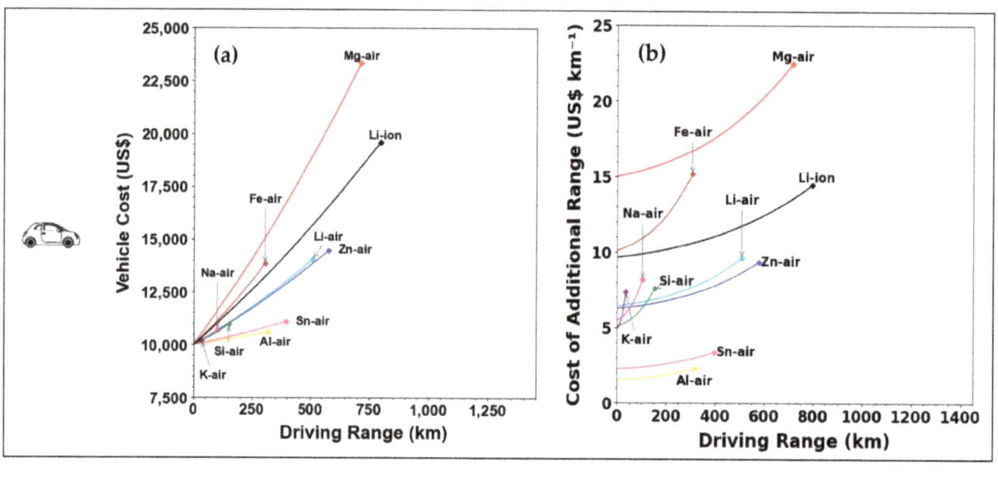

Figure 14. *Cont.*

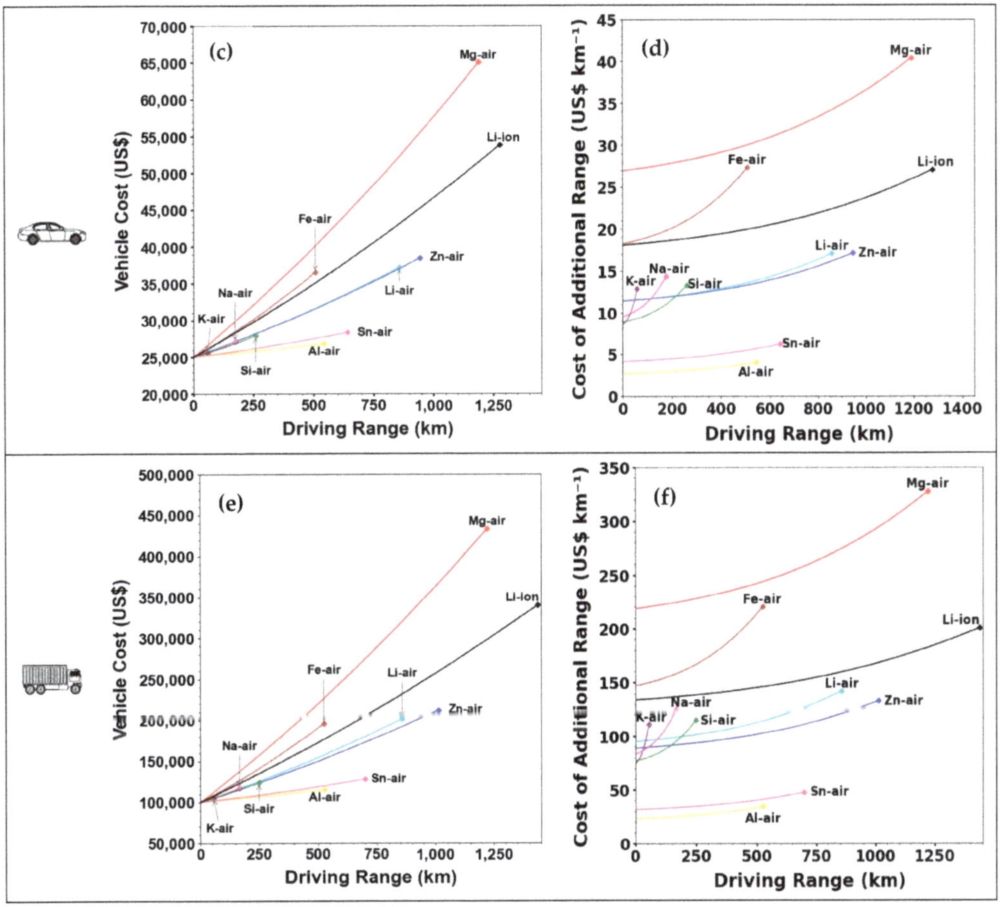

Figure 14. Vehicle cost vs. driving range and cost of additional range vs. driving range. Plots in (**a**,**c**,**e**) are computed from the equations R_V and $C_{V,T}$ (Table 1), while (**b**,**d**,**f**) are plotted from calculating the tangents of the curves of (**a**,**c**,**e**), respectively. (**a**,**b**): Mini-vehicle; (**c**,**d**): mid-size vehicle; (**e**,**f**): semi-trailer truck. Refer to Supplementary Tables S2 and S3 for further calculations and parameters.

4. Conclusions

The electrification of the transportation sector, a major contributor to climate change, necessitates the exploration of alternative battery types beyond Li-ion. MABs have emerged as a promising solution, offering the potential for extended driving ranges at reduced costs. This comparative review examines more than ten different MAB chemistries, evaluating their potential as range extenders in the EV market. The review delves into the design, functionality, and limitations of these MABs, highlighting key theoretical and practical parameters that influence their performance. This paper also shows how certain battery metrics like energy parameters and energy storage cost can be practically evaluated by estimating their effect on vehicle range, battery pack volume, and total vehicle cost. These metrics are related to the battery pack's energy. Cost considerations, particularly the relationship between vehicle cost and driving range, reveal the significant trade-offs involved in adopting these technologies. While no single battery technology can meet the diverse needs of all EV applications, certain MABs, such as Zn–air, Li–air, and Al–air, show considerable promise in addressing the specific demands of different vehicle types. Although some MABs may not be optimal for current EV applications, this review also underscores their potential significance in other domains where their unique characteristics could be highly advantageous. The viability of MABs as range extenders in EVs depends on their ability to deliver longer driving ranges and their alignment with consumer cost expectations. While Li-ion batteries are well-suited for specific electric mobility applications, broader adoption of various MAB technologies is crucial for a complete transition to clean, low-carbon transportation. This review paper lays the groundwork for researchers to develop effective, reliable, and viable MABs, paving the way for more sustainable and cost-effective solutions in the EV market.

5. Future Outlook

Though this paper reviewed MABs for EVs only, these batteries show immense potential for a wide range of applications due to their high energy density, lightweight properties, and cost-effectiveness. In the realm of light-duty transport, MABs hold promise for EVs such as passenger cars and small trucks. Li–air, Zn–air, and Al–air variants can significantly extend driving ranges compared to traditional Li-ion batteries. Their lightweight nature also makes them ideal for e-bikes and electric scooters, enhancing performance by offering longer ranges and reducing vehicle weight. Beyond transportation, MABs are already making an impact in other fields. In portable electronics, they are commonly used in hearing aids and medical devices due to their compact size and steady power output over extended periods. They are also being explored for grid storage solutions to support renewable energy from sources like solar and wind. Their cost-effectiveness and high energy density make them a viable option for large-scale energy storage, helping to stabilize power grids and ensuring reliable electricity supply. The military and aerospace sectors are investigating MABs for powering critical equipment, benefiting from their high energy density and lightweight design. Similarly, emergency backup systems in hospitals, data centers, and other critical infrastructure rely on these batteries for their long shelf life and dependable power during outages. Current implementations include Zn–air batteries in hearing aids, MABs in remote sensors deployed in harsh environments, and marine applications such as buoys and underwater sensors, where their ability to operate efficiently in wet conditions is invaluable [60]. As research continues to address challenges like improving rechargeability and enhancing performance, metal–air batteries are poised for even broader applications across multiple industries, driving innovation and supporting sustainable energy solutions.

Supplementary Materials: The following supporting information can be downloaded at https://www.mdpi.com/article/10.3390/batteries11010035/s1, Table S1: Anode Material Properties and Performance Metrics; Table S2: Electric Vehicle Specifications and Market Analysis; Table S3: Performance and Cost Metrics for Metal-Air Batteries.

Author Contributions: Conceptualization, Y.S. and S.S.M.; methodology, Y.S.; software, Y.S.; validation, Y.S., S.S.M. and S.P.; formal analysis, Y.S.; investigation, Y.S.; resources, Y.S.; data curation, Y.S.; writing—original draft preparation, Y.S.; writing—review and editing, Y.S., S.S.M., S.P., M.M. and M.F.; visualization, Y.S.; supervision, M.F.; project administration, M.F.; funding acquisition, M.F. All authors have read and agreed to the published version of the manuscript.

Funding: This research received no external funding.

Conflicts of Interest: Author Mahboubeh Mousavi was employed by the company AlumaPower Corporation, Kitchener, ON, Canada. The remaining authors declare that the research was conducted in the absence of any commercial or financial relationships that could be construed as a potential conflict of interest.

Abbreviations

Al–air	Aluminum–air batteries
BEVs	Battery electric vehicles
Ca-O_2	Calcium–oxygen battery
CNT	Carbon nanotubes
CO_2	Carbon dioxide
EVs	Electric vehicles
Fe–air	Iron–air battery
GDL	Gas diffusion layer
Ge–air	Germanium–air batteries
GPEs	Gel polymer electrolytes
HER	Hydrogen evolution reaction
ICEVs	Internal combustion engine vehicles
IEA	International Energy Agency
LEDs	Light-emitting diodes
Li–air	Lithium–air
Li-ion	Lithium-ion
LISICON	Lithium superionic conductor
MABs	Metal–air batteries
Mg–air	Magnesium–air
Na–air	Sodium–air
OCV	Open circuit voltage
OER	Oxygen-evolution reaction
ORR	Oxygen-reduction reaction
SDICH	Dual-ion-conducting hydrogels
Si–air	Silicon–air batteries
SSE	Solid-state electrolytes
UHP	Ultra-high-purity
Zn–air	Zinc–air

References

1. Østergaard, P.A.; Duic, N.; Noorollahi, Y.; Kalogirou, S. Latest progress in Sustainable Development using renewable energy technology. *Renew. Energy* **2020**, *162*, 1554–1562. [CrossRef]
2. Gaele, M.F.; Di Palma, T.M. Polymer electrolytes for Al-air batteries: Current state and future perspectives. *Energy Fuels* **2022**, *36*, 12875–12895. [CrossRef]
3. Gouveia, J.; Mendes, A.; Monteiro, R.; Mata, T.; Caetano, N.; Martins, A. Life cycle assessment of a vanadium flow battery. *Energy Rep.* **2020**, *6*, 95–101. [CrossRef]
4. Aldhafeeri, T.; Tran, M.-K.; Vrolyk, R.; Pope, M.; Fowler, M. A review of methane gas detection sensors: Recent developments and future perspectives. *Inventions* **2020**, *5*, 28. [CrossRef]
5. Shittu, E.; Suman, R.; Ravikumar, M.K.; Shukla, A.K.; Zhao, G.; Patil, S.; Baker, J. Life cycle assessment of soluble lead redox flow battery. *J. Clean. Prod.* **2022**, *337*, 130503. [CrossRef]
6. Branco, H.; Castro, R.; Lopes, A.S. Battery energy storage systems as a way to integrate renewable energy in small isolated power systems. *Energy Sustain. Dev.* **2018**, *43*, 90–99. [CrossRef]
7. Network, R.E.P. Renewables Global Status Report. Paris: REN21 Secretariat, 272p. Available online: https://www.ren21.net/gsr-2023/ (accessed on 24 June 2017).
8. Tooryan, F.; HassanzadehFard, H.; Collins, E.R.; Jin, S.; Ramezani, B. Smart integration of renewable energy resources, electrical, and thermal energy storage in microgrid applications. *Energy* **2020**, *212*, 118716. [CrossRef]
9. Madhavan, P.V.; Shahgaldi, S.; Li, X. Ex-situ Characterization of Nb-Ti Alloy/Pt Coated Stainless Steel Bipolar Plates for Proton Exchange Membrane Fuel Cells. *Energy Convers. Manag.* **2024**, *311*, 118536. [CrossRef]
10. Madhavan, P.V.; Zeng, X.; Shahgaldi, S.; Li, X. Investigation of Cr2SiC Ceramic MAX Phase Coated Metallic Bipolar Plates in Ex-situ Conditions for Proton Exchange Membrane Fuel Cells. *Int. J. Hydrogen Energy* **2024**, *96*, 1232–1242. [CrossRef]
11. Madhavan, P.V.; Shahgaldi, S.; Li, X. Modelling Anti-Corrosion Coating Performance of Metallic Bipolar Plates for PEM Fuel Cells: A Machine Learning Approach. *Energy AI* **2024**, *17*, 100391. [CrossRef]

12. Moradizadeh, L.; Madhavan, P.V.; Chellehbari, Y.M.; Gupta, A.; Li, X.; Shahgaldi, S. Porous transport layers with low Pt loading having Nb–Ta alloy as interlayer for proton exchange membrane water electrolyzers. *Int. J. Hydrogen Energy* **2024**, *94*, 1114–1129. [CrossRef]
13. Gröger, O.; Gasteiger, H.A.; Suchsland, J.-P. Electromobility: Batteries or fuel cells? *J. Electrochem. Soc.* **2015**, *162*, A2605. [CrossRef]
14. Legala, A.; Kubesh, M.; Chundru, V.R.; Conway, G.; Li, X. Machine Learning Modelling for Fuel Cell-Battery Hybrid Power System Dynamics in a Toyota Mirai 2 Vehicle Under Various Drive Cycles. *Energy AI* **2024**, *17*, 100415. [CrossRef]
15. Li, W.; Xie, Y.; Hu, X.; Zhang, B.; Fowler, M.; Panchal, S.; Fraser, R.; Zhang, Y. An efficient two-stage heating strategy for embedded heat pipe system considering power and energy requirements from battery. *Appl. Therm. Eng.* **2024**, *257*, 124499. [CrossRef]
16. Neubauer, J.; Wood, E. The impact of range anxiety and home, workplace, and public charging infrastructure on simulated battery electric vehicle lifetime utility. *J. Power Sources* **2014**, *257*, 12–20. [CrossRef]
17. Singer, M. *Consumer Views on Plug-In Electric Vehicles—National Benchmark Report*; National Renewable Energy Laboratory (NREL): Golden, CO, USA, 2016.
18. Li, W.; Long, R.; Chen, H.; Geng, J. A review of factors influencing consumer intentions to adopt battery electric vehicles. *Renew. Sustain. Energy Rev.* **2017**, *78*, 318–328. [CrossRef]
19. Tran, M.-K.; Bhatti, A.; Vrolyk, R.; Wong, D.; Panchal, S.; Fowler, M.; Fraser, R. A review of range extenders in battery electric vehicles: Current progress and future perspectives. *World Electr. Veh. J.* **2021**, *12*, 54. [CrossRef]
20. Cano, Z.P.; Banham, D.; Ye, S.; Hintennach, A.; Lu, J.; Fowler, M.; Chen, Z. Batteries and fuel cells for emerging electric vehicle markets. *Nat. Energy* **2018**, *3*, 279–289. [CrossRef]
21. Zackrisson, M.; Fransson, K.; Hildenbrand, J.; Lampic, G.; O'Dwyer, C. Life cycle assessment of lithium-air battery cells. *J. Clean. Prod.* **2016**, *135*, 299–311. [CrossRef]
22. Zhang, X.; Li, Z.; Luo, L.; Fan, Y.; Du, Z. A review on thermal management of lithium-ion batteries for electric vehicles. *Energy* **2022**, *238*, 121652. [CrossRef]
23. Pu, J.H.; Li, Y.; Li, R.C.; Hua, N.; Zhang, H.; Lu, Y.; Panchal, S.; Fraser, R.; Fowler, M.; Zhang, X.-K. Design and performance of a compact lightweight hybrid thermal management system using phase change material and liquid cooling with a honeycomb-like structure for prismatic lithium-ion batteries. *J. Power Sources* **2024**, *624*, 235632. [CrossRef]
24. Ma, W.; Xie, Y.; Guo, S.; Li, W.; Yang, R.; Panchal, S.; Zhang, Y. A mechanism-data driven resistance transfer algorithm for lithium-ion batteries and its application to thermal modeling. *J. Energy Storage* **2024**, *102*, 114066. [CrossRef]
25. O'hayre, R.; Cha, S.-W.; Colella, W.; Prinz, F.B. *Fuel Cell Fundamentals*; John Wiley & Sons: Hoboken, NJ, USA, 2016.
26. Alaswad, A.; Baroutaji, A.; Achour, H.; Carton, J.; Al Makky, A.; Olabi, A.-G. Developments in fuel cell technologies in the transport sector. *Int. J. Hydrogen Energy* **2016**, *41*, 16499–16508. [CrossRef]
27. Ramasubramanian, B.; Rao, R.P.; Chellappan, V.; Ramakrishna, S. Towards sustainable fuel cells and batteries with an AI perspective. *Sustainability* **2022**, *14*, 16001. [CrossRef]
28. Bank, D. *Welcome to the Lithium-Ion Age*; Deutsche Bank AG: Sydney, Australia, 2016.
29. Van Noorden, R. A better battery. *Nature* **2014**, *507*, 26. [CrossRef] [PubMed]
30. Placke, T.; Kloepsch, R.; Dühnen, S.; Winter, M. Lithium ion, lithium metal, and alternative rechargeable battery technologies: The odyssey for high energy density. *J. Solid State Electrochem.* **2017**, *21*, 1939–1964. [CrossRef]
31. Nayak, P.K.; Yang, L.; Brehm, W.; Adelhelm, P. From lithium-ion to sodium-ion batteries: Advantages, challenges, and surprises. *Angew. Chem. Int. Ed.* **2018**, *57*, 102–120. [CrossRef] [PubMed]
32. Li, M.; Lu, J.; Chen, Z.; Amine, K. 30 years of lithium-ion batteries. *Adv. Mater.* **2018**, *30*, 1800561. [CrossRef]
33. Popp, D. Fit for 55: MEPs Back Objective of Zero Emissions for Cars and Vans in 2035. 2022. Available online: https://www.europarl.europa.eu/news (accessed on 5 October 2024).
34. Yu, X.; Chen, R.; Gan, L.; Li, H.; Chen, L. Battery safety: From lithium-ion to solid-state batteries. *Engineering* **2023**, *21*, 9–14. [CrossRef]
35. Feng, X.; Ren, D.; He, X.; Ouyang, M. Mitigating thermal runaway of lithium-ion batteries. *Joule* **2020**, *4*, 743–770. [CrossRef]
36. Deng, D. Li-ion batteries: Basics, progress, and challenges. *Energy Sci. Eng.* **2015**, *3*, 385–418. [CrossRef]
37. Mokhtar, M.; Talib, M.Z.M.; Majlan, E.H.; Tasirin, S.M.; Ramli, W.M.F.W.; Daud, W.R.W.; Sahari, J. Recent developments in materials for aluminum–air batteries: A review. *J. Ind. Eng. Chem.* **2015**, *32*, 1–20. [CrossRef]
38. Wen, H.; Liu, Z.; Qiao, J.; Chen, R.; Zhao, R.; Wu, J.; Qiao, G.; Yang, J. High energy efficiency and high power density aluminum-air flow battery. *Int. J. Energy Res.* **2020**, *44*, 7568–7579. [CrossRef]
39. Hou, J.; Yang, M.; Wang, D.; Zhang, J. Fundamentals and challenges of lithium ion batteries at temperatures between −40 and 60° C. *Adv. Energy Mater.* **2020**, *10*, 1904152. [CrossRef]
40. Bandhauer, T.M.; Garimella, S.; Fuller, T.F. A critical review of thermal issues in lithium-ion batteries. *J. Electrochem. Soc.* **2011**, *158*, R1. [CrossRef]
41. Costa, C.M.; Barbosa, J.C.; Gonçalves, R.; Castro, H.; Del Campo, F.; Lanceros-Méndez, S. Recycling and environmental issues of lithium-ion batteries: Advances, challenges and opportunities. *Energy Storage Materials* **2021**, *37*, 433–465. [CrossRef]

42. Ma, J.; Li, Y.; Grundish, N.S.; Goodenough, J.B.; Chen, Y.; Guo, L.; Peng, Z.; Qi, X.; Yang, F.; Qie, L. The 2021 battery technology roadmap. *J. Phys. D Appl. Phys.* **2021**, *54*, 183001. [CrossRef]
43. Chen, J.; Kollmeyer, P.; Panchal, S.; Masoudi, Y.; Gross, O.; Emadi, A. Experimental results of battery power capability measurement on cells with different state of health levels. In Proceedings of the 2024 IEEE Transportation Electrification Conference and Expo (ITEC), Chicago, IL, USA, 19–21 June 2024; pp. 1–6.
44. Liu, Q.; Pan, Z.; Wang, E.; An, L.; Sun, G. Aqueous metal-air batteries: Fundamentals and applications. *Energy Storage Mater.* **2020**, *27*, 478–505. [CrossRef]
45. Li, Y.; Lu, J. Metal–air batteries: Will they be the future electrochemical energy storage device of choice? *ACS Energy Lett.* **2017**, *2*, 1370–1377. [CrossRef]
46. Li, T.; Huang, M.; Bai, X.; Wang, Y.-X. Metal–air batteries: A review on current status and future applications. *Prog. Nat. Sci. Mater. Int.* **2023**, *33*, 151–171. [CrossRef]
47. Rani, B.; Yadav, J.K.; Saini, P.; Pandey, A.P.; Dixit, A. Aluminum–air batteries: Current advances and promises with future directions. *RSC Adv.* **2024**, *14*, 17628–17663. [CrossRef] [PubMed]
48. Hardwick, L.J.; De León, C.P. Rechargeable multi-valent metal-air batteries. *Johns. Matthey Technol. Rev.* **2018**, *62*, 134–149. [CrossRef]
49. Alemu, M.A.; Getie, M.Z.; Worku, A.K. Advancement of electrically rechargeable multivalent metal-air batteries for future mobility. *Ionics* **2023**, *29*, 3421–3435. [CrossRef]
50. Fu, J.; Cano, Z.P.; Park, M.G.; Yu, A.; Fowler, M.; Chen, Z. Electrically rechargeable zinc–air batteries: Progress, challenges, and perspectives. *Adv. Mater.* **2017**, *29*, 1604685. [CrossRef]
51. Buckingham, R.; Asset, T.; Atanassov, P. Aluminum-air batteries: A review of alloys, electrolytes and design. *J. Power Sources* **2021**, *498*, 229762. [CrossRef]
52. Feng, Y.; Xiong, W.; Zhang, J.; Wang, R.; Wang, N. Electrochemical discharge performance of the Mg–Al–Pb–Ce–Y alloy as the anode for Mg–air batteries. *J. Mater. Chem. A* **2016**, *4*, 8658–8668. [CrossRef]
53. Zhang, T.; Tao, Z.; Chen, J. Magnesium–air batteries: From principle to application. *Mater. Horiz.* **2014**, *1*, 196–206. [CrossRef]
54. Arafat, Y.; Azhar, M.R.; Zhong, Y.; Tadé, M.O.; Shao, Z. Metal-free carbon based air electrodes for Zn-air batteries: Recent advances and perspective. *Mater. Res. Bull.* **2021**, *140*, 111315. [CrossRef]
55. Bi, X.; Jiang, Y.; Chen, R.; Du, Y.; Zheng, Y.; Yang, R.; Wang, R.; Wang, J.; Wang, X.; Chen, Z. Rechargeable zinc–air versus lithium–air battery: From fundamental promises toward technological potentials. *Adv. Energy Mater.* **2024**, *14*, 2302388. [CrossRef]
56. Li, L.; Chang, Z.w.; Zhang, X.B. Recent progress on the development of metal-air batteries. *Adv. Sustain. Syst.* **2017**, *1*, 1700036. [CrossRef]
57. Li, T.; Peng, X.; Cui, P.; Shi, G.; Yang, W.; Chen, Z.; Huang, Y.; Chen, Y.; Peng, J.; Zou, R. Recent progress and future perspectives of flexible metal-air batteries. *SmartMat* **2021**, *2*, 519–553. [CrossRef]
58. U.S. Geological Survey. *Mineral Commodity Summaries 2024*; U.S. Geological Survey: Reston, VA, USA, 2024. Available online: https://www.usgs.gov/publications/mineral-commodity-summaries-2024 (accessed on 5 October 2024).
59. Leimer, D.H.W. Abundances of Chemical Elements in the Earth's Crust. Available online: https://sites.tntech.edu/hwleimer/geol-1040/geol-1040-lecture/abundances-of-chemical-elements-in-the-earths-crust/ (accessed on 6 June 2024).
60. Wang, C.; Yu, Y.; Niu, J.; Liu, Y.; Bridges, D.; Liu, X.; Pooran, J.; Zhang, Y.; Hu, A. Recent progress of metal–air batteries—A mini review. *Appl. Sci.* **2019**, *9*, 2787. [CrossRef]
61. Han, X.; Li, X.; White, J.; Zhong, C.; Deng, Y.; Hu, W.; Ma, T. Metal–air batteries: From static to flow system. *Adv. Energy Mater.* **2018**, *8*, 1801396. [CrossRef]
62. Cheng, F.; Chen, J. Metal–air batteries: From oxygen reduction electrochemistry to cathode catalysts. *Chem. Soc. Rev.* **2012**, *41*, 2172–2192. [CrossRef]
63. Cheng, J.; Zhang, L.; Yang, Y.-S.; Wen, Y.-H.; Cao, G.-P.; Wang, X.-D. Preliminary study of single flow zinc–nickel battery. *Electrochem. Commun.* **2007**, *9*, 2639–2642. [CrossRef]
64. Fu, J.; Zhang, J.; Song, X.; Zarrin, H.; Tian, X.; Qiao, J.; Rasen, L.; Li, K.; Chen, Z. A flexible solid-state electrolyte for wide-scale integration of rechargeable zinc–air batteries. *Energy Environ. Sci.* **2016**, *9*, 663–670. [CrossRef]
65. Zhou, G.; Li, F.; Cheng, H.-M. Progress in flexible lithium batteries and future prospects. *Energy Environ. Sci.* **2014**, *7*, 1307–1338. [CrossRef]
66. Sumboja, A.; Ge, X.; Zong, Y.; Liu, Z. Progress in development of flexible metal–air batteries. *Funct. Mater. Lett.* **2016**, *9*, 1630001. [CrossRef]
67. Wang, Y.; Kwok, H.; Pan, W.; Zhang, H.; Leung, D.Y. Innovative paper-based Al-air batteries as a low-cost and green energy technology for the miniwatt market. *J. Power Sources* **2019**, *414*, 278–282. [CrossRef]
68. Li, F.; Zhang, T.; Zhou, H. Challenges of non-aqueous Li–O_2 batteries: Electrolytes, catalysts, and anodes. *Energy Environ. Sci.* **2013**, *6*, 1125–1141. [CrossRef]

69. Yadegari, H.; Sun, Q.; Sun, X. Sodium-oxygen batteries: A comparative review from chemical and electrochemical fundamentals to future perspective. *Adv. Mater.* **2016**, *28*, 7065–7093. [CrossRef]
70. Ren, X.; Wu, Y. A low-overpotential potassium–oxygen battery based on potassium superoxide. *J. Am. Chem. Soc.* **2013**, *135*, 2923–2926. [CrossRef] [PubMed]
71. Zhang, X.-B. *Metal-Air Batteries: Fundamentals and Applications*; John Wiley & Sons: Hoboken, NJ, USA, 2018.
72. Liu, Q.; Chang, Z.; Li, Z.; Zhang, X. Flexible metal–air batteries: Progress, challenges, and perspectives. *Small Methods* **2018**, *2*, 1700231. [CrossRef]
73. McCloskey, B.D.; Scheffler, R.; Speidel, A.; Bethune, D.S.; Shelby, R.M.; Luntz, A.C. On the efficacy of electrocatalysis in nonaqueous Li–O_2 batteries. *J. Am. Chem. Soc.* **2011**, *133*, 18038–18041. [CrossRef]
74. Peng, Z.; Freunberger, S.A.; Chen, Y.; Bruce, P.G. A reversible and higher-rate Li-O_2 battery. *Science* **2012**, *337*, 563–566. [CrossRef]
75. Chen, Y.; Freunberger, S.A.; Peng, Z.; Fontaine, O.; Bruce, P.G. Charging a Li–O_2 battery using a redox mediator. *Nat. Chem.* **2013**, *5*, 489–494. [CrossRef] [PubMed]
76. Xu, J.-J.; Wang, Z.-L.; Xu, D.; Zhang, L.-L.; Zhang, X.-B. Tailoring deposition and morphology of discharge products towards high-rate and long-life lithium-oxygen batteries. *Nat. Commun.* **2013**, *4*, 2438. [CrossRef]
77. Zhang, T.; Zhou, H. A reversible long-life lithium–air battery in ambient air. *Nat. Commun.* **2013**, *4*, 1817. [CrossRef]
78. Tu, K.-N.; Gusak, A. Mean-time-to-failure equations for electromigration, thermomigration, and stress migration. *IEEE Trans. Compon. Packag. Manuf. Technol.* **2020**, *10*, 1427–1431. [CrossRef]
79. Bard, A.J.; Faulkner, L.R.; White, H.S. *Electrochemical Methods: Fundamentals and Applications*; John Wiley & Sons: Hoboken, NJ, USA, 2022.
80. Ye, L.; Liao, M.; Zhao, T.; Sun, H.; Zhao, Y.; Sun, X.; Wang, B.; Peng, H. A sodiophilic interphase-mediated, dendrite-free anode with ultrahigh specific capacity for sodium-metal batteries. *Angew. Chem.* **2019**, *131*, 17210–17216. [CrossRef]
81. Olabi, A.G.; Sayed, E.T.; Wilberforce, T.; Jamal, A.; Alami, A.H.; Elsaid, K.; Rahman, S.M.A.; Shah, S.K.; Abdelkareem, M.A. Metal-air batteries—A review. *Energies* **2021**, *14*, 7373. [CrossRef]
82. Arai, H. Introduction—General features of metal-air batteries. In *Electrochemical Power Sources: Fundamentals, Systems, and Applications*; Elsevier: Amsterdam, The Netherlands, 2021; pp. 1–10.
83. Arai, H.; Ikezawa, A. *Electrochemical Devices | Electrochemical Power Sources: Fundamentals, Systems, and Applications*; Elsevier: Amsterdam, The Netherlands, 2021.
84. Arai, H. Metal Storage/Metal Air (Zn, Fe, Al, Mg). In *Electrochemical Energy Storage for Renewable Sources and Grid Balancing*; Elsevier: Amsterdam, The Netherlands, 2015; pp. 337–344.
85. Arai, H.; Hayashi, M. Secondary batteries–metal-air Systems | Overview (secondary and primary). *Encycl. Electrochem. Power Sources* **2009**, 347–355. [CrossRef]
86. Xu, S.; Lau, S.; Archer, L.A. CO_2 and ambient air in metal–oxygen batteries: Steps towards reality. *Inorg. Chem. Front.* **2015**, *2*, 1070–1079. [CrossRef]
87. Salado, M.; Lizundia, E. Advances, challenges, and environmental impacts in metal–air battery electrolytes. *Mater. Today Energy* **2022**, *28*, 101064. [CrossRef]
88. Park, M.; Kim, K.Y.; Seo, H.; Cheon, Y.E.; Koh, J.H.; Sun, H.; Kim, T.J. Practical challenges associated with catalyst development for the commercialization of Li-air batteries. *J. Electrochem. Sci. Technol.* **2014**, *5*, 1–18. [CrossRef]
89. Suryatna, A.; Raya, I.; Thangavelu, L.; Alhachami, F.R.; Kadhim, M.M.; Altimari, U.S.; Mahmoud, Z.H.; Mustafa, Y.F.; Kianfar, E. A Review of High-Energy Density Lithium-Air Battery Technology: Investigating the Effect of Oxides and Nanocatalysts. *J. Chem.* **2022**, *2022*, 2762647. [CrossRef]
90. Li, J.; Hou, L.; Yu, M.; Li, Q.; Zhang, T.; Sun, H. Review and Recent Advances of Oxygen Transfer in Li-air Batteries. *ChemElectroChem* **2021**, *8*, 3588–3603. [CrossRef]
91. Jilani, A.; Awan, Z.; Taqvi, S.A.A.; Khan, F.; Alshahrani, T. Recent Advances in the Development of Li-Air Batteries, Experimental and Predictive Approaches–Prospective, Challenges, and Opportunities. *ChemBioEng Rev.* **2024**, *11*, 95–114. [CrossRef]
92. Manthiram, A.; Li, L. Hybrid and Aqueous Lithium-Air Batteries. *Adv. Energy Mater.* **2015**, *5*, 1401302. [CrossRef]
93. Zhang, Z.; Xiao, X.; Zhu, X.; Tan, P. Addressing transport issues in non-aqueous Li–air batteries to achieving high electrochemical performance. *Electrochem. Energy Rev.* **2023**, *6*, 18. [CrossRef]
94. Luntz, A.C.; McCloskey, B.D. Nonaqueous Li–air batteries: A status report. *Chem. Rev.* **2014**, *114*, 11721–11750. [CrossRef] [PubMed]
95. Feng, N.; He, P.; Zhou, H. Critical challenges in rechargeable aprotic Li–O_2 batteries. *Adv. Energy Mater.* **2016**, *6*, 1502303. [CrossRef]
96. Shu, C.; Wang, J.; Long, J.; Liu, H.K.; Dou, S.X. Understanding the reaction chemistry during charging in aprotic lithium–oxygen batteries: Existing problems and solutions. *Adv. Mater.* **2019**, *31*, 1804587. [CrossRef] [PubMed]
97. Zhang, T.; Imanishi, N.; Takeda, Y.; Yamamoto, O. Aqueous lithium/air rechargeable batteries. *Chem. Lett.* **2011**, *40*, 668–673. [CrossRef]

98. Imanishi, N.; Yamamoto, O. Rechargeable lithium–air batteries: Characteristics and prospects. *Mater. Today* **2014**, *17*, 24–30. [CrossRef]
99. Imanishi, N.; Yamamoto, O. Perspectives and challenges of rechargeable lithium–air batteries. *Mater. Today Adv.* **2019**, *4*, 100031. [CrossRef]
100. Chen, P.; Bai, F.; Deng, J.W.; Liu, B.; Zhang, T. Recent progresses and challenges in aqueous lithium–air batteries relating to the solid electrolyte separator: A mini-review. *Front. Chem.* **2022**, *10*, 1035691. [CrossRef] [PubMed]
101. Imanishi, N.; Hasegawa, S.; Zhang, T.; Hirano, A.; Takeda, Y.; Yamamoto, O. Lithium anode for lithium-air secondary batteries. *J. Power Sources* **2008**, *185*, 1392–1397. [CrossRef]
102. Wang, Y.; Zhou, H. A lithium-air battery with a potential to continuously reduce O_2 from air for delivering energy. *J. Power Sources* **2010**, *195*, 358–361. [CrossRef]
103. Zhang, T.; Imanishi, N.; Hasegawa, S.; Hirano, A.; Xie, J.; Takeda, Y.; Yamamoto, O.; Sammes, N. Li/Polymer Electrolyte/Water Stable Lithium-Conducting Glass Ceramics Composite for Lithium–Air Secondary Batteries with an Aqueous Electrolyte. *J. Electrochem. Soc.* **2008**, *155*, A965. [CrossRef]
104. He, P.; Wang, Y.; Zhou, H. A Li-air fuel cell with recycle aqueous electrolyte for improved stability. *Electrochem. Commun.* **2010**, *12*, 1686–1689. [CrossRef]
105. Hasegawa, S.; Imanishi, N.; Zhang, T.; Xie, J.; Hirano, A.; Takeda, Y.; Yamamoto, O. Study on lithium/air secondary batteries—Stability of NASICON-type lithium ion conducting glass–ceramics with water. *J. Power Sources* **2009**, *189*, 371–377. [CrossRef]
106. Li, L.; Zhao, X.; Manthiram, A. A dual-electrolyte rechargeable Li-air battery with phosphate buffer catholyte. *Electrochem. Commun.* **2012**, *14*, 78–81. [CrossRef]
107. Shimonishi, Y.; Zhang, T.; Johnson, P.; Imanishi, N.; Hirano, A.; Takeda, Y.; Yamamoto, O.; Sammes, N. A study on lithium/air secondary batteries—Stability of NASICON-type glass ceramics in acid solutions. *J. Power Sources* **2010**, *195*, 6187–6191. [CrossRef]
108. Ding, F.; Xu, W.; Shao, Y.; Chen, X.; Wang, Z.; Gao, F.; Liu, X.; Zhang, J.-G. H+ diffusion and electrochemical stability of Li1+ x+ yAlxTi2−xSiyP3−yO12 glass in aqueous Li/air battery electrolytes. *J. Power Sources* **2012**, *214*, 292–297. [CrossRef]
109. Bidault, F.; Brett, D.; Middleton, P.; Brandon, N. Review of gas diffusion cathodes for alkaline fuel cells. *J. Power Sources* **2009**, *187*, 39–48. [CrossRef]
110. He, P.; Zhang, T.; Jiang, J.; Zhou, H. Lithium–air batteries with hybrid electrolytes. *J. Phys. Chem. Lett.* **2016**, *7*, 1267–1280. [CrossRef]
111. Manthiram, A.; Yu, X.; Wang, S. Lithium battery chemistries enabled by solid-state electrolytes. *Nat. Rev. Mater.* **2017**, *2*, 1–16. [CrossRef]
112. Choudhury, S.; Stalin, S.; Deng, Y.; Archer, L.A. Soft colloidal glasses as solid-state electrolytes. *Chem. Mater.* **2018**, *30*, 5996–6004. [CrossRef]
113. Wu, Y.; Li, Y.; Wang, Y.; Liu, Q.; Chen, Q.; Chen, M. Advances and prospects of PVDF based polymer electrolytes. *J. Energy Chem.* **2022**, *64*, 62–84. [CrossRef]
114. Wang, H.-F.; Wang, X.-X.; Li, F.; Xu, J.-J. Fundamental Understanding and Construction of Solid-State Li−Air Batteries. *Small Sci.* **2022**, *2*, 2200005. [CrossRef]
115. Vadhva, P.; Gill, T.E.; Cruddos, J.H.; Said, S.; Siniscalchi, M.; Narayanan, S.; Pasta, M.; Miller, T.S.; Rettie, A.J. Engineering solution-processed non-crystalline solid electrolytes for Li metal batteries. *Chem. Mater.* **2023**, *35*, 1168–1176. [CrossRef]
116. Raj, H.; Fabre, T.; Lachal, M.; Neveu, A.; Jean, J.; Steil, M.C.; Bouchet, R.; Pralong, V. Stabilizing the NASICON Solid Electrolyte in an Inert Atmosphere as a Function of Physical Properties and Sintering Conditions for Solid-State Battery Fabrication. *ACS Appl. Energy Mater.* **2023**, *6*, 1197–1207. [CrossRef]
117. Xu, A.; Wang, R.; Yao, M.; Cao, J.; Li, M.; Yang, C.; Liu, F.; Ma, J. Electrochemical properties of an Sn-doped LATP ceramic electrolyte and its derived sandwich-structured composite solid electrolyte. *Nanomaterials* **2022**, *12*, 2082. [CrossRef]
118. Zhang, H.; Yu, Z.; Cheng, J.; Chen, H.; Huang, X.; Tian, B. Halide/sulfide composite solid-state electrolyte for Li-anode based all-solid-state batteries. *Chin. Chem. Lett.* **2023**, *34*, 108228. [CrossRef]
119. Hamao, N.; Hamamoto, K. Fabrication of single-grain-layered garnet-type electrolyte sheets by a precursor method. *J. Asian Ceram. Soc.* **2022**, *10*, 1–8. [CrossRef]
120. Sun, J.; Zhao, N.; Li, Y.; Guo, X.; Feng, X.; Liu, X.; Liu, Z.; Cui, G.; Zheng, H.; Gu, L. A rechargeable Li-air fuel cell battery based on garnet solid electrolytes. *Sci. Rep.* **2017**, *7*, 41217. [CrossRef] [PubMed]
121. Lu, J.; Li, Y. Perovskite-type Li-ion solid electrolytes: A review. *J. Mater. Sci. Mater. Electron.* **2021**, *32*, 9736–9754. [CrossRef]
122. Hashishin, T.; Shimomura, H.; Kamiyama, R.; Matsuda, M. Analcime with high sodium ion conduction as a solid electrolyte. *J. Phys. Chem. C* **2022**, *126*, 19480–19486. [CrossRef]
123. Cao, W.; Yang, Y.; Deng, J.; Li, Y.; Cui, C.; Zhang, T. Localization of electrons within interlayer stabilizes NASICON-type solid-state electrolyte. *Mater. Today Energy* **2021**, *22*, 100875. [CrossRef]
124. Xu, Z.; Liu, Z.; Gu, Z.; Zhao, X.; Guo, D.; Yao, X. Polyimide-based solid-state gel polymer electrolyte for lithium–oxygen batteries with a long-cycling life. *ACS Appl. Mater. Interfaces* **2023**, *15*, 7014–7022. [CrossRef] [PubMed]

125. Hamao, N.; Yamaguchi, Y.; Hamamoto, K. Densification of a NASICON-type LATP electrolyte sheet by a cold-sintering process. *Materials* **2021**, *14*, 4737. [CrossRef] [PubMed]
126. Zhu, Q.; Ma, J.; Li, S.; Mao, D. Solid-state electrolyte for lithium-air batteries: A review. *Polymers* **2023**, *15*, 2469. [CrossRef] [PubMed]
127. Sun, Z.; Lu, H.; Fan, L.; Hong, Q.; Leng, J.; Chen, C. Performance of Al-air batteries based on Al–Ga, Al–In and Al–Sn alloy electrodes. *J. Electrochem. Soc.* **2015**, *162*, A2116. [CrossRef]
128. Zhang, Z.; Xiong, L.; Wang, S.; Xie, Y.; You, W.; Du, X. A review of the Al-gas batteries and perspectives for a "Real" Al-air battery. *J. Power Sources* **2023**, *580*, 233375. [CrossRef]
129. Elia, G.A.; Marquardt, K.; Hoeppner, K.; Fantini, S.; Lin, R.; Knipping, E.; Peters, W.; Drillet, J.F.; Passerini, S.; Hahn, R. An overview and future perspectives of aluminum batteries. *Adv. Mater.* **2016**, *28*, 7564–7579. [CrossRef] [PubMed]
130. Chen, L.; Li, B.; Zhu, L.; Deng, X.; Sun, X.; Liu, Y.; Zhang, C.; Zhao, W.; Chen, X. A PVA/LiCl/PEO interpenetrating composite electrolyte with a three-dimensional dual-network for all-solid-state flexible aluminum–air batteries. *RSC Adv.* **2021**, *11*, 39476–39483. [CrossRef] [PubMed]
131. Haynes, W.M. *CRC Handbook of Chemistry and Physics*, 95th ed.; Taylor & Francis Group, LLC: Abingdon, UK, 2014.
132. Hu, Y.; Sun, D.; Luo, B.; Wang, L. Recent progress and future trends of aluminum batteries. *Energy Technol.* **2019**, *7*, 86–106. [CrossRef]
133. Bogolowski, N.; Drillet, J.-F. An electrically rechargeable Al-air battery with aprotic ionic liquid electrolyte. *ECS Trans.* **2017**, *75*, 85. [CrossRef]
134. Carlin, R.T.; Osteryoung, R.A. Aluminum anodization in a basic ambient temperature molten salt. *J. Electrochem. Soc.* **1989**, *136*, 1409. [CrossRef]
135. Ocon, J.D.; Kim, J.W.; Abrenica, G.H.A.; Lee, J.K.; Lee, J. Quasi-perpetual discharge behaviour in p-type Ge–air batteries. *Phys. Chem. Chem. Phys.* **2014**, *16*, 22487–22494. [CrossRef]
136. Revel, R.; Audichon, T.; Gonzalez, S. Non-aqueous aluminium–air battery based on ionic liquid electrolyte. *J. Power Sources* **2014**, *272*, 415–421. [CrossRef]
137. Wang, D.; Gao, X.; Chen, Y.; Jin, L.; Kuss, C.; Bruce, P.G. Plating and stripping calcium in an organic electrolyte. *Nat. Mater.* **2018**, *17*, 16–20. [CrossRef] [PubMed]
138. Schröder, D.; Janek, J.; Adelhelm, P. Nonlithium Aprotic Metal/Oxygen Batteries Using Na, K, Mg, or Ca as Metal Anode. *Encycl. Electrochem. Online* **2007**, 1–29. [CrossRef]
139. Xiao, N.; Ren, X.; McCulloch, W.D.; Gourdin, G.; Wu, Y. Potassium superoxide: A unique alternative for metal–air batteries. *Acc. Chem. Res.* **2018**, *51*, 2335–2343. [CrossRef] [PubMed]
140. Deyab, M.; Mohsen, Q. Improved battery capacity and cycle life in iron-air batteries with ionic liquid. *Renew. Sustain. Energy Rev.* **2021**, *139*, 110729. [CrossRef]
141. McKerracher, R.; Ponce de Leon, C.; Wills, R.; Shah, A.A.; Walsh, F.C. A review of the iron–air secondary battery for energy storage. *ChemPlusChem* **2015**, *80*, 323–335. [CrossRef]
142. Zhao, X.; Xu, N.; Li, X.; Gong, Y.; Huang, K. Energy storage characteristics of a new rechargeable solid oxide iron–air battery. *RSC Adv.* **2012**, *2*, 10163–10166. [CrossRef]
143. Arai, H.; Garche, J.; Colmenares, L.C. *Electrochemical Power Sources: Fundamentals, Systems, and Applications: Metal–Air Batteries: Present and Perspectives*; Elsevier: Amsterdam, The Netherlands, 2020.
144. Sun, Q.; Dai, L.; Luo, T.; Wang, L.; Liang, F.; Liu, S. Recent advances in solid-state metal–air batteries. *Carbon Energy* **2023**, *5*, e276. [CrossRef]
145. Zhao, X.; Gong, Y.; Li, X.; Xu, N.; Huang, K. Performance of solid oxide iron-air battery operated at 550 C. *J. Electrochem. Soc.* **2013**, *160*, A1241. [CrossRef]
146. Jin, X.; Uddin, A.M.; Zhao, X.; White, R.; Huang, K. Understanding the High-Temperature Solid-Oxide Iron-Air Redox Battery Operated with an Oxygen Shuttle Mechanism: A Computational Study. *J. Electrochem. Soc.* **2015**, *162*, A1476. [CrossRef]
147. Girishkumar, G.; McCloskey, B.; Luntz, A.C.; Swanson, S.; Wilcke, W. Lithium-air battery: Promise and challenges. *J. Phys. Chem. Lett.* **2010**, *1*, 2193–2203. [CrossRef]
148. Lu, J.; Li, L.; Park, J.-B.; Sun, Y.-K.; Wu, F.; Amine, K. Aprotic and aqueous Li–O_2 batteries. *Chem. Rev.* **2014**, *114*, 5611–5640. [CrossRef]
149. Kowalczk, I.; Read, J.; Salomon, M. Li-air batteries: A classic example of limitations owing to solubilities. *Pure Appl. Chem.* **2007**, *79*, 851–860. [CrossRef]
150. Geng, D.; Ding, N.; Hor, T.A.; Chien, S.W.; Liu, Z.; Wuu, D.; Sun, X.; Zong, Y. From lithium-oxygen to lithium-air batteries: Challenges and opportunities. *Adv. Energy Mater.* **2016**, *6*, 1502164. [CrossRef]
151. Grande, L.; Paillard, E.; Hassoun, J.; Park, J.B.; Lee, Y.J.; Sun, Y.K.; Passerini, S.; Scrosati, B. The lithium/air battery: Still an emerging system or a practical reality? *Adv. Mater.* **2015**, *27*, 784–800. [CrossRef] [PubMed]
152. Staniewicz, R.J. A study of the calcium-thionyl chloride electrochemical system. *J. Electrochem. Soc.* **1980**, *127*, 782. [CrossRef]

153. Weinrich, H.; Durmus, Y.E.; Tempel, H.; Kungl, H.; Eichel, R.-A. Silicon and iron as resource-efficient anode materials for ambient-temperature metal-air batteries: A review. *Materials* **2019**, *12*, 2134. [CrossRef]
154. Xu, X.; San Hui, K.; Dinh, D.A.; Hui, K.N.; Wang, H. Recent advances in hybrid sodium–air batteries. *Mater. Horiz.* **2019**, *6*, 1306–1335. [CrossRef]
155. Faktorovich-Simon, E.; Natan, A.; Peled, E.; Golodnitsky, D. Oxygen redox processes in PEGDME-based electrolytes for the Na-air battery. *J. Solid State Electrochem.* **2018**, *22*, 1015–1022. [CrossRef]
156. Ju, H.; Lee, J. High-temperature liquid Sn-air energy storage cell. *J. Energy Chem.* **2015**, *24*, 614–619. [CrossRef]
157. Andrés, C.M.C.; Pérez de la Lastra, J.M.; Andrés Juan, C.; Plou, F.J.; Pérez-Lebeña, E. Superoxide Anion Chemistry—Its Role at the Core of the Innate Immunity. *Int. J. Mol. Sci.* **2023**, *24*, 1841. [CrossRef] [PubMed]
158. Skyllas-Kazacos, M.; Cao, L.; Kazacos, M.; Kausar, N.; Mousa, A. Vanadium electrolyte studies for the vanadium redox battery—A review. *ChemSusChem* **2016**, *9*, 1521–1543. [CrossRef]
159. Skyllas-Kazacos, M.; Grossmith, F. Efficient vanadium redox flow cell. *J. Electrochem. Soc.* **1987**, *134*, 2950. [CrossRef]
160. Wu, W.-F.; Yan, X.; Zhan, Y. Recent progress of electrolytes and electrocatalysts in neutral aqueous zinc-air batteries. *Chem. Eng. J.* **2023**, *451*, 138608. [CrossRef]
161. Li, Y.; Dai, H. Recent advances in zinc–air batteries. *Chem. Soc. Rev.* **2014**, *43*, 5257–5275. [CrossRef]
162. Xu, M.; Ivey, D.; Xie, Z.; Qu, W. Rechargeable Zn–air batteries: Progress in electrolyte development and cell configuration advancement. *J. Power Sources* **2015**, *283*, 358–371. [CrossRef]
163. Pan, J.; Xu, Y.Y.; Yang, H.; Dong, Z.; Liu, H.; Xia, B.Y. Advanced architectures and relatives of air electrodes in Zn–air batteries. *Adv. Sci.* **2018**, *5*, 1700691. [CrossRef] [PubMed]
164. Lv, X.W.; Wang, Z.; Lai, Z.; Liu, Y.; Ma, T.; Geng, J.; Yuan, Z.Y. Rechargeable zinc–air batteries: Advances, challenges, and prospects. *Small* **2024**, *20*, 2306396. [CrossRef]
165. Li, Y.; Gong, M.; Liang, Y.; Feng, J.; Kim, J.-E.; Wang, H.; Hong, G.; Zhang, B.; Dai, H. Advanced zinc-air batteries based on high-performance hybrid electrocatalysts. *Nat. Commun.* **2013**, *4*, 1805. [CrossRef] [PubMed]
166. Wang, H.F.; Tang, C.; Zhang, Q. A review of precious-metal-free bifunctional oxygen electrocatalysts: Rational design and applications in Zn–air batteries. *Adv. Funct. Mater.* **2018**, *28*, 1803329. [CrossRef]
167. Yu, P.; Wang, L.; Sun, F.; Xie, Y.; Liu, X.; Ma, J.; Wang, X.; Tian, C.; Li, J.; Fu, H. Co Nanoislands rooted on Co–N–C nanosheets as efficient oxygen electrocatalyst for Zn air batteries. *Adv. Mater.* **2019**, *31*, 1901666. [CrossRef]
168. Han, X.; Ling, X.; Wang, Y.; Ma, T.; Zhong, C.; Hu, W.; Deng, Y. Spatial isolation of zeolitic imidazole frameworks-derived cobalt catalysts: From nanoparticle, atomic cluster to single atom. *Angew. Chem. Int. Ed.* **2019**, *58*, 5359–5364. [CrossRef]
169. Shang, H.; Zhou, X.; Dong, J.; Li, A.; Zhao, X.; Liu, Q.; Lin, Y.; Pei, J.; Li, Z.; Jiang, Z. Engineering unsymmetrically coordinated Cu-S1N3 single atom sites with enhanced oxygen reduction activity. *Nat. Commun.* **2020**, *11*, 3049. [CrossRef]
170. Zhang, J.; Zhou, Q.; Tang, Y.; Zhang, L.; Li, Y. Zinc–air batteries: Are they ready for prime time? *Chem. Sci.* **2019**, *10*, 8924–8929. [CrossRef] [PubMed]
171. Deiss, E.; Holzer, F.; Haas, O. Modeling of an electrically rechargeable alkaline Zn–air battery. *Electrochim. Acta* **2002**, *47*, 3995–4010. [CrossRef]
172. Ma, H.; Wang, B.; Fan, Y.; Hong, W. Development and characterization of an electrically rechargeable zinc-air battery stack. *Energies* **2014**, *7*, 6549–6557. [CrossRef]
173. Org, W.; Hong, W.; Li, H.; Wang, B. ELECTROCHEMICAL SCIENCE A Horizontal Three-Electrode Structure for Zinc-Air Batteries with Long-Term Cycle Life and High Performance. *Int. J. Electrochem. Sci.* **2016**, *11*, 3843–3851.
174. Li, P.-C.; Chien, Y.-J.; Hu, C.-C. Novel configuration of bifunctional air electrodes for rechargeable zinc–air batteries. *J. Power Sources* **2016**, *313*, 37–45. [CrossRef]
175. Park, M.G.; Lee, D.U.; Seo, M.H.; Cano, Z.P.; Chen, Z. 3D ordered mesoporous bifunctional oxygen catalyst for electrically rechargeable zinc–air batteries. *Small* **2016**, *12*, 2707–2714. [CrossRef]
176. Nazir, G.; Rehman, A.; Lee, J.-H.; Kim, C.-H.; Gautam, J.; Heo, K.; Hussain, S.; Ikram, M.; AlObaid, A.A.; Lee, S.-Y. A Review of Rechargeable Zinc–Air Batteries: Recent Progress and Future Perspectives. *Nano-Micro Lett.* **2024**, *16*, 138. [CrossRef]
177. Liu, Q.; Liu, R.; He, C.; Xia, C.; Guo, W.; Xu, Z.-L.; Xia, B.Y. Advanced polymer-based electrolytes in zinc–air batteries. *eScience* **2022**, *2*, 453–466. [CrossRef]
178. Belete, A.S.; Worku, A.K.; Ayele, D.W.; Assegie, A.A.; Teshager, M.A. The Recent Advancement of Graphene-Based Cathode Material for Rechargeable Zinc–Air Batteries. *Processes* **2024**, *12*, 1684. [CrossRef]
179. Leong, K.W.; Wang, Y.; Ni, M.; Pan, W.; Luo, S.; Leung, D.Y. Rechargeable Zn-air batteries: Recent trends and future perspectives. *Renew. Sustain. Energy Rev.* **2022**, *154*, 111771. [CrossRef]
180. Wu, M.; Zhang, G.; Tong, H.; Liu, X.; Du, L.; Chen, N.; Wang, J.; Sun, T.; Regier, T.; Sun, S. Cobalt (II) oxide nanosheets with rich oxygen vacancies as highly efficient bifunctional catalysts for ultra-stable rechargeable Zn-air flow battery. *Nano Energy* **2021**, *79*, 105409. [CrossRef]

181. Huang, S.; Zhang, H.; Zhuang, J.; Zhou, M.; Gao, M.; Zhang, F.; Wang, Q. Redox-Mediated Two-Electron Oxygen Reduction Reaction with Ultrafast Kinetics for Zn–Air Flow Battery. *Adv. Energy Mater.* **2022**, *12*, 2103622. [CrossRef]
182. Papanikolaou, M.; Hadjithoma, S.; Keramidas, O.; Drouza, C.; Amoiridis, A.; Themistokleous, A.; Hayes, S.C.; Miras, H.N.; Lianos, P.; Tsipis, A.C. Experimental and Theoretical Investigation of the Mechanism of the Reduction of O_2 from Air to O22–by VIVO2+–N, N, N-Amidate Compounds and Their Potential Use in Fuel Cells. *Inorg. Chem.* **2024**, *63*, 3229–3249. [CrossRef] [PubMed]
183. Zhang, Y.; Deng, Y.-P.; Wang, J.; Jiang, Y.; Cui, G.; Shui, L.; Yu, A.; Wang, X.; Chen, Z. Recent progress on flexible Zn-air batteries. *Energy Storage Mater.* **2021**, *35*, 538–549. [CrossRef]
184. Fang, W.; Zhao, J.; Zhang, W.; Chen, P.; Bai, Z.; Wu, M. Recent progress and future perspectives of flexible Zn-Air batteries. *J. Alloys Compd.* **2021**, *869*, 158918. [CrossRef]
185. Liu, H.; Xie, W.; Huang, Z.; Yao, C.; Han, Y.; Huang, W. Recent advances in flexible Zn–Air batteries: Materials for electrodes and electrolytes. *Small Methods* **2022**, *6*, 2101116. [CrossRef]
186. Jiang, L.; Luo, X.; Wang, D.W. A review on system and materials for aqueous flexible metal–air batteries. *Carbon Energy* **2023**, *5*, e284. [CrossRef]
187. Liu, Y.; Sun, Q.; Li, W.; Adair, K.R.; Li, J.; Sun, X. A comprehensive review on recent progress in aluminum–air batteries. *Green Energy Environ.* **2017**, *2*, 246–277. [CrossRef]
188. Liu, X.; Jiao, H.; Wang, M.; Song, W.-l.; Xue, J.; Jiao, S. Current progresses and future prospects on aluminium–air batteries. *Int. Mater. Rev.* **2022**, *67*, 734–764. [CrossRef]
189. Goel, P.; Dobhal, D.; Sharma, R. Aluminum–air batteries: A viability review. *J. Energy Storage* **2020**, *28*, 101287. [CrossRef]
190. Mori, R. Recent developments for aluminum–air batteries. *Electrochem. Energy Rev.* **2020**, *3*, 344–369. [CrossRef]
191. Mori, R. Rechargeable aluminum–air battery using various air-cathode materials and suppression of byproducts formation on both anode and air cathode. *ECS Trans.* **2017**, *80*, 377. [CrossRef]
192. Zhao, Q.; Yu, H.; Fu, L.; Wu, P.; Li, Y.; Li, Y.; Sun, D.; Wang, H.; Tang, Y. Electrolytes for aluminum–air batteries: Advances, challenges, and applications. *Sustain. Energy Fuels* **2023**, *7*, 1353–1370. [CrossRef]
193. Xu, Y.; Zhao, Y.; Ren, J.; Zhang, Y.; Peng, H. An all-solid-state fiber-shaped aluminum–air battery with flexibility, stretchability, and high electrochemical performance. *Angew. Chem.* **2016**, *128*, 8111–8114. [CrossRef]
194. Yang, S.; Knickle, H. Design and analysis of aluminum/air battery system for electric vehicles. *J. Power Sources* **2002**, *112*, 162–173. [CrossRef]
195. Tang, Y.; Lu, L.; Roesky, H.W.; Wang, L.; Huang, B. The effect of zinc on the aluminum anode of the aluminum–air battery. *J. Power Sources* **2004**, *138*, 313–318. [CrossRef]
196. Garcia-Garcia, F.; Skeldon, P.; Thompson, G.; Smith, G. The effect of nickel on alloy microstructure and electrochemical behaviour of AA1050 aluminium alloy in acid and alkaline solutions. *Electrochim. Acta* **2012**, *75*, 229–238. [CrossRef]
197. Ryu, J.; Park, M.; Cho, J. Advanced technologies for high-energy aluminum–air batteries. *Adv. Mater.* **2019**, *31*, 1804784. [CrossRef] [PubMed]
198. Yirka, B. Phinergy Demonstrates Aluminum–air Battery Capable of Fueling an Electric Vehicle for 1000 Miles. *PHYS ORG*. 27 March 2013. Available online: https://phys.org/news/2013-03-phinergy-aluminum-air-battery-capable-fueling.html (accessed on 16 August 2024).
199. Vegh, J.M.; Alonso, J.J. Design and optimization of short-range aluminum-air powered aircraft. In Proceedings of the 54th AIAA Aerospace Sciences Meeting, San Diego, CA, USA, 4–8 January 2016; p. 1026.
200. Rewatkar, P.; Goel, S. Catalyst-mitigated arrayed aluminum-air origami fuel cell with ink-jet printed custom-porosity cathode. *Energy* **2021**, *224*, 120017. [CrossRef]
201. Huang, H.; Liu, P.; Ma, Q.; Tang, Z.; Wang, M.; Hu, J. Enabling a high-performance saltwater Al-air battery via ultrasonically driven electrolyte flow. *Ultrason. Sonochem.* **2022**, *88*, 106104. [CrossRef] [PubMed]
202. Zhang, L.; Shao, Q.; Zhang, J. An overview of non-noble metal electrocatalysts and their associated air cathodes for Mg-air batteries. *Mater. Rep. Energy* **2021**, *1*, 100002. [CrossRef]
203. Raghavan, P.; Das, A.; Jabeen Fatima, M.J. *Advanced Technologies for Rechargeable Batteries: Metal Ion, Hybrid, and Metal-Air Batteries*; CRC Press: Boca Raton, FL, USA, 2024.
204. Li, C.S.; Sun, Y.; Gebert, F.; Chou, S.L. Current progress on rechargeable magnesium–air battery. *Adv. Energy Mater.* **2017**, *7*, 1700869. [CrossRef]
205. Chen, X.; Venezuela, J.; Shi, Z.; Wang, L.; Dargusch, M. Ultra-high-purity Mg-Ge anodes enable a long-lasting, high energy-density Mg-air battery. *Nano Energy* **2024**, *122*, 109269. [CrossRef]
206. Li, L.; Chen, H.; He, E.; Wang, L.; Ye, T.; Lu, J.; Jiao, Y.; Wang, J.; Gao, R.; Peng, H. High-energy-density magnesium-air battery based on dual-layer gel electrolyte. *Angew. Chem.* **2021**, *133*, 15445–15450. [CrossRef]
207. Huang, X.; Liu, Y.; Park, W.; Zhao, Z.; Li, J.; Lim, C.K.; Wong, T.H.; Yiu, C.K.; Gao, Y.; Zhou, J. Stretchable magnesium–air battery based on dual ions conducting hydrogel for intelligent biomedical applications. *InfoMat* **2023**, *5*, e12388. [CrossRef]

208. Guo, Z.; Zhao, S.; Li, T.; Su, D.; Guo, S.; Wang, G. Recent advances in rechargeable magnesium-based batteries for high-efficiency energy storage. *Adv. Energy Mater.* **2020**, *10*, 1903591. [CrossRef]
209. Peled, E.; Golodnitsky, D.; Mazor, H.; Goor, M.; Avshalomov, S. Parameter analysis of a practical lithium-and sodium-air electric vehicle battery. *J. Power Sources* **2011**, *196*, 6835–6840. [CrossRef]
210. Qin, L.; Xiao, N.; Zhang, S.; Chen, X.; Wu, Y. From K-O2 to K-Air batteries: Realizing superoxide batteries on the basis of dry ambient air. *Angew. Chem. Int. Ed.* **2020**, *59*, 10498–10501. [CrossRef] [PubMed]
211. Mao, P.; Arandiyan, H.; Mofarah, S.S.; Koshy, P.; Pozo-Gonzalo, C.; Zheng, R.; Wang, Z.; Wang, Y.; Bhargava, S.K.; Sun, H. A comprehensive review of cathode materials for Na–air batteries. *Energy Adv.* **2023**, *2*, 465–502. [CrossRef]
212. Qin, L.; Ao, H.; Wu, Y. Feasibility of achieving two-electron K–O_2 batteries. *Faraday Discuss.* **2024**, *248*, 60–74. [CrossRef]
213. Yu, W.; Lau, K.C.; Lei, Y.; Liu, R.; Qin, L.; Yang, W.; Li, B.; Curtiss, L.A.; Zhai, D.; Kang, F. Dendrite-free potassium–oxygen battery based on a liquid alloy anode. *ACS Appl. Mater. Interfaces* **2017**, *9*, 31871–31878. [CrossRef] [PubMed]
214. Wang, T.; Han, Y.; Xu, P.; Feng, X.; Ji, W.; Arandiyan, H. In-situ generating highly efficient sites for multifunctional water splitting/Zn-air battery over hierarchically architectured Fe_2Ni_3 nanoalloy/N-incorporated carbon matrix. *Chem. Eng. J.* **2022**, *450*, 138245. [CrossRef]
215. Lutz, L.; Alves Dalla Corte, D.; Tang, M.; Salager, E.; Deschamps, M.; Grimaud, A.; Johnson, L.; Bruce, P.G.; Tarascon, J.-M. Role of electrolyte anions in the Na–O_2 battery: Implications for NaO_2 solvation and the stability of the sodium solid electrolyte interphase in glyme ethers. *Chem. Mater.* **2017**, *29*, 6066–6075. [CrossRef]
216. Yadegari, H.; Sun, X. Sodium–oxygen batteries: Recent developments and remaining challenges. *Trends Chem.* **2020**, *2*, 241–253. [CrossRef]
217. Kang, Y.; Wang, S.; Zhu, S.; Gao, H.; San Hui, K.; Yuan, C.-Z.; Yin, H.; Bin, F.; Wu, X.-L.; Mai, W. Iron-modulated nickel cobalt phosphide embedded in carbon to boost power density of hybrid sodium–air battery. *Appl. Catal. B Environ.* **2021**, *285*, 119786. [CrossRef]
218. Xu, C.; Zhang, K.; Zhang, D.; Chang, S.; Liang, F.; Yan, P.; Yao, Y.; Qu, T.; Zhan, J.; Ma, W. Reversible hybrid sodium-CO2 batteries with low charging voltage and long-life. *Nano Energy* **2020**, *68*, 104318. [CrossRef]
219. Enterría, M.; Reynaud, M.; Paredes, J.I.; Medinilla, L.; Younesi, R.; Ortiz-Vitoriano, N. Driving the sodium-oxygen battery chemistry towards the efficient formation of discharge products: The importance of sodium superoxide quantification. *J. Energy Chem.* **2022**, *68*, 709–720. [CrossRef]
220. Duan, C.; Li, X.; Wang, D.; Wang, Z.; Sun, H.; Zheng, R.; Liu, Y. Nanosized high entropy spinel oxide $(FeCoNiCrMn)_3O_4$ as a highly active and ultra-stable electrocatalyst for the oxygen evolution reaction. *Sustain. Energy Fuels* **2022**, *6*, 1479–1488. [CrossRef]
221. Cohn, G.; Starosvetsky, D.; Hagiwara, R.; Macdonald, D.D.; Ein-Eli, Y. Silicon–air batteries. *Electrochem. Commun.* **2009**, *11*, 1916–1918. [CrossRef]
222. Zhong, X.; Zhang, H.; Liu, Y.; Bai, J.; Liao, L.; Huang, Y.; Duan, X. High-Capacity Silicon–Air Battery in Alkaline Solution. *ChemSusChem* **2012**, *5*, 177–180. [CrossRef]
223. Park, D.-W.; Kim, S.; Ocon, J.D.; Abrenica, G.H.A.; Lee, J.K.; Lee, J. Controlled electrochemical etching of nanoporous Si anodes and its discharge behavior in alkaline Si–air batteries. *ACS Appl. Mater. Interfaces* **2015**, *7*, 3126–3132. [CrossRef] [PubMed]
224. Durmus, Y.E.; Aslanbas, Ö.; Kayser, S.; Tempel, H.; Hausen, F.; De Haart, L.; Granwehr, J.; Ein-Eli, Y.; Eichel, R.-A.; Kungl, H. Long run discharge, performance and efficiency of primary Silicon–air cells with alkaline electrolyte. *Electrochim. Acta* **2017**, *225*, 215–224. [CrossRef]
225. Schalinski, R.; Schweizer, S.L.; Wehrspohn, R.B. The Role of Silicate Enrichment on the Discharge Duration of Silicon-Air Batteries. *ChemSusChem* **2023**, *16*, e202300077. [CrossRef]
226. Montiel Guerrero, S.S.; Durmus, Y.E.; Tempel, H.; Roth, C.; Kungl, H.; van Waasen, S.; Ein-Eli, Y.; Eichel, R.A. Unveiling the Potential of Silicon-Air Batteries for Low-Power Transient Electronics: Electrochemical Insights and Practical Application. *Batter. Supercaps* **2024**, *7*, e202300573. [CrossRef]
227. Yang, W.-T.; Yu, J.; Li, D.-X.; Hu, S.-Q.; Ding, X.; Chen, F.-Y.; Qiao, C. Performance of silicon-air batteries using industrial silicon as anode. *J. Mater. Sci. Mater. Electron.* **2023**, *34*, 800. [CrossRef]
228. Lu, Y.-T.; Neale, A.R.; Hu, C.-C.; Hardwick, L.J. Divalent nonaqueous metal-air batteries. *Front. Energy Res.* **2021**, *8*, 602918. [CrossRef]
229. Ocon, J.D.; Kim, J.W.; Uhm, S.; Mun, B.S.; Lee, J. An etched nanoporous Ge anode in a novel metal–air energy conversion cell. *Phys. Chem. Chem. Phys.* **2013**, *15*, 6333–6338. [CrossRef]
230. Ye, L.; Liao, M.; Zhang, K.; Zheng, M.; Jiang, Y.; Cheng, X.; Wang, C.; Xu, Q.; Tang, C.; Li, P. A rechargeable calcium–oxygen battery that operates at room temperature. *Nature* **2024**, *626*, 313–318. [CrossRef]
231. Fang, C.; Tang, X.; Wang, J.; Yi, Q. Performance of iron-air battery with iron nanoparticle-encapsulated CN composite electrode. *Front. Energy* **2024**, *18*, 42–53. [CrossRef]

232. Global EV Outlook 2024, Paris. 2024. Available online: https://www.iea.org/reports/global-ev-outlook-2024 (accessed on 15 October 2024).
233. Gallagher, K.G.; Goebel, S.; Greszler, T.; Mathias, M.; Oelerich, W.; Eroglu, D.; Srinivasan, V. Quantifying the promise of lithium–air batteries for electric vehicles. *Energy Environ. Sci.* **2014**, *7*, 1555–1563. [CrossRef]

batteries

MDPI

Article

Engineering Hierarchical Porous Electrodes Integrated with Conformal Ultrathin Nanosheets for Achieving Rapid Kinetics in High-Power Microbatteries

Xin Chen [1,†], Minjian Gong [2,†], Jiantao Li [2,*], Wei Yang [1,*] and Xu Xu [2,3,*]

[1] School of Materials Science and Engineering, Wuhan University of Technology, Wuhan 430070, China; 331251@whut.edu.cn

[2] State Key Laboratory of Advanced Technology for Materials Synthesis and Processing, International School of Materials Science and Engineering, Wuhan University of Technology, Wuhan 430070, China; gongminjian@whut.edu.cn

[3] Zhongyu Feima New Material Technology Innovation Center (Zhengzhou) Co., Ltd., High Technology Industrial Development Zone, Zhengzhou 450001, China

* Correspondence: jiantao_work@126.com (J.L.); yangwei830@whut.edu.cn (W.Y.); xuxu@whut.edu.cn (X.X.)

† These authors contributed equally to this work.

Abstract: With the rapid development of the Internet of Things (IoT), there is an increasing demand for batteries with high energy and power densities. Three-dimensional microstructures present a promising approach for achieving high areal mass loading and an expanded electrochemical reaction surface. However, their high cost and complexity have hindered their widespread adoption. In this study, hierarchical porous electrodes integrated with conformal ultrathin nanosheets were fabricated to enhance reaction kinetics. The hierarchical porous skeleton provides a continuous pathway for electron transport and electrolyte diffusion, while the amorphous vanadium oxide (α-VO_x) nanosheets offer short ion diffusion channels and a large electrochemical surface area. Additionally, the internal space of the hierarchical structure accommodates substantial growth of the α-VO_x nanosheets, thereby supporting high mass loading and preserving areal capacity. The resulting hierarchical electrode structure demonstrates a high energy density of 0.49 mAh cm^{-2} at 1 mA cm^{-2} and an ultrahigh power density of 410 mW cm^{-2} at 250 mA cm^{-2}. The assembled microbattery, using lithium metal as the anode, is encapsulated with a novel packaging process. This microbattery can power an electronic clock for up to 18 h on a single charge, retaining 75% of its initial capacity after 180 cycles.

Keywords: keyword hierarchical porous; ultrahigh power density; micro lithium battery

1. Introduction

With the rapid development of mass consumer electronics and the Internet of Things (IoT) [1–3], there is an increasing demand for energy storage devices with higher performance, particularly in terms of energy and power density [4,5]. As electronic devices continue to miniaturize, they rely heavily on small button batteries. Conventional button batteries, which use densely stacked active materials for high areal energy density, suffer from limited electron transport and ion diffusion paths, resulting in low output currents that fail to meet the power requirements of small microelectronic devices. Consequently, the development of microbatteries with both high energy and power densities has become urgent [6]. Meanwhile, there is growing interest in aqueous microbatteries [7–9], particularly those based on zinc electrodes, such as Mn-Zn [10,11], I_2-Zn [12], and PBA-Zn [13] batteries.

While these devices can deliver relatively high power densities, their limited operating voltage window leads to low energy densities. As a result, micro lithium batteries, which offer both a high voltage window and high energy density, are emerging as the preferred choice.

Researchers are currently employing strategies such as anode-free [14,15] designs and 3D printing [16,17] to maximize energy density. However, achieving high power densities at high active mass loadings [18] remains a significant challenge. The simultaneous attainment of high energy and power densities in conventional batteries is difficult, as increased material loading reduces the contact area between the electrode material and the electrolyte, increases ion transport distances, and disrupts electron conduction paths. To address this, strategies such as gradient [19,20] and three-dimensional electrode structures [21–24] are often used to enhance the power density. However, the 3D structuring of electrodes is frequently associated with high cost and complexity. Therefore, developing low-cost, low-complexity electrode fabrication methods is essential for advancing microbatteries. Electrochemical deposition, which is highly controllable, cost-effective, environmentally friendly, and easy to implement, is commonly used to prepare materials such as transition metal oxides and hydroxides [25].

In this work, a hierarchical porous electrode with ultrathin α-VO_x nanosheets was fabricated. The hierarchical porous current collector was prepared using a hydrogen templating method on nickel foam, followed by the conformal construction of ultrathin α-VO_x nanosheets within this framework to create high-performance electrodes. This hierarchical structure provides ample space for the growth of active materials, continuous pathways for electron conduction, and efficient electrolyte diffusion. Additionally, the high specific surface area ensures a large electrochemical interface, while the ultrathin nanosheets offer short ion diffusion paths. The resulting electrode demonstrates a high energy density of 0.49 mAh cm^{-2} and a power density of 410 mW cm^{-2}. Furthermore, a lithium microbattery based on this high-performance electrode was developed, incorporating a novel packaging strategy that achieves a cycle life of 180 cycles and powers an electronic clock for over 18 h on a single charge.

2. Materials and Methods

Chemicals: Nickel sulfate hexahydrate ($NiSO_4{\cdot}6H_2O$), ammonium sulfate ($(NH_4)_2SO_4$), Vanadium (IV) oxide sulfate hydrate ($VOSO_4{\cdot}xH_2O$), sodium acetate (CH_3COONa). All the chemicals were used directly after purchase.

Fabrication of 3D hierarchical porous current collector: A 3D hierarchical porous current collector was fabricated on commercial nickel foam using the electrodeposition process to construct an electrode current collector. First, an electrolyte with a concentration of 0.1 M $NiSO_4$ and 0.75 M $(NH_4)_2SO_4$ was configured. A two-electrode electrodeposition process was carried out using a DC-regulated power supply. A $2.5 \times 0.5 \times 0.01$ cm^3 nickel foam was used as the working electrode and a Pt foil was used as the counter electrode. Then, the typical electrodeposition was performed with a constant current of 1–3 mA cm^{-2} for 1–3 min. The 3D hierarchical porous current collector was rinsed with deionized (DI) water, followed by drying at 80 °C for 30 min.

Active materials deposition: Once the porous nickel was fabricated on the nickel foam to form a hierarchical porous current collector structure, the VO_x cathode material was deposited on the current collector by cyclic voltammetry. Among them, a 3D hierarchical porous current collector, Pt foil, and a saturated Ag/AgCl electrode were used as the working electrode, counter electrode, and reference electrode, respectively. The electrodeposition process was carried out at a sweep rate of 50mV s^{-1} in the operating voltage range of

0–1.2 V for 300 cycles. The completed hierarchical porous VO_x electrode was rinsed with DI water, followed by drying at 80 °C for 6 h.

Assembly of HP-VO_x//Li MB: The HP-VO_x microelectrodes and lithium–copper composite tape microelectrodes were affixed to commercial nano-adhesives, and the microelectrodes were covered with a box-type nano-adhesive. Subsequently, the nano-adhesive fit should be coated with TB-6201 UV curing adhesive. Once the light-curing adhesive has undergone the curing process, the electrolyte should be added to the electrode. The PE film should then be pasted on the top side of the frame-type nano-adhesive. Finally, the light-curing adhesive should be used to coat the PE film and nano-adhesive fit.

Material characterization and electrochemical testing: The coin-type cells with hierarchical porous electrodes as positive electrodes and lithium metal as negative electrodes were investigated using a 2500 separator (Celgard, NC, USA) and 1 M LiTFSI in DME—DOL = 1:1 Vol%—with 1% $LiNO_3$ as electrolyte. Scanning electron microscopy (SEM) images were obtained with a JEOL JSM-7100F SEM (JEOL Ltd., Tokyo, Japan) at an acceleration voltage of 15.0 kV. The microscope pictures were taken using a light microscope with a 20× objective lens. Energy dispersive X-ray spectroscopy (EDS) results were recorded by an Oxford IE250 system (Oxford Instruments Co., Oxford, UK). X-ray photoelectron spectroscopy (XPS) results were recorded by the AXIS SUPRA (Shimadzu Group Co., Kyoto, Japan). Transmission electron microscope (TEM) results were obtained with a JEM-1400Plus (JEOL Ltd., Tokyo, Japan). High-resolution transmission electron microscopy (HRTEM) results were obtained with a JEM-2100F (JEOL Ltd., Tokyo, Japan). Atomic force microscope (AFM) (Dimension ICON-IR, Bruker Ltd., Karlsruhe, Germany) results were obtained with a Dimension ICON-IR. X-ray diffractometer result (XRD) was obtained by the D8 Discover (Bruker Ltd., Karlsruhe, Germany). Constant current charge/discharge curve results were tested through the Neware BTS83 battery test system (Neware Ltd., Shenzhen, China) and the long cycle result was tested by the Land T2001A battery testing station (LAND Ltd., Wuhan, China). Cyclic voltammetry curve (CV) results were tested within a voltage window of 1.8–3.8 V and the electrochemical impedance spectroscopy (EIS) results were measured in a frequency range of 0.01–100,000 Hz by the CHI760D. ECSA results were obtained from the electrochemical workstation CHI760D (Chenhua Ltd., Shanghai, China). GITT results were conducted on Land T2001A battery testing stations with a voltage window range of 1.8–3.8 V. The current pulse of 1 mA cm^{-2} for 1 min and the following relaxation for 30 min were applied in the GITT test.

3. Results and Discussion

As schematically illustrated in Figure 1, hierarchical porous electrodes were fabricated via electrodeposition. The hierarchical porous current collector was first created by introducing secondary pores into commercial nickel foam using the hydrogen bubble templating method. Subsequently, α-VO_x nanosheets were deposited onto the hierarchical porous current collector through an electrodeposition process (Figure 1a,b). For microbattery packaging with a lithium metal anode, the organic electrolyte leakage, which is closely linked to battery performance and safety, has been a persistent issue. To address this, we employed a reproducible, highly efficient, and cost-effective double-stacked encapsulation process. This method ensures that the electrolyte remains within the interdigital electrodes over extended periods (Figure 1c). As shown in Figure 1d, excellent electronic conductivity is provided by a continuous structure composed of nickel foam and porous nickel, and a large reactive area and short ion-transfer path are realized by ultrathin nanosheets, which ingeniously achieve excellent electron-ion bicontinuity.

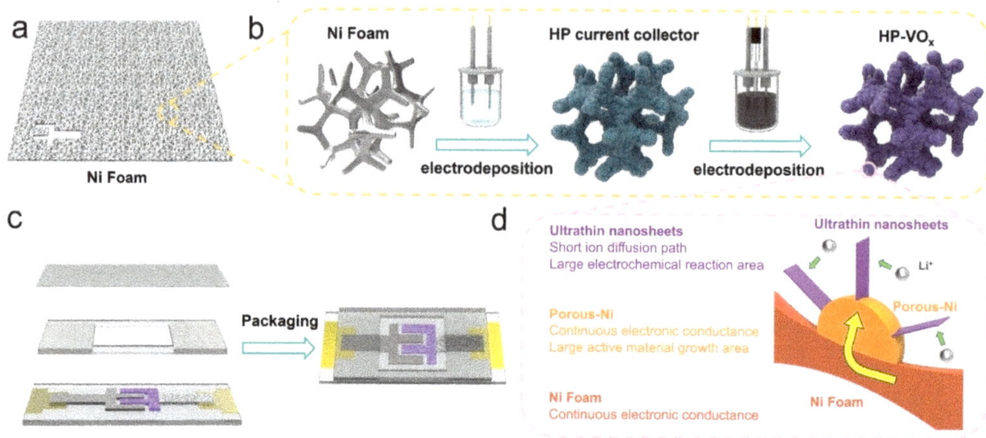

Figure 1. Fabrication process of hierarchical porous electrodes. (**a**) The diagram of the interdigital microelectrode from the commercial Ni foam. (**b**) Fabrication of hierarchical porous electrodes. (**c**) Package for lithium microbattery. (**d**) Reaction mechanism diagram of microelectrode.

Suitable parameters for preparing hierarchical porous electrodes were determined based on the results shown in Figures S1–S5 (Supporting Information). A microscope image of the hierarchical porous current collector (Figure S6a, Supporting Information) reveals that the original pores of the nickel foam remain intact, with secondary pores uniformly distributed throughout the structure. Electrochemical active areas of the nickel foam and hierarchical porous current collectors were subsequently evaluated (Figure S7, Supporting Information), showing that the hierarchical porous current collector has an active area six times larger than the original. This enhancement enables higher active material loading and promotes efficient ion and electron transport. Given the significant influence of electrodeposition solution concentration on the structure and properties of the electrodes, and consequently on their ion and electron transport capabilities, tests were conducted at various concentrations. Electrodeposition solutions of 20, 40, 50, 75, and 100 mM were prepared and applied via cyclic voltammetry over 300 cycles, with a voltage range of 0–1.2 V and a sweep rate of 50 mV s^{-1} (denoted as HP-VO_x-20, 40, 50, 75, and 100, respectively). Additionally, α-VO_x was directly electrodeposited onto flat nickel foam, designated as P-VO_x-50. Comparative SEM images of these electrodes are provided in the Supporting Information (Figures S8–S12).

The morphology of the HP-VO_x-50 electrode was analyzed using scanning electron microscopy (SEM) (Figure 2a and Figure S13, Supporting Information). Unlike the material structure of nickel foil, [26] the active material and nickel particles feature large pores distributed across the nickel foam's cross-structure. Optical microscopy (Figure S6b, Supporting Information) confirms that the primary and secondary pores of the nickel foam retain their original structure, indicating the successful construction of α-VO_x nanosheets through electrodeposition. Energy-dispersive spectroscopy (EDS) mapping (Figure 2b) demonstrates the uniform distribution of V and O elements, verifying consistent α-VO_x loading across the electrode. Similar results from other electrodes (Figures S14–S18, Supporting Information) further confirm the homogeneity of α-VO_x during electrodeposition. This hierarchical structure enhances surface area and active sites, facilitating the growth of active materials and providing ion transport pathways. Atomic force microscopy (AFM) results (Figure 2c) reveal that the electrodeposited nanosheets are 5–7 nm thick, supporting efficient ion diffusion. Additionally, the HP-VO_x-50 electrode surface exhibits greater roughness (Figure S19a, Supporting Information) than the standard hierarchical porous electrode (Figure S19b, Supporting Information). Transmission electron microscopy (TEM) and high-resolution TEM (HRTEM) further characterize the α-VO_x nanosheets. Figure 2d shows that individual nanosheets are 6–8 μm long and about 500–800 nm wide. The absence of clear lattice fringes in the HRTEM image (Figure 2e) confirms the amorphous nature of the α-VO_x nanosheets, which provide abundant reaction sites. EDS results (Figure 2f) reinforce the uniform distribution of V and O elements. X-ray diffraction (XRD) analysis (Figure 2g) reveals peaks at $2\theta = 44°$, $51°$, and $76°$, corresponding to the nickel substrate, with no visible crystalline peaks, consistent with the amorphous nature observed in Figure 2e. This structural feature ensures efficient ion transport and ample electrochemical reaction interfaces. X-ray photoelectron spectroscopy (XPS) characterization (Figure 2h,i) of the HP-VO_x-50 electrode identifies V $2p_{3/2}$ peaks at 517.7 eV and 516.2 eV, corresponding to V^{5+} and V^{4+}, respectively, indicating the presence of high-valence vanadium species. Raman spectroscopy (Figure S20, Supporting Information) exhibits distinct peaks at 153, 262, and 700 cm^{-1}. The peak at 153 cm^{-1} is attributed to backbone bending vibrations: 262 cm^{-1} to O-V-O bond bending vibrations, Ref. [27], and 700 cm^{-1} to V-O bond stretching vibrations, Ref. [28], confirming the material as vanadium oxide.

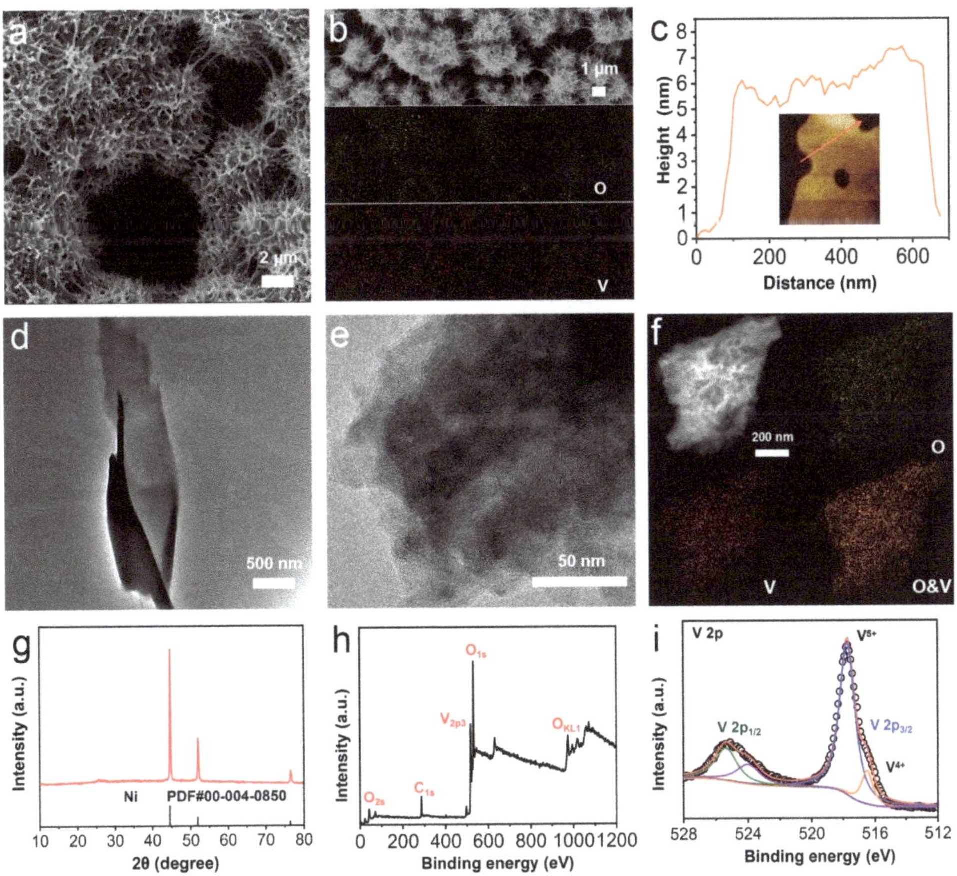

Figure 2. Structural characterization of hierarchical porous electrodes. (**a**) SEM images of HP-VO_x-50 electrodes. (**b**) EDS mapping images of V and O elements of HP-VO_x-50. (**c**) AFM image of HP-VO_x-50 electrode. (**d**) TEM image of VO_x. (**e**) HRTEM image of VO_x. (**f**) EDS mapping images of V and O elements of VO_x. (**g**) XRD image of HP-VO_x-50 electrode. (**h**) XPS full spectrum and (**i**) V 2p spectrum of HP-VO_x-50 electrode.

Coin-type batteries were assembled using these electrodes as cathodes to evaluate the electrochemical performance of the hierarchical porous electrodes. Figure 3a presents the galvanostatic charge/discharge (GCD) curves of the HP-VO_x-20, 40, 50, 75, and 100 electrodes at a current density of 1 mA cm^{-2}. As the electrodeposition solution concentration increases (20–100 mM), the energy density of the electrodes exhibits a general upward trend, maintaining a consistent voltage plateau (2.3–3.0 V), which indicates sufficient specific surface area within the hierarchical porous structure to accommodate the active material. Figure 3b displays the capacities of these electrodes at different current densities. At 1 mA cm^{-2}, the capacities are 0.05, 0.21, 0.49, 0.40, and 1.12 mAh cm^{-2} for the HP-VO_x-20, 40, 50, 75, and 100 electrodes, respectively. However, at higher current densities (10 mA cm^{-2}), the capacities decrease to 0.03, 0.10, 0.37, 0.25, and 0.39 mAh cm^{-2}, respectively. The HP-VO_x-75 and HP-VO_x-100 electrodes exhibit significant capacity loss due to excess VO_x material clogging the porous structure, which impedes ion diffusion. Notably, the HP-VO_x-50 electrode retains an energy density of 0.22 mAh cm^{-2} even at 50 mA cm^{-2}, demonstrating its superior ion transport capabilities. Figure 3c shows the Ragone plot, where it is evident that while the HP-VO_x-75 and HP-VO_x-100 electrodes achieve high energy densities, their power densities are poor. Conversely, the HP-VO_x-50 electrode exhibits an optimal balance of energy and power densities, achieving a high energy density of 1.27 mWh cm^{-2} at 1 mA cm^{-2} and an ultrahigh power density of 410 mW cm^{-2} at 250 mA cm^{-2}. Figure 3d highlights the GCD curves of the HP-VO_x-50 electrode across a current density range of 1–50 mA cm^{-2} with a voltage window of 1.8–3.8 V. These curves maintain a stable voltage plateau and high Coulombic efficiency, even as the current density increases. GCD curves for the other electrodes under varying current densities are included in Figures S21–S24 (Supporting Information). Figure 3e compares the performance of the HP-VO_x-50 electrode with the P-VO_x-50 electrode, which lacks secondary pore structures. The P-VO_x-50 electrode shows a substantial decrease in energy density, attributed to insufficient active sites for VO_x growth and an uneven material stacking structure, leading to pronounced capacity decay (Figure S25, Supporting Information). The HP-VO_x-50 electrode, on the other hand, benefits from its hierarchical structure, maintaining higher energy density and improved performance. The cycling performance of the HP-VO_x-50 electrode was tested at a high current density of 20 mA cm^{-2} (Figure 3f,g). Remarkably, the electrode retains nearly 100% of its capacity after 700 cycles, demonstrating excellent long-term stability. The capacity values remain consistent with those observed in Figure 3d. Finally, the Ragone plot in Figure 3h compares the energy and power densities of the HP-VO_x-50 electrode with those of previously reported micro lithium batteries [29–33]. This comparison underscores the superior performance of the HP-VO_x-50 electrode in balancing high energy and power densities. In addition, the HP-VO_x-50 electrode was characterized after cycling. The XRD and SEM results shown in Figures S26 and S27 (Supporting Information) illustrate that the electrode maintains the hierarchical porous structure of amorphous vanadium oxides after long cycling. Similarly, the EDS results of Figure S28 demonstrate the uniform distribution of the V and O elements.

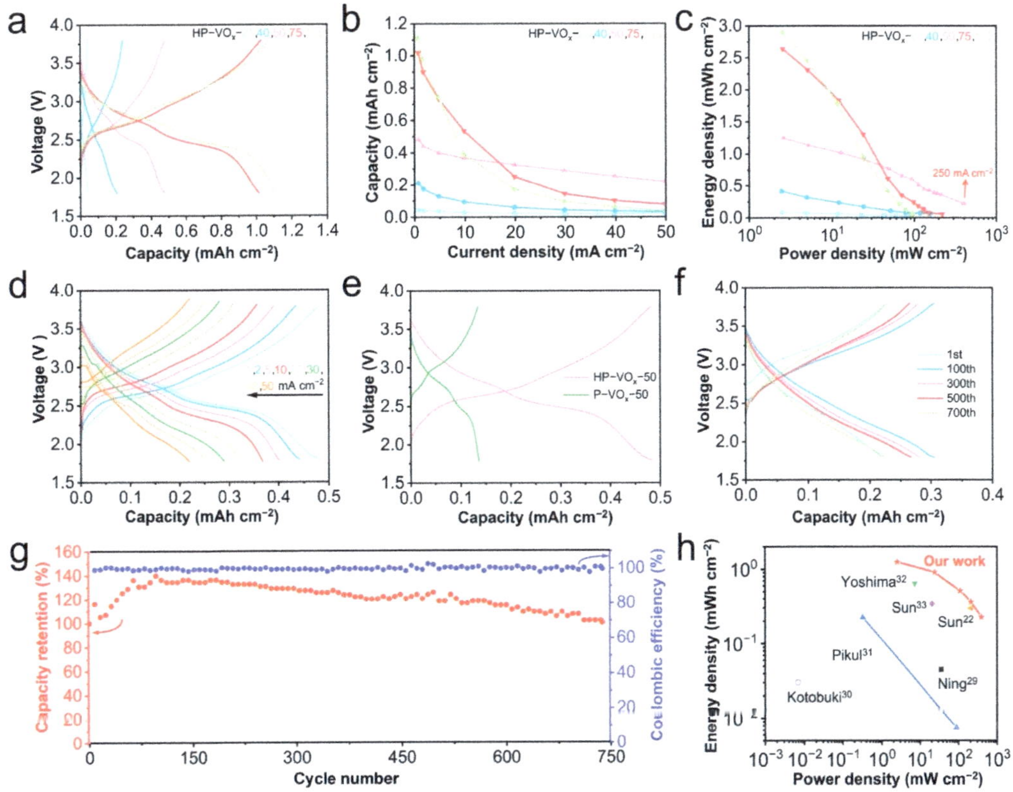

Figure 3. Electrochemical properties of HP-VO_x-20, 40, 50, 75, 100, and P-VO_x-50 electrodes. (**a**) GCD curves and (**b**) capacities at various current densities of different HP-VO_x electrodes. (**c**) Ragone plot of each electrode. (**d**) GCD curves of HP-VO_x-50 electrode at various current densities. (**e**) Capacity comparison of HP-VO_x-50 and P-VO_x-50 electrodes. (**f**) Comparison of the capacity of HP-VO_x-50 electrode at different number of cycles. (**g**) Cycling performance of HP-VO_x-50 electrode. (**h**) Ragone plot showing the performance of our battery and other reported batteries.

To evaluate the electrochemical behavior of the HP-VO_x-50 electrode and its ability to maintain high power density, cyclic voltammetry (CV) tests were conducted at sweep rates of 1 to 5 mV s^{-1}, with results shown in Figure 4a. The CV curves remained nearly identical across the varying sweep rates, indicating stable electrochemical performance. Two reduction peaks at 2.4 V and 2.8 V in the cathodic sweep at 1 mV s^{-1} were observed, consistent with the GCD results. To further investigate the reaction kinetics and charge storage mechanisms of the HP-VO_x-50 electrode, a power–law relationship between the peak current (i) and sweep rate (v) was analyzed, as shown in Equation (1) [34]:

$$i = av^b \tag{1}$$

where a and b are constants related to the peak current and sweep rate. The value of b is typically between 0.5 and 1, representing diffusion-controlled and capacitance-controlled behavior, respectively. The log(i)-log(v) plots in Figure 4b were used to calculate the b-values for the oxidation and reduction peaks of the HP-VO_x-50 electrode, yielding values of 0.87 and 0.65, respectively. This indicates a combination of diffusion-controlled and capacitive behavior. Figure 4c further illustrates that as the sweep rate increases, surface capacitive behavior becomes more pronounced, with a b-value of 0.73 for both the oxidation and reduction peaks. To quantify the contributions of diffusion-controlled and capacitance-controlled behaviors, the total current response at a given potential was separated into diffusion-controlled (k_1v) and capacitive ($k_2v^{1/2}$) components, as shown in Equation (2) [35]:

$$i = k_1 v + k_2 v^{\frac{1}{2}} \quad (2)$$

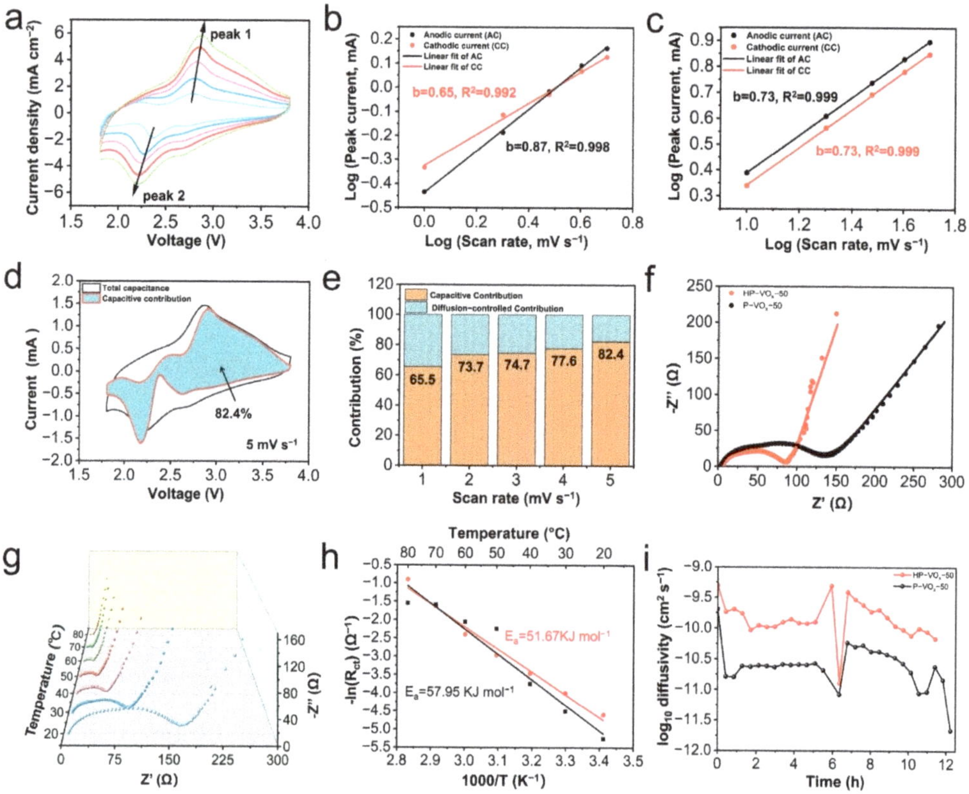

Figure 4. Electrochemical properties of HP-VO_x-50 electrode. (**a**) CV curves at sweep rates of HP-VO_x-50 electrode. (**b**) Determination of the b-value of the redox peaks by evaluating the relationship between the peak current and the sweep rates of the HP-VO_x-50 electrode from 1 to 5 mV s^{-1} and (**c**) from 10 to 50 mV s^{-1}. (**d**) The cyan region from 1.8 to 3.8V shows capacitive contribution. (**e**) Separations of capacitive and diffusion-controlled contribution at different sweep rates for HP-VO_x-50 electrode. (**f**) EIS curves for HP-VO_x-50 and P-VO_x-50 electrodes. (**g**) 3D EIS plots of HP-VO_x-50 electrode at different temperatures. (**h**) Fitting and activation energy plots of $-ln(R_{ct})$ at about $1000/T$ for HP-VO_x-50 electrode. (**i**) Comparison of Li^+ diffusion rates for HP-VO_x-50 and P-VO_x-50 electrodes.

Figure 4d presents the contributions of these behaviors for the HP-VO_x-50 electrode at a sweep rate of 5 mV s^{-1}, where the capacitive contribution (cyan region) accounts for 82.4% of the total performance. Additionally, the capacitive contribution at various sweep rates (1–4 mV s^{-1}) is shown in Figure S29 (Supporting Information), with values of 65.5%, 73.7%, 74.7%, and 77.6%, respectively. As shown in Figure 4e, the capacitive contribution increases steadily with the sweep rate, reaching 82.4% at 5 mV s^{-1}, suggesting that the electrode predominantly undergoes surface capacitive reactions with faster reaction kinetics at higher sweep rates.

To investigate the impact of the secondary structure on the electrochemical reaction mechanism of the electrode, we performed electrochemical impedance spectroscopy (EIS) on both the HP-VO_x-50 and P-VO_x-50 electrodes, with results shown in Figure 4f. The impedance of the electrode with the hierarchical porous structure is significantly lower, indicating improved ion and electron transport.

Next, the electrochemical impedance of the HP-VO_x-50 electrode was evaluated at varying temperatures, with 3D impedance spectra between 20 and 80 °C shown in Figure 4g. The EIS curve remains consistent across the temperature range, suggesting that the electrode maintains stable electrochemical performance at elevated temperatures. In contrast, the P-VO_x-50 electrode exhibits relatively higher impedance, as shown in Figure S30 (Supporting Information). Figures S31 and S32 (Supporting Information) further confirm that the charge transfer resistance (R_{ct}) of P-VO_x-50 is higher at all temperatures, suggesting that hierarchical porous electrodes with secondary pore structures provide more electrochemical reaction surface area, significantly enhancing reaction kinetics. Finally, the reaction activation energy was calculated using the Arrhenius equation, as detailed below [36]:

$$lnk = \frac{E_a}{RT} + lnA \tag{3}$$

where A is the prefinger factor, T is the thermodynamic temperature, R is the gas constant, and E_a is the activation energy. Figure 4h shows the linear fit of $-ln(R_{ct})$-1000/T, which further yields a reaction activation energy of 51.67 kJ mol^{-1} for HP-VO_x-50 compared to 57.95 kJ mol^{-1} for P-VO_x-50, suggesting that the hierarchical structure can accelerate the reaction kinetics to some extent. Furthermore, the galvanostatic intermittent titration technique was employed to determine the Li^+ diffusion coefficient in HP-VO_x-50 and P-VO_x-50 electrodes, utilizing the following equation [37]:

$$D = \frac{4}{\tau\pi}\left(\frac{m_B V_m}{M_B S}\right)^2\left(\frac{\Delta E_s}{\Delta E_t}\right)^2\left(\tau \ll \frac{L^2}{D}\right) \tag{4}$$

where τ is the time of applying constant current, m_B is the mass of the active substance (g), V_m is the molar volume of the sample (cm^3 mol^{-1}), M_B is the relative molecular mass of the material (g mol^{-1}), S is the contact area of the electrode/electrolyte (cm^2), ΔE_s is the change in the cell voltage during the charging (discharging) process, ΔE_t the change in voltage during resting to equilibrium, and L is the thickness of the electrode. Figure S33a (Supporting Information) shows that the diffusion coefficient of the HP-VO_x-50 electrode is about 10^{-9} to 10^{-11}, which is enhanced by an average of 1–2 orders of magnitude (Figure 4i) at different charging and discharging voltages when compared with P-VO_x-50 (Figure S33b, Supporting Information). This also demonstrates that the HP-VO_x-50 electrode exhibits fast electrochemical reaction kinetics during both charging and discharging.

Next, we assembled a HP-VO_x//Li interdigital microbattery using the HP-VO_x-50 electrode and lithium–copper composite tape. The electrochemical performance of the HP-VO_x//Li microbattery (MB) was evaluated, and the cyclic voltammetry (CV) results, shown in Figure 5a, maintain a similar profile to that of the individual electrode. Figure 5b presents the galvanostatic charge/discharge (GCD) curves of the HP-VO_x//Li MB at current densities ranging from 0.5 to 2.0 mA cm^{-2}, displaying a discharge plateau between 2.3 and 3.0 V, consistent with the CV results. At a current density of 0.5 mA cm^{-2}, the HP-VO_x//Li MB demonstrates an area capacity of 0.25 mAh cm^{-2}, with a capacity retention of 60% at 2.0 mA cm^{-2}. After 180 cycles, the capacity retention remains at 75%, as shown in Figure 5c. To evaluate the practical application potential of the HP-VO_x//Li MB, we powered an electronic timer continuously for more than 18 h, as shown in Figure 5d. This result highlights the effectiveness of our double-stacked encapsulation process. The simple and cost-effective electrode preparation method, combined with the encapsulation strategy, presents a promising approach for future practical applications.

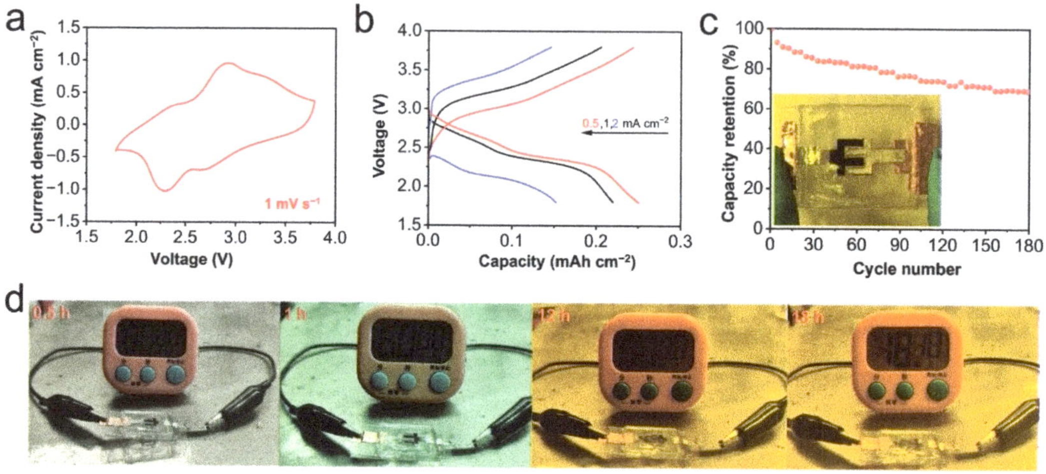

Figure 5. Schematic and performance diagram of a microbattery fabricated from a HP-VO_x electrode. (**a**) CV curves of microbattery at 1 mV s^{-1}. (**b**) GCD curves of microbattery at different current densities. (**c**) Cycling performance of microbattery. (**d**) Application of microbattery to electronic clocks.

4. Conclusions

In conclusion, hierarchical porous current collectors were fabricated on commercial nickel foam using a simple, low-cost hydrogen templating method. Ultrathin α-VO_x nanosheets were then conformally integrated onto the collectors using the same deposition method to create the hierarchical porous current electrodes. These electrodes exhibit high porosity, a dense network of active sites, and retain their porous structure even after the active material is loaded. The resulting electrode achieves a high energy density of 0.49 mAh cm^{-2} at 1 mA cm^{-2} and an ultrahigh power density of 410 mW cm^{-2} at 250 mA cm^{-2}. These remarkable performance characteristics are attributed to the dominance of capacitive behavior at the surface interface and the fast Li^+ diffusion rate (~10^{-9}–10^{-11} cm^2 s^{-1}), facilitated by the hierarchical porous structure and ultrathin α-VO_x nanosheets (5–7 nm). Furthermore, a novel double-layer encapsulation method was employed to prevent electrolyte leakage and Li-metal electrode oxidation, significantly improving the microbattery's stability. This approach allowed the microbattery to cycle more than 180 times and power an electronic clock for over 18 h on a single charge.

Supplementary Materials: The following supporting information can be downloaded at: https://www.mdpi.com/article/10.3390/batteries11020081/s1, Figure S1: SEM images of the hierarchical porous current collector at 1 A cm^{-2} -2 min; Figure S2: SEM images of the hierarchical porous current collector at 2 A cm^{-2} -2 min; Figure S3: SEM images of the hierarchical porous current collector at 3 A cm^{-2} -2 min; Figure S4: SEM images of the hierarchical porous current collector at 2 A cm^{-2} -1 min; Figure S5: SEM images of the hierarchical porous current collector at 2 A cm^{-2} -3 min; Figure S6: Microscopic images of (a) the hierarchical porous current collector, (b) HP-VO_x-50 electrode; Figure S7: CV curves of (a) Ni foam and (b) hierarchical porous electrodes. (c) Capacitive current densities of the above electrodes versus the scan rates and corresponding C_{dl} values were estimated through the linear fitting of the plots; Figure S8: SEM images of P-VO_x-50 porous electrode; Figure S9: SEM images of HP-VO_x-20 hierarchical porous electrode; Figure S10: SEM images of HP-VO_x-40 hierarchical porous electrode; Figure S11: SEM images of HP-VO_x-75 hierarchical porous electrode; Figure S12: SEM images of HP-VO_x-100 hierarchical porous electrode; Figure S13: SEM images of HP-VO_x-50 hierarchical porous electrode; Figure S14: EDS mapping images of V and O elements of HP-VO_x-20; Figure S15: EDS mapping images of V and O elements of HP-VO_x-40; Figure S16: EDS mapping images of V and O elements of HP-VO_x-75; Figure S17: EDS mapping images of V and O elements of HP-VO_x-100; Figure S18: EDS mapping images of V and O elements of P-VO_x-50; Figure S19: AFM image of (a) hierarchical porous current collector and (b) HP-VO_x electrode; Figure S20: Raman scattering spectra of HP-VO_x-50; Figure S21: GCD curves of HP-VO_x-20 electrode at various current densities; Figure S22: GCD curves of HP-VO_x-40 electrode at various current densities; Figure S23: GCD curves of HP-VO_x-75 electrode at various current densities; Figure S24: GCD curves of HP-VO_x-100 electrode at various current densities; Figure S25: GCD curves of P-VO_x-50 electrode at various current densities; Figure S26: XRD spectra of HP-VO_x-50 electrode after long cycling; Figure S27: SEM images of HP-VO_x-50 electrode after long cycling; Figure S28: EDS spectra of HP-VO_x-50 electrode after long cycling; Figure S29: The 3D plot of the contribution to the capacitive behavior of the HP-VO_x-50 electrode from 1 to 4 mV s^{-1} sweep rates; Figure S30: 3D EIS plots of P-VO_x-50 electrode at different temperatures; Figure S31: EIS curves at different temperatures for HP-VO_x-50 electrode. (a) 20 °C, (b) 30 °C, (c) 40 °C, (d) 50 °C, (e) 60 °C, (f) 70 °C and (g) 80 °C; Figure S32: EIS curves at different temperatures for P-VO_x-50 electrode. (a) 20 °C, (b) 30 °C, (c) 40 °C, (d) 50 °C, (e) 60 °C, (f) 70 °C and (g) 80 °C; Figure S33: GITT curves and Li^+ diffusion rate plots for (a) HP-VO_x-50 and (b) P-VO_x-50 electrode.

Author Contributions: Conceptualization, X.X.; methodology, X.C.; formal analysis, X.C., M.G. and J.L.; investigation, X.C. and M.G.; writing—original draft preparation, X.C.; writing—review and editing, W.Y. and X.X.; supervision, J.L.; funding acquisition, W.Y. All authors have read and agreed to the published version of the manuscript.

Funding: This research was funded by the Fundamental Research Funds for the Central Universities (WUT:104972024JYS0048) and the Postdoctoral Fellowship Program of CPSF (GZB20240581).

Data Availability Statement: The raw data supporting the conclusions of this article will be made available by the authors on request.

Conflicts of Interest: Author Xu Xu was employed by the company Zhongyu Feima New Material Technology Innovation Center (Zhengzhou) Co., Ltd. The remaining authors declare that the research was conducted in the absence of any commercial or financial relationships that could be construed as a potential conflict of interest.

References

1. Zhu, M.; Schmidt, O.G. Tiny Robots and Sensors Need Tiny Batteries—Here's How to Do It. *Nature* **2021**, *589*, 195–197. [CrossRef]
2. Zhu, J.; Ni, J.; Li, L. Harnessing In Situ Iodine Ionic Liquids for High Areal Capacity and Long Cycle Life Zn//I_2 Solid Microbatteries. *Adv. Funct. Mater.* **2024**, 2416799. [CrossRef]
3. Li, S.; Zhao, T.; Wang, H.; Wang, Z.; Zhang, M.; Shao, A.; Liu, J.; Wang, Z.; Ma, Y. High-Areal-Capacity/Power Lithium Metal Microbattery Configuration Based on the Mechanically Flexible, Ultra-Lightweight, Nanocellulose Framework. *Nano Res.* **2024**, *17*, 8155–8162. [CrossRef]
4. Choi, J.M.; Saroha, R.; Kim, J.S.; Jang, M.R.; Cho, J.S. Porous Nanofibers Comprising VN Nanodots and Densified N-Doped CNTs as Redox-Active Interlayers for Li–S Batteries. *J. Power Sources* **2023**, *559*, 232632. [CrossRef]
5. Liu, S.; Liu, W.; Zhao, Y.; Ye, Y.; Li, Y.; Liu, J. Filler-Integrated Composite Polymer Electrolyte for Solid-State Lithium Batteries. *Adv. Mater.* **2023**, *35*, 2110123. [CrossRef] [PubMed]
6. Li, Y.; Xiao, S.; Qiu, T.; Lang, X.; Tan, H.; Wang, Y.; Li, Y. Recent Advances on Energy Storage Microdevices: From Materials to Configurations. *Energy Storage Mater.* **2022**, *45*, 741–767. [CrossRef]
7. Peng, J.; Zhang, W.; Wang, S.; Huang, Y.; Wang, J.; Liu, H.; Dou, S.; Chou, S. The Emerging Electrochemical Activation Tactic for Aqueous Energy Storage: Fundamentals, Applications, and Future. *Adv. Funct. Mater.* **2022**, *32*, 2111720. [CrossRef]
8. Wang, X.; Wu, Z. Zinc Based Micro-electrochemical Energy Storage Devices: Present Status and Future Perspective. *EcoMat* **2020**, *2*, e12042. [CrossRef]
9. Jiang, M.; Hou, Z.; Ren, L.; Zhang, Y.; Wang, J.-G. Prussian Blue and Its Analogues for Aqueous Energy Storage: From Fundamentals to Advanced Devices. *Energy Storage Mater.* **2022**, *50*, 618. [CrossRef]
10. Zhang, X.; Hu, L.; Zhou, K.; Zhang, L.; Zeng, X.; Shi, Y.; Cai, W.; Wu, J.; Lin, Y. Fully Printed and Sweat-Activated Micro-Batteries with Lattice-Match Zn/MoS_2 Anode for Long-Duration Wearables. *Adv. Mater.* **2024**, *36*, 2412844. [CrossRef] [PubMed]

11. Li, X.; Jin, X.; Wang, Y.; Zhang, X.; Li, D.; Wang, J.; Yuan, M.; Liu, J.; Zhao, Y. All-Direct Laser Patterning Zinc-Based Microbatteries. *Adv. Funct. Mater.* **2024**, *34*, 2314060. [CrossRef]
12. Jin, X.; Song, L.; Dai, C.; Xiao, Y.; Han, Y.; Li, X.; Wang, Y.; Zhang, J.; Zhao, Y.; Zhang, Z.; et al. A Flexible Aqueous Zinc–Iodine Microbattery with Unprecedented Energy Density. *Adv. Mater.* **2022**, *34*, 2109450. [CrossRef]
13. Yang, W.; Xu, L.; Luo, W.; Huang, M.; Fu, K.; Song, R.; Han, C.; Tu, R.; Shi, J.; Mai, L. Rechargeable Zinc-Ammonium Hybrid Microbattery with Ultrahigh Energy and Power Density. *Matter* **2023**, *6*, 3006–3020. [CrossRef]
14. Tang, M.; Dong, S.; Wang, J.; Cheng, L.; Zhu, Q.; Li, Y.; Yang, X.; Guo, L.; Wang, H. Low-Temperature Anode-Free Potassium Metal Batteries. *Nat. Commun.* **2023**, *14*, 6006. [CrossRef] [PubMed]
15. Liang, P.; Sun, H.; Huang, C.; Zhu, G.; Tai, H.; Li, J.; Wang, F.; Wang, Y.; Huang, C.; Jiang, S.; et al. A Nonflammable High-Voltage 4.7 V Anode-Free Lithium Battery. *Adv. Mater.* **2022**, *34*, 2207361. [CrossRef] [PubMed]
16. Ma, J.; Zheng, S.; Chi, L.; Liu, Y.; Zhang, Y.; Wang, K.; Wu, Z. 3D Printing Flexible Sodium-Ion Microbatteries with Ultrahigh Areal Capacity and Robust Rate Capability. *Adv. Mater.* **2022**, *34*, 2205569. [CrossRef]
17. Lu, Y.; Wang, Z.; Li, M.; Li, Z.; Hu, X.; Xu, Q.; Wang, Y.; Liu, H.; Wang, Y. 3D Printed Flexible Zinc Ion Micro-Batteries with High Areal Capacity Toward Practical Application. *Adv. Funct. Mater.* **2024**, *34*, 2310966. [CrossRef]
18. Zhang, Y.; Wang, Y.; Zhao, W.; Zuo, P.; Tong, Y.; Yin, G.; Zhu, T.; Lou, S. Delocalized Electronic Engineering of $TiNb_2O_7$ Enables Low Temperature Capability for High-Areal-Capacity Lithium-Ion Batteries. *Nat. Commun.* **2024**, *15*, 6299. [CrossRef] [PubMed]
19. Yang, J.; Li, Y.; Mijailovic, A.; Wang, G.; Xiong, J.; Mathew, K.; Lu, W.; Sheldon, B.W.; Wu, Q. Gradient Porosity Electrodes for Fast Charging Lithium-Ion Batteries. *J. Mater. Chem. A* **2022**, *10*, 12114–12124. [CrossRef]
20. Zhang, X.; Hui, Z.; King, S.T.; Wu, J.; Ju, Z.; Takeuchi, K.J.; Marschilok, A.C.; West, A.C.; Takeuchi, E.S.; Wang, L.; et al. Gradient Architecture Design in Scalable Porous Battery Electrodes. *Nano Lett.* **2022**, *22*, 2521–2528. [CrossRef]
21. Liu, X.; Liu, Y.; Zhang, Q.; Zhou, S.; Li, X.; Yi, B.; Liu, Y.; Liu, X.; Ding, Y. Integrating Aperiodic 3D Porous Electrodes into 3D Batteries through Spray-Deposited Polymer Electrolytes. *Adv. Energy Mater.* **2024**, *14*, 2401330. [CrossRef]
22. Sun, P.; Li, X.; Shao, J.; Braun, P.V. High-Performance Packaged 3D Lithium-Ion Microbatteries Fabricated Using Imprint Lithography. *Adv. Mater.* **2021**, *33*, 2006229. [CrossRef]
23. Hur, J.I.; Smith, L.C.; Dunn, B. High Areal Energy Density 3D Lithium-Ion Microbatteries. *Joule* **2018**, *2*, 1187–1201. [CrossRef]
24. Hao, Z.; Xu, L.; Liu, Q.; Yang, W.; Liao, X.; Meng, J.; Hong, X.; He, L.; Mai, L. On-Chip Ni–Zn Microbattery Based on Hierarchical Ordered Porous Ni@$Ni(OH)_2$ Microelectrode with Ultrafast Ion and Electron Transport Kinetics. *Adv. Funct. Mater.* **2019**, *29*, 1808470. [CrossRef]
25. Lee, S.A.; Yang, J.W.; Choi, S.; Jang, H.W. Nanoscale Electrodeposition: Dimension Control and 3D Conformality. *Exploration* **2021**, *1*, 20210012. [CrossRef] [PubMed]
26. Yang, W.; Xu, L.; Luo, W.; Li, M.; Hu, P.; Dai, Y.; Ye, F.; Han, C.; Zhou, M.; Tu, R.; et al. 3D Macroporous Frame Based Microbattery with Ultrahigh Capacity, Energy Density, and Integrability. *Adv. Energy Mater.* **2023**, *13*, 2300574. [CrossRef]
27. Li, Z.; Xu, Y.; Wu, L.; Dou, H.; Zhang, X. Microstructural Engineering of Hydrated Vanadium Pentoxide for Boosted Zinc Ion Thermoelectrochemical Cells. *J. Mater. Chem. A* **2022**, *10*, 21446–21455. [CrossRef]
28. Hevia, S.A. High Performance of V_2O_5 Thin Film Electrodes for Lithium-Ion Intercalation. *Appl. Surf. Sci.* **2022**, *576*, 151710. [CrossRef]
29. Ning, H.; Pikul, J.H.; Zhang, R.; Li, X.; Xu, S.; Wang, J.; Rogers, J.A.; King, W.P.; Braun, P.V. Holographic Patterning of High-Performance on-Chip 3D Lithium-Ion Microbatteries. *Proc. Natl. Acad. Sci. USA* **2015**, *112*, 6573–6578. [CrossRef]
30. Kotobuki, M.; Suzuki, Y.; Munakata, H.; Kanamura, K.; Sato, Y.; Yamamoto, K.; Yoshida, T. Effect of Sol Composition on Solid Electrode/Solid Electrolyte Interface for All-Solid-State Lithium Ion Battery. *Electrochim. Acta* **2011**, *56*, 1023–1029. [CrossRef]
31. Pikul, J.H.; Gang Zhang, H.; Cho, J.; Braun, P.V.; King, W.P. High-Power Lithium Ion Microbatteries from Interdigitated Three-Dimensional Bicontinuous Nanoporous Electrodes. *Nat. Commun.* **2013**, *4*, 1732. [CrossRef]
32. Yoshima, K.; Munakata, H.; Kanamura, K. Fabrication of Micro Lithium-Ion Battery with 3D Anode and 3D Cathode by Using Polymer Wall. *J. Power Sources* **2012**, *208*, 404–408. [CrossRef]
33. Sun, K.; Wei, T.; Ahn, B.Y.; Seo, J.Y.; Dillon, S.J.; Lewis, J.A. 3D Printing of Interdigitated Li-Ion Microbattery Architectures. *Adv. Mater.* **2013**, *25*, 4539–4543. [CrossRef] [PubMed]
34. Acharya, J.; Ko, T.H.; Seong, J.-G.; Seo, M.-K.; Khil, M.-S.; Kim, H.-Y.; Kim, B.-S. Hybrid Electrodes Based on Zn–Ni–Co Ternary Oxide Nanowires and Nanosheets for Ultra-High-Rate Asymmetric Supercapacitors. *ACS Appl. Nano Mater.* **2020**, *3*, 8679–8690. [CrossRef]
35. Brezesinski, T.; Wang, J.; Tolbert, S.H.; Dunn, B. Ordered Mesoporous α-MoO_3 with Iso-Oriented Nanocrystalline Walls for Thin-Film Pseudocapacitors. *Nat. Mater* **2010**, *9*, 146–151. [CrossRef] [PubMed]

36. Tu, S.; Zhang, B.; Zhang, Y.; Chen, Z.; Wang, X.; Zhan, R.; Ou, Y.; Wang, W.; Liu, X.; Duan, X.; et al. Fast-Charging Capability of Graphite-Based Lithium-Ion Batteries Enabled by Li_3P-Based Crystalline Solid–Electrolyte Interphase. *Nat. Energy* **2023**, *8*, 1365–1374. [CrossRef]
37. Gong, S.; Sun, K.; Cao, N.; Chao, D.; Jia, X.; Wang, C. Potassium Phthalimide Doped with Delta-Site Structures to Construct Ultra-Long Cycle Life Aqueous Zn-Polymer Batteries. *Adv. Funct. Mater.* **2024**, *34*, 2409805. [CrossRef]

batteries

MDPI

Article

Interactions Between Trivalent Elements Enable Ultrastable LDH Cathode for High-Performance Zinc Battery

Junhua Zeng, Jinlei Gao, Wenyao Lu, Jiashuo Feng and Ting Deng *

Key Laboratory of Automobile Materials of MOE, School of Materials Science and Engineering, and Jilin Provincial International Cooperation Key Laboratory of High-Efficiency Clean Energy Materials, Jilin University, Changchun 130012, China; zengjh1621@mails.jlu.edu.cn (J.Z.); gaojl1621@mails.jlu.edu.cn (J.G.); luwy1621@mails.jlu.edu.cn (W.L.); fengjs1621@mails.jlu.edu.cn (J.F.)
* Correspondence: tdeng@jlu.edu.cn

Abstract: Layered double hydroxides (LDHs) are one class of two-dimensional materials, with tunable chemical composition and large interlayer spacing, that is a potential cathode material candidate for aqueous zinc-ion batteries (AZIBs). Nevertheless, the low conductivity and fragile structure of LDH have impeded their practical application in AZIBs. Herein, a ternary CoMnAl LDH is synthesized via the facile coprecipitation method as the cathode material for AZIB. The interaction between trivalent Al^{3+} and Mn^{3+} not only lowers the redox energy barrier but also enhances the electronic structure, as proved by EIS analysis and DFT simulation. As a result, the synthesized CoMnAl LDH displays a high specific capacity of 238.9 mAh g^{-1} at 0.5 A g^{-1}, an outstanding rate performance (138.8 mAh g^{-1} at 5 A g^{-1}), and a stable cyclability (92% capacity retention after 2000 cycles).

Keywords: layered double hydroxide; cathode material; electron interaction; aqueous zinc-ion battery

1. Introduction

As lithium ores become strategic resources, developing alternative high-performance electrochemical energy storage devices (EESDs) with high safety and low cost in the post-lithium age is of great significance for environmental and resource sustainability [1,2]. Aqueous zinc-ion batteries (AZIBs) have become a hot spot in EESD research due to the high specific capacity (820 mAh g^{-1}) and high crustal content [3–5]. However, besides the dendrite issue, the application of AZIBs also suffers from the unsatisfying electrochemical performance of cathode materials, such as the low redox potential and unstable cycling life [6,7]. Therefore, investigations on cathode materials with high specific capacity, fast rate capability, and long cyclability are imperative for high-performance AZIBs.

Layered double hydroxides (LDHs) are a class of 2D layered materials, which can be represented by the general formula $[M^{2+}_{1-x}M^{3+}_{x}(OH)_2]^{x+}[A^{n-}]_{x/n}\cdot yH_2O$ [8,9]. In this formula, M^{2+} and M^{3+} stand for divalent and trivalent metal cations occupying the octahedral holes in the brucite-like layer, while A^{n-} stands for anions from the precursor located in the hydrated interlayer galleries (Figure 1A). It is explicit that the electrochemical performance of LDHs originates from the redox reaction of transition metal atoms during the charge/discharge process. The unique 2D morphology can expose massive atoms on the LDH's surface, which provide sufficient active sites for the electrochemical reaction [10,11]. And the tunable composition makes LDH quite adaptable in different EESDs as an electrode material. Co and Mn species have shown promising properties, indicating that CoMn LDH is one potential electrode material for use in AZIBs [12–15].

However, general issues of weak electronic conductivity and structural collapse still restrict CoMn LDH from real application. LDHs face large polarization and redox irreversibility due to their semi-conductor nature [16,17]. Moreover, Mn^{3+} species on 2D surfaces can accelerate the dissolution into electrolytes, compromising the structural stability. Very recently, ternary LDHs with a third transition metal element have exhibited outstanding electrochemical performance in supercapacitors, such as NiCoAl, NiCoMn, CoZnAl, and CoNiFe. These LDHs usually consist of two divalent transition metal elements and one trivalent transition metal element. Compared to divalent metal ions, trivalent metal ions show higher reactivity and act as a bridge to conduct electrons. Thus, doping a second trivalent metal element in CoMn LDH can effectively modify the electronic structure of ambient atoms and regulate the band-gap structure, thus not only increasing the carrier in the system and promoting electron conduction but also creating strong interaction with soluble species to stabilize the structure (Figure 1B).

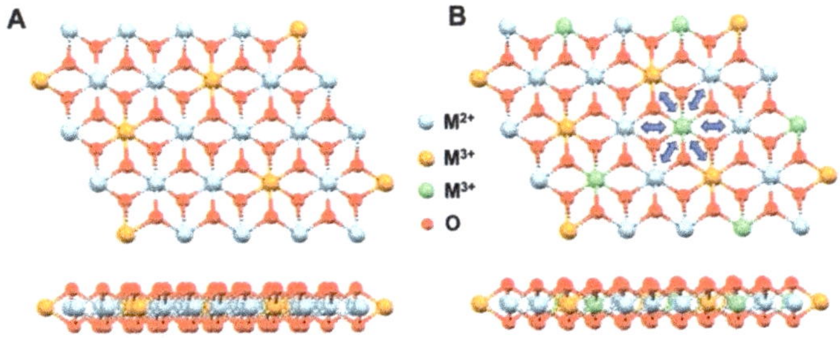

Figure 1. (**A**) An illustration of the LDH structure. (**B**) The structure of the LDH with one divalent and two trivalent transition metal elements.

Herein, a tenary CoMnAl LDH was synthesized via a coprecipitation method. As a AZIB cathode, CoMnAl LDH exhibited a high specific capacity of 238.9 mAh g^{-1} at 0.5 A g^{-1} and an outstanding rate performance (138.8 mAh g^{-1} at 5 A g^{-1}). X-ray photoemission spectroscopy (XPS) revealed that the valence of Mn species was partially increased by Al^{3+} introduction, while Co^{2+} remained, demonstrating that the interaction was between trivalent Mn^{3+} and Al^{3+}. Density functional theory (DFT) calculations confirmed the formation of a conductive CoMnAl-O network, which aligned with electrochemical impedance spectroscopy (EIS). In addition, both the overlap and the intensity of the density of state indicate that the structure of CoMnAl LDH was more stable. As a result, the performance of CoMn LDH decayed quickly within 500 cycles at 0.5 A g^{-1}. As a comparison, CoMnAl LDH showed a stable cycle life of 2000 cycles at the same current density with a capacity retention of 92%, which demonstrated the prospect of LDH for use as the cathode material of AZIBs.

2. Experimental Section

2.1. Synthesis of CoMn, CoAl, and CoMnAl LDHs

$Co(NO_3)_2$, $Mn(NO_3)_2$, $Al_2(SO_4)_3$, and NH_3H_2O were all purchased from Aladdin Reagent Co., LTD (Shanghai, China). **CoMnAl:** $Co(NO_3)_2$, $MnSO_4$ and $Al_2(SO_4)_3$ were dissolved in deionized water with a mole ratio of 6:2:2. Then, NH_3H_2O was dropwise added into the solution until no more sediment appeared. After a 30 min stirring period, the sediment was filtered and washed with deionized water, then put into a 60 °C oven overnight. **CoMn** and **CoAl** were synthesized under the same procedure with a mole ratio of 6:4 (Co:Mn and Co:Al).

2.2. Physical Characterization

The X-ray diffraction tests of LDHs were conducted by a Bruker D8 powder diffractometer with a LynxEye XE-T detector. X-ray photoelectron spectroscopy, Raman spectroscopy and Fourier transform infrared spectroscopy were conducted by a Thermo ESCALAB 250Xi tool, a Thermo DXR2 instrument, and a Perkin Elmer Spectrum One B device. The BET specific area was analyzed by N_2 adsorption measurements via an auto-JW-BK132F instrument from JWGB SCI. & TECH instruments. Scanning electron microscopy and transmission electron microscopy were performed by a Hitachi SU8000 scanning electron microscope and JEM ARM 1300S.

2.3. Electrochemical Evaluation

The electrochemical performances of LDHs were assessed in both a three-electrode system and a CR2032 coin cell. In the three-electrode system, a platinum plate and a saturated calomel electrode (SCE) served as the counter and reference electrodes. In the coin cell, Zn foil and a porous non-woven cloth served as the anode and separator. A total of 1 M KOH solution acted as the electrolyte in both the three-electrode system and coin cell. The electrolyte was prepared by dissolving 56 g of KOH pellets in 1000 mL of distilled water. The LDH cathode for AZIBs was fabricated as follows: LDH, acetylene black, and nafion (wt%: 8:1:1) were mixed in a small amount of ethanol for 12 h to make a homogeneous paste, which was coated onto carbon fiber paper as the current collector. Then, the electrode was put into a 60 °C oven overnight. The loading mass was ~2 mg/cm^2. The charge/discharge profiles and cycling performance were determined by the Neware battery testing system, and cycling voltammetry and a Nyquist plot were carried out by a CHI660e potentiostat.

2.4. DFT Calculations

All theoretical calculations were performed by the use of density functional theory (DFT) with CASTEP code. The exchange/correlation energies were calculated using the generalized gradient approximation introduced by Perdew, Burke, and Ernzerhof. Ultrasoft pseudopotentials were conducted to describe the coupling between the ionic core and valence electrons, and the spin-polarized calculations were considered. The cutoff energy for the plane–wave basis set was set at 380.0 eV. For optimizing the cell, the convergence tolerance was 5.0×10^{-6} eV/atom to energy, 0.01 eV/Å to maximum force, 0.02 GPa to maximum stress, and 5.0×10^{-4} Å to maximum displacement. For geometric optimization and energy calculations, the separation of k-point sampling was adopted with 0.07/Å and 0.06/Å, respectively. A convergence accuracy of 5.0×10^{-7} eV/atom was set for the self-consistent field calculation. In our work, the mixtures of different elements were created. Mixture atoms can simulate disorder in a structure, such as in the case where one atomic site is randomly occupied by two or more different types of atoms.

3. Results and Discussion

CoMnAl LDH was synthesized by a facile coprecipitation method, as shown in Figure 2A, in which Co, Mn, and Al hydroxides shared edges and octahedral sheets bounded together. The crystalline structure of CoMnAl LDH was characterized by X-ray diffraction (XRD). Figure 2B displays the XRD patterns of CoMnAl LDH and CoMn counterpart, both of which show the typical (006) peak located around 20.6° [18]. The intensity of the (006) peak was enhanced after Al^{3+} introduction, while (100) and (110) peaks were reduced. This was probably due to the layer charge density being affected by the charge and size of the metal ions in LDH. The calculated d_{006} was 0.435 nm, which decreased to 0.433 nm after Al^{3+} introduction (Table S1). To compensate these positive charges, anions from precursor solutions usually absorb and occupy the interlayer spacing between LDH layers. Figure 2C depicts the Fourier transform infrared spectroscopy (FTIR) spectra of CoMnAl and CoMn LDHs. In both spectra, one can see that peaks located around 1500 cm^{-1} are due to the N-O stretching mode and the peak centered at 1120 cm^{-1} is due to the S-O stretching vibration, indicating NO_3^- and SO_4^{2-}, intercalated into the layer structure [19,20]. The board band at 3470 cm^{-1} can be assigned to the O-H vibration from the OH group and interlayer water. The peaks at around 600 cm^{-1} are the characteristic of Co/Mn/Al-O and Co/Mn/Al-OH bending vibrations, respectively.

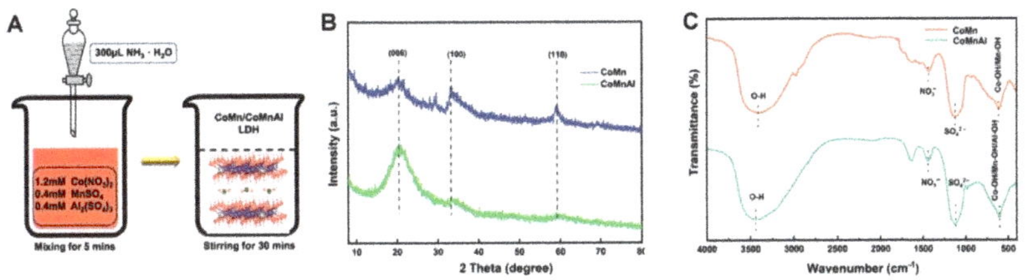

Figure 2. (**A**) An illustration of LDH synthesis. (**B**,**C**) The XRD and FTIR spectra of CoMn and CoMnAl.

Scanning electron spectroscopy (SEM) and transmission microscopy (TEM) were conducted to analyze the morphologies of CoMnAl and CoMn LDHs. Figure 3A shows that CoMn LDH displayed a plicated and porous particle morphology in nano-scale. Figure 3B confirms the character-layered structure of LDH. The lattice fringe was calculated to be 2.71 Å, corresponding to d_{100}-spacing. In the LDH layer, transparent regions are observed, indicating that pores with an average diameter of 20–30 nm (P stands for pore structure) exist in these LDH layers. The specific surface area of CoMn LDH was examined by nitrogen adsorption (Figure 3C). The Brunauer–Emmett–Teller (BET) surface area of the CoMn LDH was calculated to be 47.260 m^2g^{-1}. Figure 3D exhibits the SEM image of CoMnAl LDH, in which more pores were detected. The layered structure of the CoMnAl LDH was also confirmed by HRTEM, but the pore diameter was increased, with an average size of 10 nm (Figure 3E). Accordingly, the BET surface area increased to 106.530 m^2g^{-1} (Figure 3F). The increased BET surface and porous structure were attributed to Al^{3+}, with a smaller size and high charge strongly affecting the charge density of the LDH layers.

To further probe the influence of Al^{3+} introduction to the LDH surface state, X-ray photoemission spectroscopy (XPS) measurement was carried out. Figure 4A shows the Co 2p XPS spectrum in CoMnAl LDH. The binding energies of 797.1 eV and 781.5 eV for $2p_{1/2}$ and $2p_{3/2}$ were observed as photoemission maxima, with two satellite structures. In the fitted curve, the main peak of Co $2p_{3/2}$ centering at 780.5 eV is typical evidence of Co^{2+} species [21]. The valence of the Co element was also demonstrated to be +2 in CoMn (Figure 4B) and CoAl LDHs (Figure 4C), suggesting that Mn and Al had no effects on the electronic structure of Co element. The Mn 2p XPS spectra of CoMn and CoMnAl LDHs are displayed in Figure 4D,E. In the fitted curve of CoMn LDH, the main peak of Mn $2p_{2/3}$ is located at 642.3 eV, implying that the valence state of the Mn species was +3 [22]. But the increased content of Mn^{4+} species was observed in CoMnAl LDH, while the valence of Co element remained. In CoAl LDH, the valence state of Al was evidenced to be +3, since the main peak was located at 75 eV (Figure 4F). Nevertheless, the main peak of Al 2p negatively shifted by 0.47 eV, which indicates the decrease in Al valence in CoMnAl. It seems that partial Al^{3+} substitution with high charge density caused stronger electronic interaction with ambient trivalent Mn than bivalent Co species.

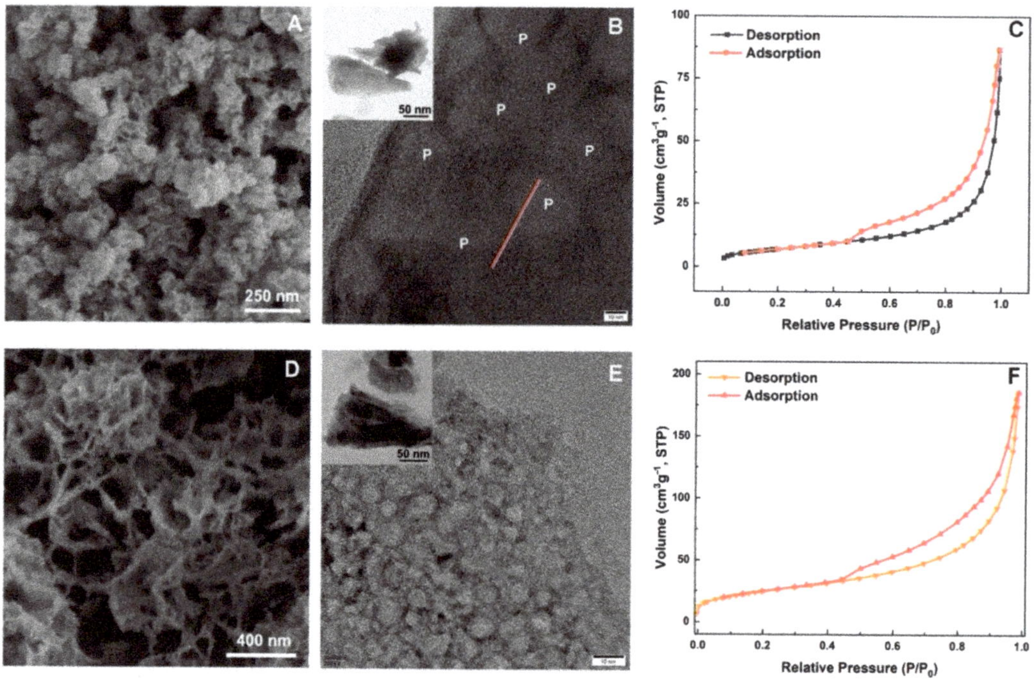

Figure 3. (**A–C**) SEM, TEM, and HRTEM images and N_2 isotherm of CoMn LDH. (**D–F**) SEM, TEM, and HRTEM images and N_2 isotherm of CoMnAl LDH.

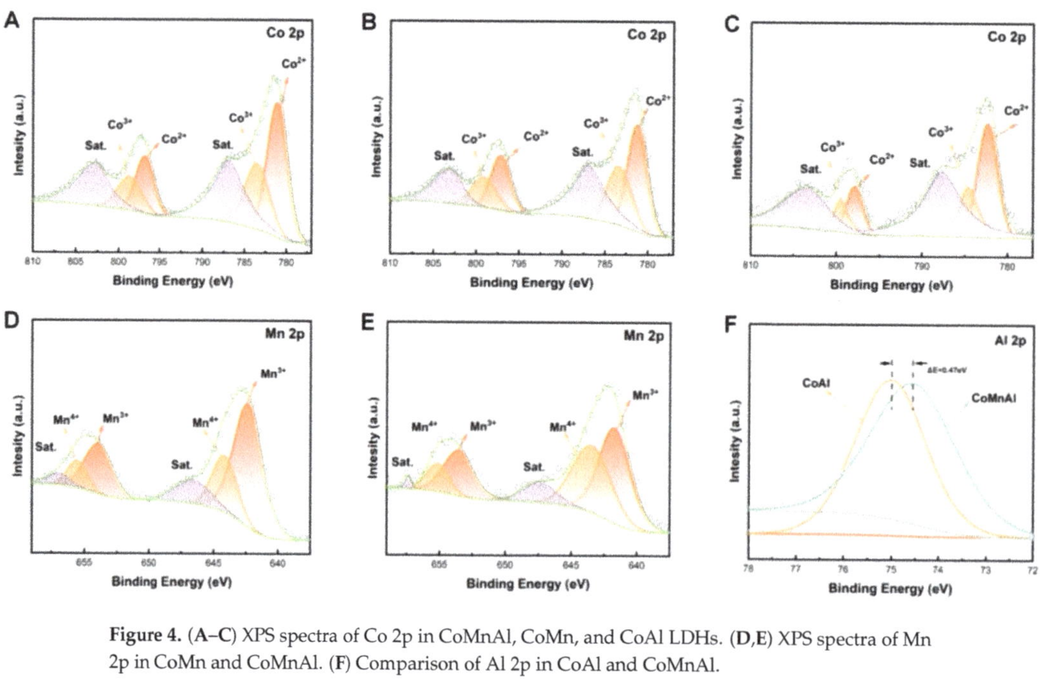

Figure 4. (**A–C**) XPS spectra of Co 2p in CoMnAl, CoMn, and CoAl LDHs. (**D**,**E**) XPS spectra of Mn 2p in CoMn and CoMnAl. (**F**) Comparison of Al 2p in CoAl and CoMnAl.

Further, we employed DFT to calculate their electronic structures, and the results are shown in Figure 5. It was found that their bands consist of three parts in general, which are marked by dashed rectangles. The bands in the top part are nearly empty, and the bands in the bottom part are full of electrons. In contrast, the Fermi level goes through the middle part containing the bands composed of mainly d orbitals and a small number of p orbitals. Noticeably, the bands in the top part are the most dispersed, and the bands in the middle part are the most localized. Alpha and beta represent the bands of electrons with upward and downward spin directions, respectively. For $Co(OH)_2$, the top part overlaps slightly with the middle part, implying that they are partially hybrid. The gap is about 2.05 eV between the middle and bottom parts. In addition, the spin polarization clearly appears near the Fermi level due to the presence of the Co atom, which mainly occurs in the middle part. When 40% of the Mn element is introduced into $Co(OH)_2$, the bimetallic component CoMn LDH is formed. It was observed that the bands near the Fermi level are degenerate. Such degenerate orbitals cause the energy region in the middle part to become narrow compared with $Co(OH)_2$. Accordingly, the density of states (DOS) near the Fermi level also obviously increases. The gap is widened to 3.28 eV between the middle and bottom parts. When Al is further introduced, the trimetallic component of CoMnAl LDH can be also synthesized. It was found that the bands in the top and bottom parts shift upwards due to electron deficiency. The elevation of the bands in the top part leads to overlap decrease between the top and middle parts, which means less hybridization. Sequentially, the bands in the middle part further narrow and the DOS appears sharply. The gap is narrowed to 1.68 eV between the middle and bottom parts. Due to fewer electrons being present, the spin polarization reappears near the Fermi level. When one component of the bimetallic component of hydroxide is Al (CoAl LDH), the bands in the top and bottom parts further shift upwards. A gap is formed between the top and middle parts. In contrast, the bands in the bottom part enter into the middle part, resulting in significant orbital hybridization and some spin polarization. The dispersed behavior widens the bands in the middle part, and the corresponding DOS is less sharp. The strength of the M-O bond (M: metals atoms) can be evaluated by the bands in the bottom part, and activity can be assessed by the bands in the middle part. They are analyzed from their corresponding DOSs. It was observed that the peaks in the bottom part are mainly hybridized by the states of metals and oxygen for CoMn, CoAl, and CoMnAl LDHs. In particular, for the top peak, its overlap and intensity suggest that the crystalline structure of CoMnAl LDH is the most stable among them. For the CoMnAl LDH, a sharp peak at the Fermi level means that metals have the very localized electrons. Such nonbonding electrons are conducive to the activation of metals. In addition, it was observed that, for CoMn LDH, the electrons of CoAl LDH at the Fermi level are more localized than those of $Co(OH)_2$, and the bands of CoAl LDH near the Fermi level are more dispersed than those of $Co(OH)_2$, indicating that CoAl LDH is more conductive.

The electrochemical performances of CoMn and CoMnAl LDHs were examined in a coin cell (Figure 6A), in which LDHs, Zn foil, and 1 M KOH solution were utilized as the cathode, anode, and electrolyte, respectively. The open-circuit voltages of Zn-CoMn and Zn-CoMnAl cells were 1.15 and 1.21 V. Accordingly, the Zn-CoMnAl cell displayed a better electrochemical performance. Figure 6B exhibits the charge/discharge profiles of the Zn-CoMnAl cell at different current densities. Two discharge plateaus indicate a stepwise reaction mechanism, corresponding to Co and Mn redox. At a current density of 0.5 A g^{-1}, the Zn-CoMnAl cell deliver a high specific capacity of 238.9 mAh g^{-1}. On the other hand, the Zn-CoMn cell displays an inferior electrochemical performance (Figure 6C). Two discharge plateaus remain, but the discharge time is much shorter. The Zn-CoMn cell only showed a capacity of 159 mAh g^{-1} at 0.5 A g^{-1}, and the capacity was 81.2, 68.6, and 51.2 mAh g^{-1} as the current density increased to 1, 2, and 5 A g^{-1}, respectively. In contrast, an excellent rate capability was obtained by the Zn-CoMnAl cell at the same current density (194.75, 179.05, and 138.75 mAh g^{-1} at 1, 2, and 5 A g^{-1}, respectively; Figure 6D).

The cyclic voltammetry (CV) curves of Zn-CoMnAl and Zn-CoMn cells are depicted in Figure 6E. The CV test starts from 0.6 V to a high voltage, and the arrows represent the directions of oxidation and reduction scans. The CV curve of Zn-CoMnAl shows a higher peak current and larger enclosed area than that of Zn-CoMn, implying higher activity after Al^{3+} introduction. To gain more insight into the effect of Al^{3+} introduction in CoMnAl LDH, electrochemical impedance spectroscopy (EIS) was conducted in a three-electrode system. Figure 6F shows the Nyquist plots of the CoMnAl and CoMn LDHs, obtained at ambient temperature. R_{ct} in electrical equivalent circuit (Figure 6F inset) stands for charge-transfer resistance. The calculated R_{ct} of CoMnAl LDH is 2.5 Ω, while the R_{ct} of CoMn is 3.6 Ω, indicative of a low resistance of electron transfer from the interface between the LDH and the electrolyte and a higher energy storage efficiency resulting from Al^{3+} introduction. Furthermore, the activation energy was calculated as 1.4 V, at which potential the charge process initiated. The activation energy was examined based on a series of EIS tests at different temperatures, representing the energy barrier of Co and Mn redox reactions (Figure 6G,H). The activation energy of the Zn-CoMnAl cell is 16.28 kJ mol^{-1} was lower than that of the Zn-CoMn cell (28.56 kJ mol^{-1}) and could be attributed to a highly conductive CoMnAl-O network. The lower energy barrier of the CoMnAl LDH could also lead to a high conversion efficiency and structural stability. In accordance with DFT simulation, the capacity of the Zn-CoMn cell dropped quickly within 500 cycles at 0.5 A g^{-1} (76% retention), while the Zn-CoMnAl cell showed a stable cycle life of 2000 cycles at the same current density with a capacity retention of 92%.

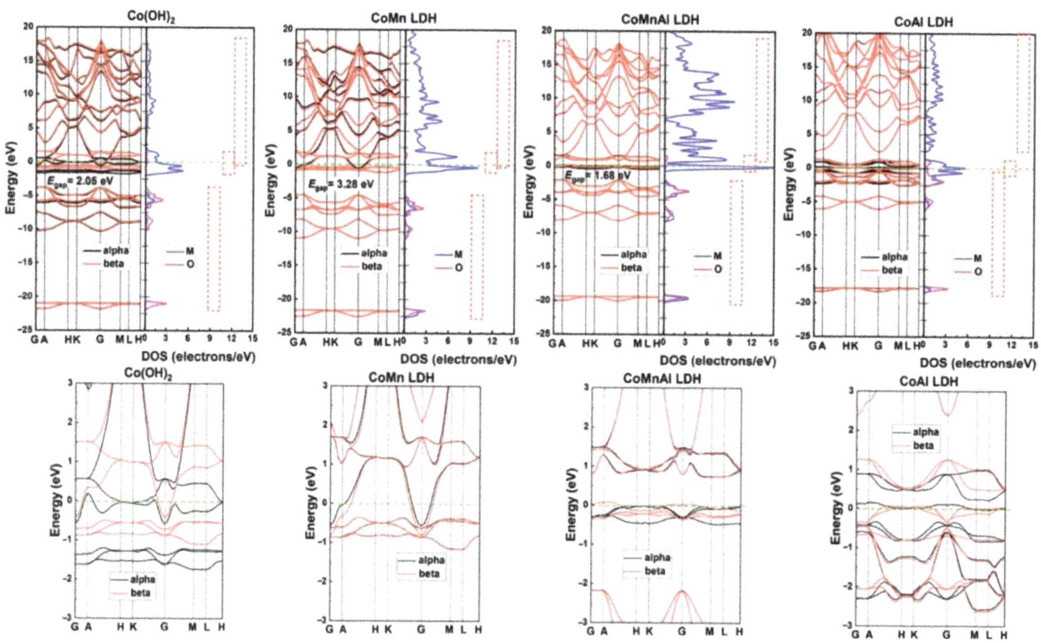

Figure 5. The energy bands (**left**) and density of states (**right**) for different structures. M and O denote metal and oxygen atoms, respectively. The spectra in the bottom panel show the magnified images of the band dispersion near the Fermi level. The Fermi level is set to 0 eV. Inset: the dashed rectangles schematically indicate different band parts.

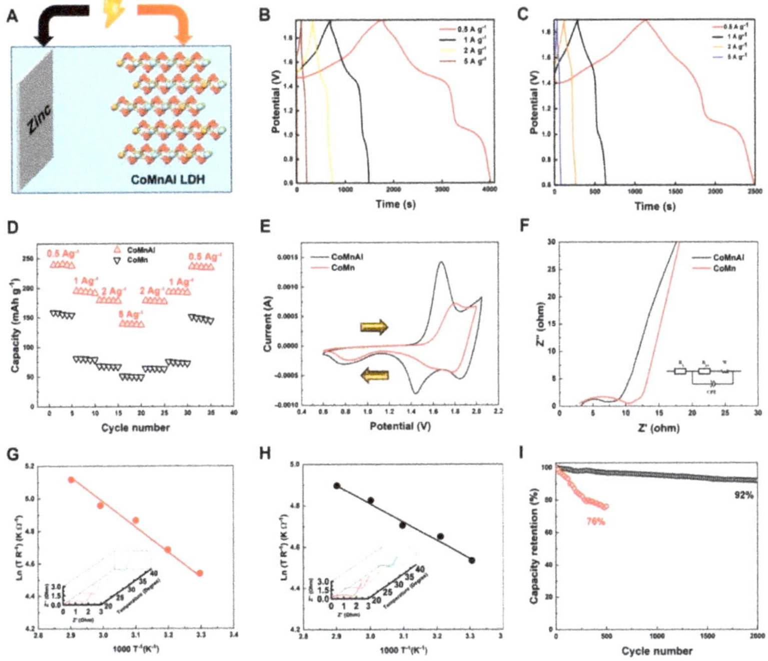

Figure 6. (**A**) The AZIB with CoMnAl LDH cathode and zinc anode. (**B**,**C**) The charge/discharge profiles of Zn-CoMnAl and Zn-CoMn AZIB cells at different current densities. (**D**) The rate capability of Zn-CoMnAl and Zn-CoMn AZIB cells. (**E**) A CV comparison of Zn-CoMnAl and Zn-CoMn AZIB cells at a scan rate of 4 mVs^{-1}. (**F**) The Nyquist plots and the equivalent electrical circuit. (**G**,**H**) The EIS curves of Zn-CoMn and Zn-CoMnAl AZIB cells at different temperatures. (**I**) The cyclability of Zn-CoMn and Zn-CoMnAl AZIB cells.

4. Conclusions

In summary, CoMnAl LDH is synthesized via a facile coprecipitation method, and it has proved to be promising as a cathode material for AZIB. The interaction between trivalent Mn and Al element not only lowers the intrinsic resistance and redox energy barrier but also strengthens the electronic structure and the electrochemical stability of LDH, as demonstrated by the DFT simulation and cycling performance. All these results demonstrate the prospects of LDH as a cathode material for AZIBs.

Supplementary Materials: The following supporting information can be downloaded at: https://www.mdpi.com/article/10.3390/batteries11050170/s1, Table S1: Interlayer spacings of CoMn and CoMnAl LDHs.

Author Contributions: Conceptualization, T.D.; methodology, T.D.; software, J.Z., J.G., W.L. and J.F.; validation, J.Z., J.G., W.L. and J.F.; formal analysis, J.Z. and J.G.; investigation, J.Z., J.G., W.L. and J.F.; resources, T.D.; data curation, T.D.; writing—original draft preparation, J.Z. and J.G.; writing—review and editing, T.D.; visualization, T.D.; supervision, T.D.; project administration, T.D. All authors have read and agreed to the published version of the manuscript.

Funding: This work was supported by the International Collaboration Program of the Jilin Provincial Department of Science and Technology (No. 20230402051GH), the 2024 Innovation-High-tech Industry Program of Jilin Province Development and Reform Commission (No. 2024C018-8), the Fundamental Research Funds for the Central Universities, JLU, and the "Double-First Class" Discipline for Materials Science & Engineering.

Data Availability Statement: The raw data supporting the conclusions of this article will be made available by the authors on request.

Conflicts of Interest: The authors declare no conflicts of interest.

References

1. Ma, W.; Wang, S.; Wu, X.; Liu, W.; Yang, F.; Liu, S.; Jun, S.C.; Dai, L.; He, Z.; Zhang, Q. Tailoring desolvation strategies for aqueous zinc-ion batteries. *Energy Environ. Sci.* **2024**, *17*, 4819–4846. [CrossRef]
2. Zheng, X.; Han, C.; Lee, C.-S.; Yao, W.; Zhi, C.; Tang, Y. Materials challenges for aluminum ion based aqueous energy storage devices: Progress and prospects. *Prog. Mater. Sci.* **2024**, *143*, 101253. [CrossRef]
3. Li, C.; Hu, L.; Ren, X.; Lin, L.; Zhan, C.; Weng, Q.; Sun, X.; Yu, X. Asymmetric Charge Distribution of Active Centers in Small Molecule Quinone Cathode Boosts High-Energy and High-Rate Aqueous Zinc-Organic Batteries. *Adv. Funct. Mater.* **2024**, *34*, 2313241. [CrossRef]
4. Guo, X.; Zhang, S.; Hong, H.; Wang, S.; Zhu, J.; Zhi, C. Interface regulation and electrolyte design strategies for zinc anodes in high-performance zinc metal batteries. *iScience* **2025**, *28*, 111751. [CrossRef]
5. Peng, Z.; Shen, X.; Li, B.; Cheng, J.; He, Z.; Sun, Z.; Li, B.; Zhang, Z.; Zhuang, Z.; Wu, X.; et al. Comprehensive crystallographic engineering for high-efficiency and durable zinc metal anodes. *Prog. Mater. Sci.* **2025**, *152*, 101453. [CrossRef]
6. Cao, J.; Wu, H.; Zhang, D.; Luo, D.; Zhang, L.; Yang, X.; Qin, J.; He, G. In-Situ Ultrafast Construction of Zinc Tungstate Interface Layer for Highly Reversible Zinc Anodes. *Angew. Chem. Int. Ed.* **2024**, *63*, e202319661. [CrossRef]
7. Cao, J.; Sun, M.; Zhang, D.; Zhang, Y.; Yang, C.; Luo, D.; Yang, X.; Zhang, X.; Qin, J.; Huang, B.; et al. Tuning Vertical Electrodeposition for Dendrites-Free Zinc-Ion Batteries. *ACS Nano* **2024**, *18*, 16610–16621. [CrossRef]
8. Chen, F.; Li, j.; Shao, Y.; Zhu, Z.; Shen, T.; Chen, K.; Chen, Y.; Chen, Y. ZIF-67 wraps Ni-Mn LDHs nanosheets to enhance the capacitive contribution of supercapacitors. *Chem. Eng. J.* **2025**, *507*, 160454. [CrossRef]
9. Zhang, S.; Zhao, Y.; Gao, S.; Liu, C.; Chen, X. Design and synthesis of nickel-cobalt-aluminum hydroxide battery electrode materials for high-performance hybrid supercapacitors. *J. Power Sources* **2025**, *632*, 236370. [CrossRef]
10. Wang, C.; Wang, D.; Lv, D.; Peng, H.; Song, X.; Yang, J.; Qian, Y. Interface Engineering by Hydrophilic and Zincophilic Aluminum Hydroxide Fluoride for Anode-Free Zinc Metal Batteries at Low Temperature. *Adv. Energy Mater.* **2023**, *13*, 2204388. [CrossRef]
11. Zhao, Y.; Zhang, P.; Liang, J.; Xia, X.; Ren, L.; Song, L.; Liu, W.; Sun, X. Unlocking Layered Double Hydroxide as a High-Performance Cathode Material for Aqueous Zinc-Ion Batteries. *Adv. Mater.* **2022**, *34*, e2204320. [CrossRef] [PubMed]
12. Li, Q.; Wang, C.; Zhu, Y.; Du, W.; Liu, W.; Yao, M.; Wang, Y.; Qian, Y.; Feng, S. Unlocking the critical role of Mg doping in α-MnO_2 cathode for aqueous zinc ion batteries. *Chem. Eng. J.* **2024**, *485*, 150077. [CrossRef]
13. Zhao, X.; Zhang, F.; Li, H.; Dong, H.; Yan, C.; Meng, C.; Sang, Y.; Liu, H.; Guo, Y.-G.; Wang, S. Dynamic heterostructure design of MnO_2 for high-performance aqueous zinc-ion batteries. *Energy Environ. Sci.* **2024**, *17*, 3629–3640. [CrossRef]
14. Liang, X.; Liu, X.; Wang, P.; Guo, Z.; Chen, X.; Yao, J.; Li, J.; Gan, Y.; Lv, L.; Tao, L.; et al. Ion-exchange induced Ni doping of α-MnO_2 cathode with structural modification for aqueous zinc ion batteries. *J. Power Sources* **2025**, *635*, 236518. [CrossRef]
15. Fu, H.; Wang, X.; Ye, L.; Wu, Z.; Yang, J.; Shi, M.; Ang, E.H. Optimizing Fe in Mn-based Prussian blue analogs with dual redox-active sites to enhance operating voltage and durability in Zn-ion batteries. *Chem. Eng. J.* **2025**, *506*, 160308. [CrossRef]
16. Zhang, Y.; Hu, S.; Li, C.-e.; Yan, X.-h.; Zhang, Y.-y.; Yin, R.-b.; Wei, Y.-f.; Gao, K.-z.; Gao, H.-l. Advanced strategies for enhancing electrochemical performance of NiAl LDH electrodes in supercapacitors. *Coord. Chem. Rev.* **2025**, *531*, 216497. [CrossRef]
17. Bai, S.; Liu, G.; Shen, T.; Wu, Z.; Chen, W.; Song, Y.-F. Controllable synthesis of layered double hydroxides: From macroscopic morphology to microscopic coordination at the atomic level. *Coord. Chem. Rev.* **2025**, *529*, 216437. [CrossRef]
18. Bujdosó, T.; Patzkó, Á.; Galbács, Z.; Dékány, I. Structural characterization of arsenate ion exchanged MgAl-layered double hydroxide. *Appl. Clay Sci.* **2009**, *44*, 75–82. [CrossRef]
19. Wang, H.; Lin, T.; Song, Z.; Huang, M.; Chai, R.; An, S.; Song, Y.-F. Simultaneous mineralization of Cd(II), Pb(II) and As(V) using MgAl-NO3: Performance and mechanism. *Sep. Purif. Technol.* **2025**, *362*, 131853. [CrossRef]
20. Zhang, F.; Liu, L.; Zhang, C.; Shang, D.; Wu, L. Preparation of ZnAl layered double hydroxides supported by silica for the treatment of Cr(VI) and Cu(II) in aqueous solution. *Sci. Rep.* **2025**, *15*, 2522. [CrossRef]
21. Wang, C.; Han, L.; Yang, S.; Liu, Z.; Liu, M.; Li, B. Nanosheet-structured ZnCo-LDH microsphere as active material for rechargeable zinc batteries. *J. Colloid Interface Sci.* **2024**, *659*, 119–126. [CrossRef] [PubMed]
22. Moradi, M.; Afkhami, A.; Madrakian, T.; Moazami, H.R. Electrosynthesis of Co Mn layered-double-hydroxide as a precursor for Co-Mn-MOFs and subsequent electrochemical sulfurization for supercapacitor application. *J. Energy Storage* **2023**, *71*, 108177. [CrossRef]

batteries

MDPI

Article

Electrochemical–Thermal Modeling of Lithium-Ion Batteries: An Analysis of Thermal Runaway with Observation on Aging Effects

Milad Tulabi and Roberto Bubbico *

Department of Chemical Materials and Environmental Engineering, "Sapienza" Rome University, Via Eudossiana 18, 00184 Rome, Italy; milad.tulabi@uniroma1.it

* Correspondence: roberto.bubbico@uniroma1.it

Abstract: The increasing demand for energy storage solutions, particularly in electric vehicles and renewable energy systems, has intensified research on lithium-ion (Li-ion) battery safety and performance. A critical challenge is thermal runaway (TR), a highly exothermic sequence of reactions triggered by mechanical, electrical, or thermal abuse, which can lead to catastrophic failures. While most TR models focus on fresh cells, aging significantly impacts battery behavior and safety. This study develops an electrochemical–thermal coupled model that incorporates aging effects to better predict thermal behavior and TR initiation in cylindrical Li-ion batteries. The model is validated against experimental data for fresh NMC and aged NCA cells, and statistical analysis is conducted to identify key factors influencing TR ($p < 0.05$). A full factorial design evaluates the effects of internal resistance (10, 20, 30, and 40 mΩ), capacity (1, 2, 3, and 5 Ah), and current rate (1C, 3C, 6C, and 8C) on temperature evolution. Additionally, a machine learning algorithm (logistic regression) is employed to identify an internal resistance threshold, beyond which thermal runaway (TR) becomes highly probable, and to predict TR probability based on key battery parameters. The model achieved a high prediction accuracy of 95% on the test dataset. Results indicate that aging affects thermal stability in complex ways. The increased internal resistance exacerbates heating rates, while capacity fade reduces stored energy, mitigating TR risk. These findings provide a validated framework for enhancing battery thermal management and predictive safety mechanisms, which contributed to the development of safer, more reliable Li-ion energy storage systems.

Keywords: thermal runaway; Li-ion battery; modeling; aging; battery safety

1. Introduction

The energy sector's reliance on fossil fuels contributes significantly to global greenhouse gas emissions and air pollution annually [1]. Rising demand, resource depletion, and geopolitical instability also induce a shift toward low-carbon energy sources such as wind and solar power. To balance the variability of renewable energy, efficient energy storage solutions like secondary (rechargeable) batteries and supercapacitors are critical. The ongoing transition to fully electric and hybrid vehicles to reduce transportation-related emissions already represents an example of this paradigm shift.

Li-ion batteries have become a cornerstone of low-carbon energy applications due to their high energy density and compact design, particularly in electric vehicles (EVs) and portable devices. However, their safety has gained attention due to the risks of catastrophic failures like thermal runaway [2–5]. Thermal runaway (TR) occurs when an uncontrollable

 https://doi.org/10.3390/batteries11050178

rise in temperature leads to strongly exothermic side reactions, such as those associated with the decomposition of the solid electrolyte interface (SEI) of the anode, cathode, and electrolyte, which can result in fires or explosions [6–9].

Lithium-ion battery modeling has evolved significantly, beginning with the foundational work by Doyle and Newman [10,11]. Their "pseudo-2D" (P2D) model incorporates one-dimensional cell layers and solid lithium diffusion in electrode particles, which require about 40 parameters for accurate representation. While highly detailed, this complexity limits its practicality in engineering. For low to moderate currents, the model is often simplified using a single-particle (SP) approach, which assumes uniform conditions across porous electrodes and uses lumped resistance terms to enhance computational efficiency [12]. Different studies have utilized the single-particle (SP) model and the more comprehensive pseudo-2D (P2D) model as foundational frameworks to investigate the thermal behavior of lithium-ion batteries. These models serve as critical tools in analyzing heat generation mechanisms, temperature distribution, and their effects on battery performance and safety [13–17].

Thermal runaway in Li-ion cells involves a chain reaction mechanism triggered by abnormal temperature increases under abusive conditions. This process is fueled by the Heat–Temperature–Reaction (HTR) loop. Initially, abnormal heat generation raises the cell's temperature, initiating side reactions like the decomposition of the solid electrolyte interface (SEI). These reactions release additional heat, perpetuating the HTR loop until the cell undergoes thermal runaway [4]. There are three types of abuse that trigger thermal runaway: mechanical, electrical, and thermal abuse. Mechanical abuse refers to physical damage to the battery, such as crushing and penetration, which can compromise the battery's structural integrity and lead to ignition. Examples of electrical abuse are high currents, overcharging, over-discharging, or short-circuiting. Thermal abuse occurs when a Li-ion battery is exposed to external temperatures outside its safe operating range, which can happen due to environmental conditions, poor thermal management, or external heat sources.

Modeling thermal runaways in lithium-ion batteries is a critical area of research aimed at understanding and mitigating safety risks associated with battery operation. One common strategy involves the development of electrochemical–thermal models, such as those based on the Newman framework [10,11], which integrate electrochemical reactions with heat generation and transport mechanisms within the battery cell. These models typically consider factors like heat generation from electrochemical reactions, ohmic heating due to internal resistance, and heat dissipation through conduction, convection, and radiation. Thermal runaway modeling has also advanced, with Hatchard et al. [14] developing a model focusing on side reactions during runaway events, which are validated through oven tests. Kim et al. [9] expanded this to a 3D framework, incorporating additional reactions, thus inspiring further research into the thermal behavior and safety of Li-ion batteries. Another significant aspect is the incorporation of thermal abuse models, which simulate extreme conditions such as overcharging or external heating to predict how these conditions can trigger thermal runaway [15–17]. Talele et al. [15] used a thermal runaway numerical model to study the trigger point evolution in prismatic cells with different cathode chemistries (NMC, LFP, and LCO) under varying ambient temperatures. They employed nail penetration-assisted thermal abuse testing and empirical correlations to simulate the source term. By ANOVA-based statistical evaluation, they found ambient temperature to be the most influential factor on trigger time for both natural and forced convection conditions. Jelly roll thickness and heat transfer coefficient (HTC) were identified as additional critical parameters. Azuaje–Berbecí et al. [16] proposed a multilayered electrochemical–thermal model designed to predict heat generation, battery temperature, voltage, and the potential for thermal runaway during lithium-ion battery charge/discharge cycles under various operating conditions. Their findings revealed that cyclic charging and discharging can elevate the battery temperature to levels capable of triggering thermal runaway. Özdemir et al. [17] investigated the thermal behavior of lithium-ion cells under both normal and abusive operating conditions through experimental and numerical approaches. They developed an electrochemical–thermal model to simulate temperature and voltage variations during discharge processes. The model was enhanced to incorporate the effects of temperature-dependent heat generation by utilizing Arrhenius equations.

Most battery modeling for thermal runaway focuses on a fresh battery and does not contain any aging-related mechanism, which is an important contributing factor that may heighten the risk [18,19]. Aging refers to the degradation of the cell over time, which is evidenced by battery performance. To understand the aging of a Li-ion battery, three main questions need to be answered: What factors cause battery aging? What happens to a battery during aging? And what are the consequences of the aging process?

There are two main categories of aging: cycling aging and calendar aging. During the operation of a battery in an electrical application such as in an electric vehicle, aging occurs primarily due to the repeated charging and discharging phases called "cycling aging". This type of lithium-ion battery aging is significantly affected by key operating conditions such as temperature, charge/discharge current rates (C-rate), depth of discharge (DOD), and mechanical stress [20]. Battery aging can also occur while the battery is at rest in a process known as calendar aging. During this period, the battery is exposed to ambient temperatures without active thermal management, potentially leading to more aggressive aging due to exposure to higher or lower ambient temperatures than those reached during cycling when the thermal management system can better control the temperature. This rate of degradation is influenced by factors such as the battery's state of charge (SOC) and the duration of storage [21]. Lithium-ion battery degradation typically follows four key modes: loss of Lithium inventory, loss of active material in the negative and positive electrodes [22], and conductivity loss [20]. Lithium ions can be consumed inside reactions such as solid electrolyte interface (SEI) layer growth, electrolyte decomposition, or lithium plating, causing lithium shortage for cycling. The active material in the anode and cathode becomes unavailable for lithium insertion due to mechanisms such as particle cracking, loss of electrical contact, or blockage by resistive surface layers. It is worth mentioning that these mechanisms have different paths and effects depending on battery chemistry, and they can affect the performance of the cell through three main factors—capacity, power, and internal resistance—which have been the preferred aging indicators of lithium-ion batteries. Structural disorder, loss of active material, crystal distortion, and mechanical stress contribute to capacity and power fading, while surface layer formation, such as in lithium plating and SEI layer growth, increases impedance and resistance [21]. These aging effects alter the physical and chemical properties of cell components, significantly influencing the thermal behavior of lithium-ion batteries. However, there are fewer studies on the thermal safety of aged lithium-ion cells, especially those using numerical methods, compared to fresh cells.

In the study by Fleischhammer et al. [23], the impact of cyclic aging on the thermal behavior of Li-ion cells was investigated, and the results revealed that cells subjected to high-rate cycling exhibited the same thermal behavior as fresh cells. In contrast, Wu et al. [18] conducted an experimental study using ARC tests to investigate how low-temperature aging affects the thermal behavior of Li-ion batteries. The results showed that low-temperature cycling significantly influences TR performance, leading to a faster onset of thermal runaway and increased self-heating rates at the same temperature. Another study [24], conducted to investigate the influence of calendar aging on the thermal stability of a commercial Li-ion cell, revealed that the self-heating onset temperature of aged cells was much lower than that of fresh cells. Garcia et al. [19] examined the thermal runaway (TR) behavior of fresh and aged NMC 811 batteries using ARC testing at different SOCs. Their findings showed that higher SOC led to earlier exothermic reaction detection, highlighting the rapid escalation risk at high SOC conditions. Guo et al. [25] investigated the effects of aging on the safety performance of commercial 18650 lithium-ion batteries by subjecting them to fast charge/discharge cycles and conducting thermal abuse tests. The results revealed that aged cells exhibited improved thermal stability compared to fresh cells and had significantly higher thermal runaway temperatures. This study suggests a greater likelihood of thermal runaway propagation in the module of aged cells used in applications such as electric vehicles due to the higher temperature of the thermal runaway. Zhang et al. [26,27] conducted a comprehensive investigation into the thermal safety evolution of lithium-ion batteries during high-temperature aging. Their study revealed that both high-temperature cyclic aging and calendar aging result in similar thermal hazard and degradation mechanisms. The findings showed that high-temperature aging significantly reduces the thermal stability of the cells. The study observed that the gas released during the aging process reduces the amount of gas available for release during a potential thermal runaway event. Although aging decreases the cells' thermal stability, it also mitigates the overall thermal hazards posed by the batteries through this gas release.

Abada et al. [28] developed a 3D model for an aged battery. They considered the SEI growth as the main cause of calendar aging in Li-ion cells and incorporated this concept into the 3D thermal model to investigate the thermal behavior of a cell subjected to external heating. The results showed that SEI growth hinders the anode decomposition and consequently increases the onset temperature of thermal runaway. However, their model did not consider the effects of SEI growth on the increase in the internal resistance of the cell. Grandjacques et al. [29] developed a semi-empirical model for predicting the thermal behavior of an aged lithium-ion cell based on an Arrhenius kinetic equation. Zhao et al. [30] developed a thermal runaway model for lithium-ion batteries subjected to low-temperature cycling. Their model incorporates an exothermic reaction between metallic lithium and the electrolyte to account for aging effects. The results indicated that aging slightly increases the total heat generated during thermal runaway, but the impact remains relatively small. In the study by Pastor et al. [31], the impact of aging on thermal runaway behavior was experimentally investigated for lithium-ion batteries with NMC and LFP cathode chemistries using Accelerating Rate Calorimetry (ARC) tests. Aged batteries were subjected to external heating abuse to evaluate thermodynamic parameters related to thermal runaway. A computational model was also developed using GT-Autolion software for further analysis. Results indicated that aged batteries released less heat during thermal runaway when triggered by external heating. This reduced heat generation was attributed to SEI growth, a key aging mechanism causing capacity to fade, which decreases the availability of lithium and electrolyte in the anode, thus diminishing the exothermic reactions during thermal runaway. Kabra et al. [32] utilized a combination of experimental and modeling approaches to explore the impact of aging mechanisms on the thermal stability of Li-ion cells. They developed a physics-based aging model to study degradation processes such as lithium plating, SEI layer growth, and active material loss during cycling. By integrating insights from this aging model with Accelerating Rate Calorimetry (ARC) experiments, they created a degradation-aware thermal stability model capable of predicting the onset and intensity of thermal runaway (TR) in Li-ion cells. The findings revealed that while SEI decomposition contributes to earlier TR onset, additional SEI growth can mitigate anode–electrolyte reactions, delaying TR. In contrast, for cells cycled at low temperatures, TR was predominantly driven by lithium plating and the subsequent reaction between plated lithium and the electrolyte. These studies mainly consider the external heat as a trigger for thermal runaway, but they did not consider the effect of high current or other electrical abuse conditions, possibly resulting in a thermal runaway.

This study focuses on developing an electrochemical–thermal coupled model that accounts for side reactions, which lead to thermal runaway, while also trying to include aging effects. The model is validated using experimental data for fresh NMC Li-ion batteries and for aged NCA Li-ion batteries [33]. It aims to simulate the thermal behavior of aged cells and to explore the impact of external factors, such as high current rates, and internal factors, including impedance and capacity variations involved in aging effects, on thermal runaway initiation. Statistical analysis is used to evaluate the influence of these conditions on the thermal behavior of an NMC cell. The study focuses on cylindrical NMC and NCA Li-ion batteries due to their widespread applications, and it consists of three parts:

- Development of the electrochemical–thermal coupled model;
- Validation of the model;
- Statistical analysis for finding the most influential factors on an aged cell's thermal behavior.

2. Methods

2.1. Geometry

The battery under study is a cylindrical lithium-ion battery with a length of 65 mm, a radius of 9 mm, a container thickness of 0.25 mm, and a mandrel thickness of 2 mm and it has been represented by using a 2D geometry with axial symmetry feature. Figure 1 represents the schematic of the geometry. The three main parts of the cell, represented by the adopted geometry, are as follows:

- The container: it is wrapped around the active volume of the cell; the heat-transfer between the battery and the external ambient happens through this part, made of steel.
- The battery's active material: the different parts where the electrochemical reactions occur are assumed to be lumped in this part.
- Mandrel: it is the isolator around which the battery cell sheets are wound.

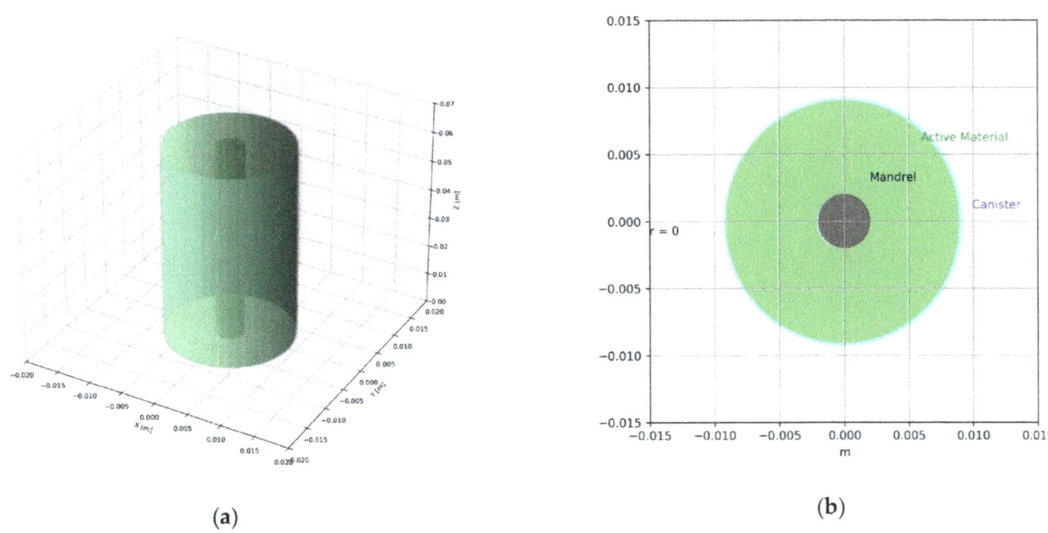

Figure 1. Schematic of Li-ion Battery: (**a**) 3D view; (**b**) Top cross section radial view.

2.2. Mass Balance and Electrochemical Reactions

In this work, the electrochemical behavior of the Li-ion battery is modeled using the single-particle (SP) approach, a simplified lumped model derived from the more detailed pseudo-2D (P2D) model. In the selected model, each electrode is represented by spherical intercalation particles of uniform size, and the current distribution is assumed to be uniform across the electrode thickness. This simplification captures the essential physics of lithium intercalation while significantly reducing computational complexity, making it well-suited for coupled thermal simulations. The mass balance is represented in Equations (1) and (2).

$$\frac{\partial c_{a,j}}{\partial t} + \frac{15D_{1,j}}{R_j^2}\left(c_{s,j} - \bar{c}_{1,j}\right) = 0 \tag{1}$$

$$\frac{15D_{1,j}}{R_j}\left(\bar{c}_{1,j} - c_{s,j}\right) + \frac{J_j}{A_j F} = 0 \tag{2}$$

where $c_{s,j}$ is average concentration and $\bar{c}_{1,j}$ is the concentration at the surface of the sphere; the initial condition is given by Equation (3):

$$c_{s,j}\big|_{t=0} = c_{1,j,0} \tag{3}$$

where $c_{1,j,0}$ denotes the concentration of lithium inside electrode 'j' at the beginning of charge or discharge. The surface concentration is then used in place of $c_{1,j}\big|_{r=R_j}$ in the flux expressions. The battery cell voltage E_{cell} is defined as in Equation (4) [34]

$$\mathrm{E}_{cell} = \mathrm{E}_{OCV}\left(SOC_{average}\right) + \eta_{IR} + \eta_{act} + \eta_{conc} \tag{4}$$

Using spherical symmetry for spherical particles, the diffusion process is defined by Equation (5)

$$\tau\frac{\partial SOC}{\partial t} = -\nabla.(-\nabla SOC) \tag{5}$$

The diffusion time constant, τ (in seconds), characterizes the timescale for diffusion within the particle. The spatial domain represents an average particle in the electrode, with $X = 0$, $X = 1$ corresponding to the particle center and surface, respectively. The boundary conditions at these locations are specified as follows:

$$\nabla \mathrm{SOC} = 0|_{X=0} \tag{6}$$

$$\nabla \mathrm{SOC} = \frac{\tau I_{cell}}{N_{shape} Q_{cell,0}}\bigg|_{X=1} \tag{7}$$

For spherical particles, the parameter N_{shape} is set to 3. The initial state of charge (SOC) of the cell is defined by SOC_0. The surface SOC ($\mathrm{SOC}_{surface}$) is determined at the particle surface (X = 1), while the average SOC ($\mathrm{SOC}_{average}$) is obtained by integrating over the particle volume, with appropriate consideration for spherical coordinates, as described in Equation (8).

$$\mathrm{SOC}_{average} = \frac{\int_0^1 SOC4\pi X^2 dX}{\int_0^1 4\pi X^2 dX} = 3\int_0^1 SOC{\cdot}X^2 dX \tag{8}$$

The lumped voltage loss due to ohmic processes within the electrolyte and electrodes used in Equation (4) is expressed by Equation (9)

$$\eta_{IR} = \eta_{IR,1c}\frac{I_{cell}}{I_{1C}} = R_{ohm}I_{cell} \tag{9}$$

where $\eta_{IR,1C}$ (V) is the ohmic overpotential at 1C, I_{cell} is the applied current, and the 1C current, I_{1C} (A), is defined as Equation (10) where $Q_{cell,0}$ (C) is the battery cell capacity. This equation highlights the role of ohmic resistance Rohm in determining the internal resistance voltage drop η_{IR}. Rohm (ohmic resistance) is a specific type of resistance that obeys Ohm's Law and excludes any reactive components. It is generally constant across frequencies and provides a purely resistive behavior. This factor, the internal resistance, and capacity are the main indicators of the aged cell.

$$I_{1C} = \frac{Q_{cell,0}}{3600} \tag{10}$$

The dimensionless charge exchange current, J_0, is utilized to characterize the lumped voltage loss linked to charge transfer reactions occurring on the surfaces of both the positive and negative electrodes. This relationship is expressed in Equation (11), where R represents the molar gas constant, T is the temperature, and F is Faraday's constant.

$$\eta_{act} = \frac{2RT}{F}\sinh^{-1}\left(\frac{I_{cell}}{2J_0 I_{1c}}\right) \tag{11}$$

Finally, the lumped voltage loss associated with concentration overpotential is defined as in Equation (12)

$$\eta_{conc} = E_{OCV}\left(SOC_{surface}\right) - E_{OCV}\left(SOC_{average}\right) \tag{12}$$

2.3. Energy Balance

The energy balance governing the thermal behavior of the single cell is represented in Equation (13), where ρ is the density of the system, c_p is the specific heat capacity, k is the thermal conductivity

$$\frac{\partial(\rho c_p T)}{\partial t} = -\nabla(k\nabla T) + Q_{TR} + I(\eta_{IR} + \eta_{act} + \eta_{conc}) \tag{13}$$

The term $I(\eta_{IR} + \eta_{act} + \eta_{conc})$ in Equation (13) refers to activation, ohmic, and concentration (entropic) heat generation rate in batteries, which also can be calculated as defined by the literature [35,36]; the latter one is neglected by the uniform concentration of electrolyte in the lumped assumption. At an elevated temperature, side reactions will happen, so that Q_{TR} represents the overall heat generated by these side reactions which can be calculated by Equation (14), where S_{SEI}, S_{ne}, S_{pe}, S_{ed} and S_{bd} are the heat sources for SEI decomposition, the negative electrode-electrolyte reaction (NE), the positive electrode-electrolyte reaction (PE), the electrolyte decomposition reaction (ED), and binder reaction (BD), respectively [9]:

$$Q_{TR} = S_{SEI} + S_{ne} + S_{pe} + S_{ed} + S_{bd} \tag{14}$$

At high current densities, concentration gradients in the electrolyte become significant. To reduce the calculations, the real sandwich-type structure of the cell has been replaced, assuming a pseudo-homogeneous active material; because of this assumption, the thermal properties of the model have to be calculated based on their specific values, according to the following equations [37]:

$$k_{T,ang} = \frac{\sum L_i k_{T,i}}{\sum L_i} \quad (15)$$

$$k_{T,r} = \frac{\sum L_i}{\sum L_i / k_{T,i}} \quad (16)$$

where L_i is the thickness of each layer of the cell, and $k_{T,i}$ the thermal conductivity of the materials constituting these layers. In particular, $k_{T,ang}$ and $k_{T,r}$ are the thermal conductivities in the cylinder length direction and radial direction, respectively. For the other physical properties, similar equations can be used.

As mentioned before, at increasing temperatures, exothermic side reactions will happen (SEI decomposition, anode–electrolyte reaction, and cathode conversion) so that additional heat will be released. The kinetics of these reactions will be represented by the following Arrhenius-type equation [38]:

$$R_i = A_i exp\left(-\frac{E_{a,i}}{RT}\right) c_i^{ni,1} (1 - c_i)^{ni,2} \quad (17)$$

where c_i is the dimensionless concentration calculated as

$$\frac{dc_i}{dt} = \begin{cases} R_i & for\ PE \\ -R_i & for\ others \end{cases} \quad (18)$$

The heat generated by each reaction will be subsequently calculated as follows:

$$S_i = H_i W_i R_i \quad (19)$$

where S_i is the heat release rate for the ith side reaction (W/m^3), H_i is the reaction heat for the ith side reaction (J/kg), W_i is the specific active substance content of the ith side reaction (kg/m^3).

For solving the energy balance, the following boundary conditions are needed:

- At t = 0 (beginning of the simulation), the battery (all parts) and the surrounding area are at a steady state at the same constant ambient temperature (298.15 K).
- At r = 0, a symmetrical boundary condition is used, i.e., the derivation of the temperature with respect to the radius equals zero.
- On the external surface, i.e., at r = R, at z = 0, and z = L, the heat conducted through the inner volumes of the cell is transferred to the external ambient air by natural convection, and the corresponding heat transfer coefficient has been assumed equal to 10 $W/m^2{*}K$ (this factor is mainly used for fitting purposes).

The simulations will be carried out in Comsol Multiphysics 6.3 and conducted in two modes: charging and charge/discharge cycle. The charging and cycling will be performed using a constant current. As mentioned, two battery chemistries have been used for this study, and the related model parameters are extracted from the literature [22,39,40] and shown in Tables 1 and 2.

Table 1. Physical properties of different components of a Li-ion cell [39] used in Energy balance.

ent	Thermal Conductivity (W/m K)	Heat Capacity (J/kg K)	Density (kg/m^3)	Thickness of Component Layer (m)
urrent or	398	385	8933	7×10^{-6}
e	1.04	1437	2660	5.5×10^{-5}
tor	0.33	1978	492	3×10^{-5}
NMC)	1.5	710	2380	5.5×10^{-5}
NCA)	6.10	651	4581	5.5×10^{-5}
urrent tor	238	903	1500	1×10^{-5}

Table 2. Physical and kinetic parameters adopted in the side reactions [22].

ith Reaction	A_i (s^{-1})	$E_{a,i}$ (J/mol)	H_i (J/kg)	$c_{i,0}$	$n_{i,1}$	$n_{i,2}$	W_i (kg/m^3)
SEI	1.667×10^{15}	1.35×10^5	2.57×10^5	0.15	1	0	875
NE	2.5×10^{13}	1.35×10^5	1.714×10^6	0.75	1	0	875
PE (NMC)	2.25×10^{14}	1.54×10^5	7.9×10^5	0.04	1	1	1293
PE (NCA)	7.25×10^{16}	1.3×10^5	2.18×10^5	0.04	1	1	1274

2.4. Aging

Internal resistance is a well-established and effective indicator for battery aging because it encapsulates the cumulative impact of various degradation mechanisms. As a battery ages, processes such as SEI layer growth, active material loss, electrolyte decomposition, and structural changes within the electrodes contribute to an increase in internal resistance. This rise in resistance not only reflects these individual degradation phenomena but also directly affects battery performance by reducing power output and increasing voltage drops. Therefore, by monitoring changes in internal resistance, one can obtain an integrated measure of battery health that is sensitive to a range of aging mechanisms. This makes internal resistance a practical and reliable parameter for assessing battery aging and predicting the remaining useful life of the cell.

2.5. Statistical Analysis

To identify the most influential factors affecting the thermal behavior of the cell, the p-value will be analyzed using an analysis of variance (ANOVA). The p-value is commonly used to test the null hypothesis (H_0), which assumes no effect or no difference. It is determined based on the F- or t-score and is compared against a known probability distribution to assess the likelihood of obtaining the observed results if H_0 were true. It is important to note that the p-value does not represent the probability of a specific factor but rather the likelihood of rejecting or failing to reject a hypothesis. In this study, a low p-value (≤ 0.05) indicates a statistically significant result, leading to the rejection of H_0, suggesting that the factor has a meaningful impact on thermal behavior.

The ANOVA in this study is performed using a design of experiments (DOE). A full factorial design is used with three four-level factors to investigate the effects of capacity, ohmic resistance, and charging rate on thermal behavior, which consist of 64 simulations reported in Table 3. To analyze thermal behaviour, two key outputs have been chosen: the average cell temperature and the temperature derivative over time, both in a battery charged at 10, 50, and 100% state of charge (SOC). The cell's initial SOC is set near zero (0.05), and the initial (ambient) temperature is considered 298.15 K. The ANOVA will be carried out in Python v3.9 and the open library pyDOE2 v1.3.0, and *p*-values with confidence levels of 95% will be used to find the best features presenting the related results.

Table 3. Selected parameters used in the simulations of the Full Factorial design.

Run Order	Internal Resistance (mOhm)	Capacity (Ah)	Current C-Rate (A)	Run Order	Internal Resistance (mOhm)	Capacity (Ah)	Current C-Rate (A)	Run Order	Internal Resistance (mOhm)	Capacity (Ah)	Current C-Rate (A)
1	10	1	1	**26**	20	3	3	**51**	30	1	8
2	20	1	1	**27**	30	3	3	**52**	40	1	8
3	30	1	1	**28**	40	3	3	**53**	10	2	8
4	40	1	1	**29**	10	5	3	**54**	20	2	8
5	10	2	1	**30**	20	5	3	**55**	30	2	8
6	20	2	1	**31**	30	5	3	**56**	40	2	8
7	30	2	1	**32**	40	5	3	**57**	10	3	8
8	40	2	1	**33**	10	1	6	**58**	20	3	8
9	10	3	1	**34**	20	1	6	**59**	30	3	8
10	20	3	1	**35**	30	1	6	**60**	40	3	8

Table 3. *Cont.*

Run Order	Internal Resistance (mOhm)	Capacity (Ah)	Current C-Rate (A)	Run Order	Internal Resistance (mOhm)	Capacity (Ah)	Current C-Rate (A)	Run Order	Internal Resistance (mOhm)	Capacity (Ah)	Current C-Rate (A)
11	30	3	1	**36**	40	1	6	**61**	10	5	8
12	40	3	1	**37**	10	2	6	**62**	20	5	8
13	10	5	1	**38**	20	2	6	**63**	30	5	8
14	20	5	1	**39**	30	2	6	**64**	40	5	8
15	30	5	1	**40**	40	2	6				
16	40	5	1	**41**	10	3	6				
17	10	1	3	**42**	20	3	6				
18	20	1	3	**43**	30	3	6				
19	30	1	3	**44**	40	3	6				
20	40	1	3	**45**	10	5	6				
21	10	2	3	**46**	20	5	6				
22	20	2	3	**47**	30	5	6				
23	30	2	3	**48**	40	5	6				
24	40	2	3	**49**	10	1	8				
25	10	3	3	**50**	20	1	8				

2.6. Machine Learning Analysis

To find a correlation between exercised features in analytical analysis, a linear regression model will be fitted for different sets of results. The basic model is presented in Equation (20).

$$a_0 + a_1x_1 + a_2x_2 + \cdots + a_nx_n = \hat{y} \quad (20)$$

where the a_0, a_1 to an are coefficients for x_1 to x_n features and $\hat{y}$ is the predicted output. If the relationship between features and output is not linear, y_{actual} and $\hat{y}$ have different values, so the difference between these two values needs to be minimized. The objective function is presented in Equation (21). To find the best fit, the coefficients are updated in an iterative approach to satisfy the objective function.

$$minimize\sum_{i=1}^{n}(\hat{y} - y_{actual})^2 \quad (21)$$

As a battery ages, its internal resistance increases, and if it exceeds a critical threshold, it can trigger rapid temperature escalation and increase the risk of thermal runaway. Identifying this threshold is crucial for ensuring battery safety. By defining this limit, proactive measures such as thermal management and early fault detection can be implemented to prevent hazardous conditions, extend battery lifespan, and maintain safe operation in applications like electric vehicles and energy storage systems. To determine this threshold, a classification task will be performed using internal resistance, current, and capacity as input features, with thermal runaway occurrence as the target variable. A logistic regression model is selected for its versatility and strong predictive capabilities. The logistic regression function is presented in Equation (21), requiring an iterative process to satisfy Equation (20) for the predicted $\hat{y}$ in Equation (21), which is the probability. The dataset will be divided into training and testing sets to ensure proper model evaluation and prevent overfitting. Once trained, the model can be used to determine the internal resistance threshold by fixing the output and other feature values, leading to the derivation of Equation (23) from Equation (22) for this specific task.

$$\hat{y} = \frac{e^{b_0+b_1x_1+\ldots+b_nx_n}}{1+e^{b_0+b_1x_1+\ldots+b_nx_n}} \tag{22}$$

$$x_1 = -\frac{1}{b_1}\left(\ln\left(\frac{1}{\hat{y}}-1\right)+(b_0+b_2x_2+b_3x_3)\right) \tag{23}$$

These tasks will be carried out in Python with the open library scikit-learn. To check the performance of the fitted values, R-squared for linear regression and confusion matrix for logistic regression will be used.

2.7. Validation

2.7.1. Charge Testing: Voltage Comparison

The proposed model has been validated using experimental data from the literature [41], referred to as a commercial 26,650 NCM/graphite cell with a capacity of 5 Ah, with cell dimensions of 26 mm in diameter and 65 mm in height. The study evaluated the cell's performance at charge rates of 0.2C, 0.5C, and 1C, with a cut-off voltage set at 4.2 V. Figure 2 illustrates the validation of the electrochemical model, confirming the cell's performance during the charging process without triggering side reactions. The results demonstrate that the simulated cell voltage closely aligns with the experimental data throughout the charging period, highlighting the model's accuracy. Additionally, as shown in Figure 2, the voltage derivation increases over time with higher charging currents. This trend is accurately captured by the model. The small discrepancies between the experimental and simulated results are attributed to the simplifying assumptions of a single-particle model, which does not fully account for the complexities of a real cell.

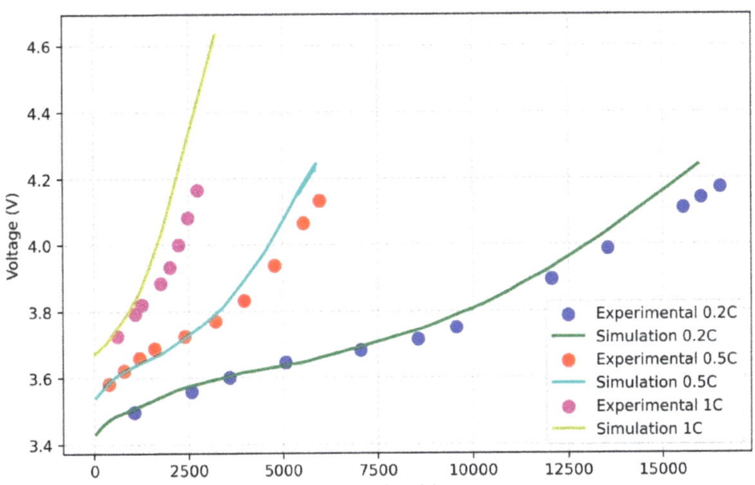

Figure 2. Charging test for 0.2C, 0.5C, and 1C simulation vs. experiment [41].

2.7.2. Charge Testing: Temperature Comparison

The proposed model was then validated using experimental data to check the temperature behavior at constant current charging rates of 1C and 8C. For this purpose, the study by Dong et al. [42] was chosen, as it provided temperature data at various charging and discharging rates. In the experiments, the temperature monitoring point was positioned at the center of the outer surface, and in the simulation, the same point on the surface was adopted. The cell under investigation is a cylindrical spiral 18650 battery with an NCM cathode, a radius of 9 mm, a height of 65 mm, and a nominal capacity of 1.5 Ah. Figure 3 compares the simulated and experimental cell temperatures during the 1C and 8C charging processes, starting with an initial SOC of approximately 0.05. The results demonstrate that at a constant SOC, the cell's temperature increases more significantly at higher charging rates: as expected, the 8C charge rate leads to a greater temperature rise and higher temperature derivation over time compared to the 1C charge rate. The data show a strong correlation between the simulation and experimental results, indicating good agreement. By comparing the proposed numerical model with experimental results from the literature, it can be concluded that a single-particle electrochemical model coupled with a heat balance is an effective approach to understanding the thermal behavior of lithium-ion batteries.

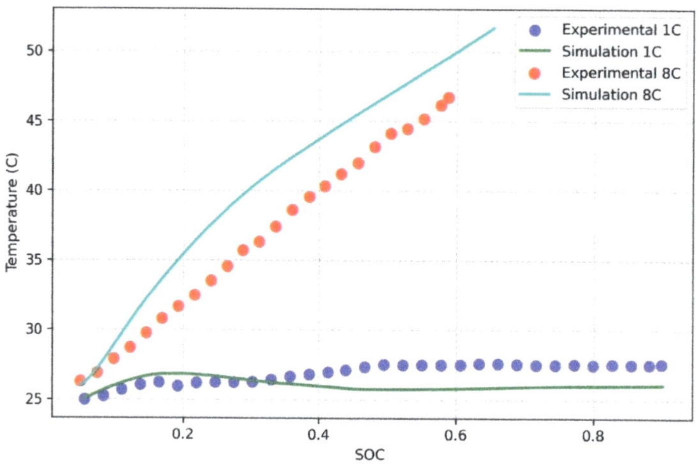

Figure 3. Temperature vs. SOC in Charging test for 1C and 8C simulation vs. experiment [42].

2.7.3. Overcharging and Thermal Runaway

The third validation test for the proposed model focused on the thermal behavior of a cell during overcharging. The model has been validated against experimental data to analyze the temperature behavior during constant current charging at rates of 1C, 2C, and 3C, as well as under overcharging conditions beyond full cell capacity. The study by Liu et al. [43] was selected due to its comprehensive temperature data for different overcharging rates. Their experiments utilized 18650 NCM523 lithium-ion batteries with a capacity of 2.6 Ah, a radius of 9 mm, and a height of 65 mm. Testing was conducted at an ambient temperature of about 298.15 K under natural ventilation. Throughout the experiments, and as reported in the literature, the surface temperature of the battery was monitored, and the simulation recorded temperatures at the center of the outer surface of the cell. Figure 4 shows that overcharging beyond a certain point can lead to thermal runaway. This phenomenon is attributed to electrochemical and side reactions, as discussed previously, causing a catastrophic increase in the heat released by the cell. The figure clearly illustrates that higher charging rates significantly reduce the time to thermal runaway. One limitation of the model is its omission of gas release and ignition processes, which contributes to discrepancies between the simulated and experimental results. These mechanisms are more prominent at lower C-rates, where the system has more time to dissipate heat, resulting in less severe temperature buildup, particularly evident under 1C overcharge conditions. Interestingly, the model demonstrates better agreement with experimental data at higher current rates. Given this observation and the relatively lower error margins at moderate currents, the study places greater emphasis on moderate C-rate scenarios, which will be further discussed in the following sections. Overall, Figure 4 supports the model's effectiveness in analyzing thermal runaway behavior in lithium-ion batteries, showing good alignment with the experimental trends.

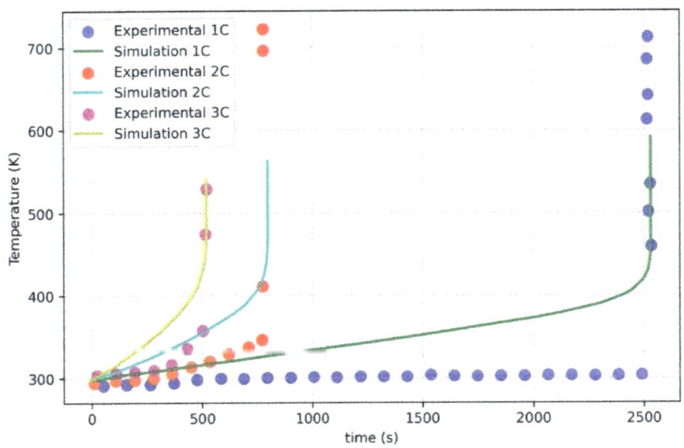

Figure 4. Overcharging test for 1C, 2C, and 3C simulation vs. experiment [43].

2.7.4. Aged Cell

The fourth set of validation runs was performed to assess the voltage and temperature trend predictions for aged cells. The dataset [33] was collected from a custom-built battery prognostics testbed at the NASA Ames Prognostics Center of Excellence (PCoE), and the study involved testing Li-ion batteries under three different operational profiles: charge, discharge, and Electrochemical Impedance Spectroscopy (EIS). The data used for the validation refer to cell No. 7 in the NASA dataset, which is an NCA 18650-sized rechargeable battery with 2 Ah nominal capacity, discharged with -1.98 A from a fully charged state. The ambient temperature was set at 297.5 K, and the battery cycled 168 times. Figure 5 represents the cell's thermal behavior during discharge at the 19th cycle and 168th cycle; both the experimental data and the simulations use the fitted internal resistance value, which shows a relation to the aged battery trend. Simulations with fitted internal resistance resulted in 0.4257689 and 0.4653298 RMSE for the 19th and 168th cycles, respectively. The figure shows a very good agreement between simulations and experimental data, so it can be concluded that the developed model is capable of properly predicting the thermal behavior of an aged cell.

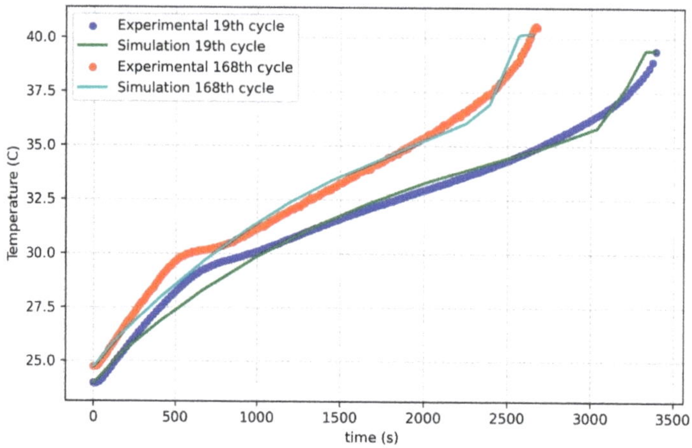

Figure 5. Fresh and aged cell temperature profile [33].

3. Results and Discussion

Figure 6 illustrates the results of the first simulation during charging. Initially, the cell's state of charge (SOC) was near zero, and the cell was charged using a 1C current to reach a fully charged state (Figure 6a). As shown in Figure 6b, the cell temperature increases steadily due to the heat generated by electrochemical reactions. In Figure 6b, at the start of the charging, heat dissipation to ambient is poor due to the fact that surface temperature is close to ambient temperature. After a while, the temperature increment reaches the surface of the cell, heat dissipation becomes larger, and the temperature almost reaches a steady state. Figure 6c shows that the temperature in the middle of the cell is higher compared to other regions. This phenomenon is related to the geometry of the cell, as heat accumulates more at the core due to limited pathways for thermal dissipation in cylindrical battery designs.

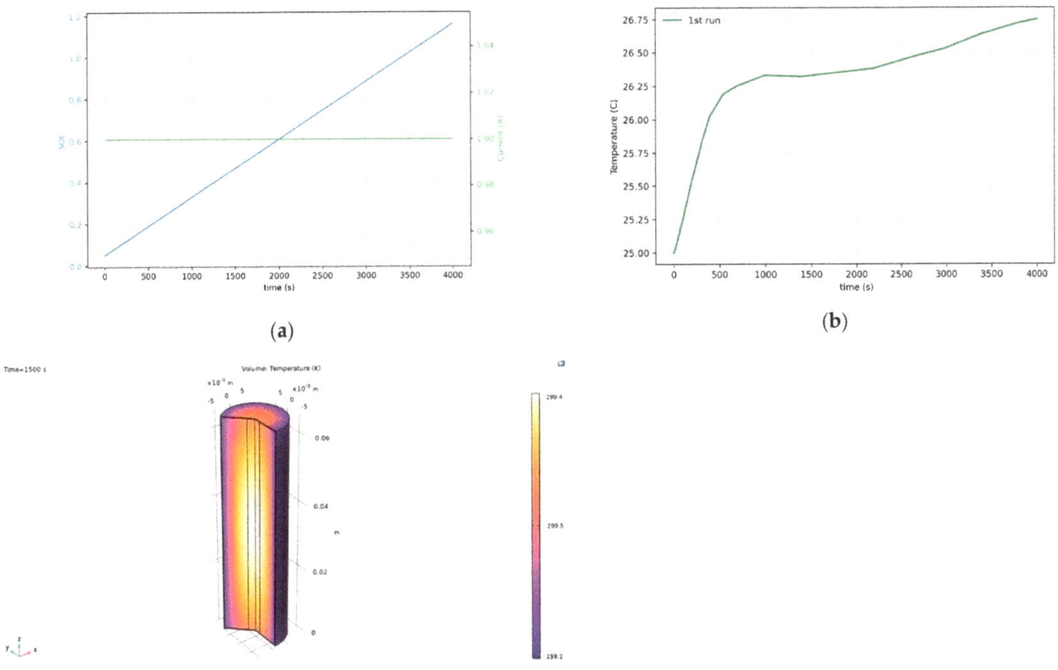

(a) (b) (c)

Figure 6. Simulation of the cell number 1 (1 Ah, 1C and 10 mOhm): (**a**) state of charge and current vs. time; (**b**) mean temperature derivation of cell vs. time; (**c**) temperature distribution in the cell at t = 1500 s.

Figure 7 shows the results of the charging/discharging cycles of simulation no. 33 in which a cell with 1 Ah capacity and 10 mOhm internal resistance was charged/discharged using a constant current of 6C. Figure 7a illustrates the voltage and current of the cell. In the charging and discharging phases, the current is positive and negative, respectively, and voltage increases from around 2.7 V to 4.2 V in charging and vice versa in discharging. Figure 7b presents the thermal behavior of the cell during this process: at the beginning and after a few cycles, temperature increases due to heat generation by reversible entropy losses [44]; after the temperature has reached a certain point, heat dissipation to the environment and heat generation within the cell due to charging/discharging almost balance each other and the temperature trend becomes nearly constant.

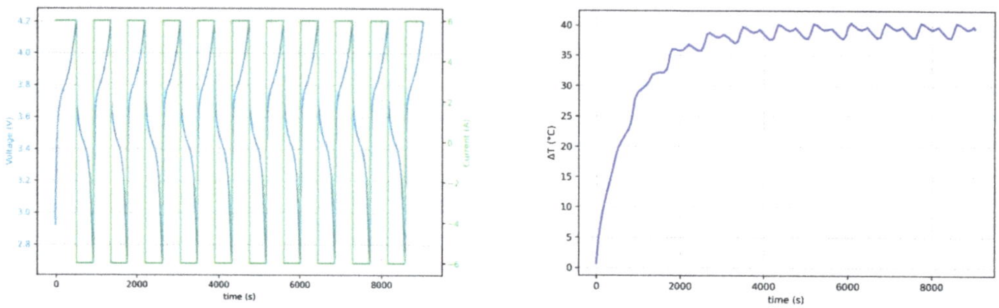

(a) (b)

Figure 7. Simulation n. 33 during cycling (1 Ah, 6C, and 10 mOhm): (**a**) voltage and current vs. time; (**b**) mean temperature increase vs. time.

3.1. Influence of Charging Rate on Thermal Behavior

Figure 8 illustrates the temperature trends for four simulations with 1C, 3C, 6C, and 8C charging rates and fixed resistance and cell capacity. As shown, the charging rate has a significant impact on the temperature behavior of lithium-ion cells, with higher charging rates resulting in increased thermal effects. This trend is expected and aligns with the theoretical understanding of heat generation in electrochemical systems. As illustrated by Equation (13), the rate of heat generation is directly influenced by the charging current, which is multiplied by the internal overpotentials within the cell. These overpotentials, including both ohmic and reaction-related components, increase as the current rises, leading to greater irreversible heat production. Consequently, at higher charging rates, the cell experiences more intense thermal buildup, which can elevate the risk of thermal instability if not properly managed. This emphasizes the importance of thermal modeling and management, particularly in high-rate charging applications where temperature control is critical for both performance and safety.

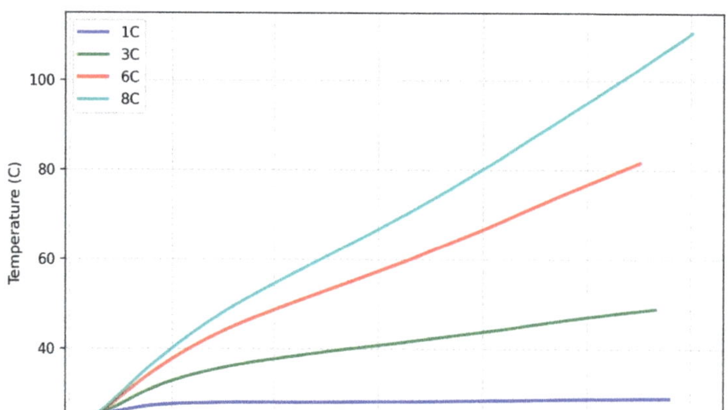

Figure 8. Temperature against SOC for four simulations with 1 Ah capacity, 10 mOhm internal resistance, and with charging rates of 1C, 3C, 6C, and 8C.

3.2. *Effect of Internal Resistance on Thermal Runaway*

To examine the impact of internal resistance on thermal behavior, simulations were conducted on NMC cells with identical capacities (5 Ah) and charging rates (3C) and different internal resistances (10, 20, 30, and 40 mΩ). Figure 9 shows the mean cell temperature (Average Cell Temperature) over time. The results reveal that a higher internal resistance correlates with increased cell temperature over the same time frame, leading to a higher likelihood of thermal runaway. This effect can be attributed to the additional heat generated by ohmic losses in the cell, and these findings agree with the results of Yoo et al. [45]. Since the internal resistance increase is often associated with battery aging, this means that aged cells are progressively more prone to thermal runaway events due to their increased thermal sensitivity, which is opposite of the study of Fleischhammer et al. [23] but in agreement with the study of Han et al. [46].

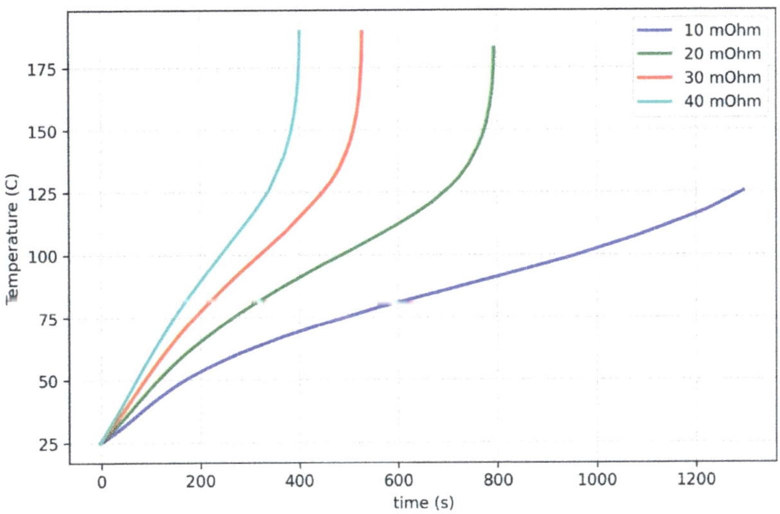

Figure 9. Temperature against time for simulations with 5 Ah capacity at 3C charging rate and with internal resistance of 10, 20, 30, and 40 mOhm.

3.3. *Effect of Capacity on Thermal Runaway*

The influence of cell capacity on thermal behavior was analyzed by comparing simulations with capacities of 1, 2, 3, and 5 Ah under a fixed internal resistance (30 mΩ) and a charging rate 3C. Figure 10 presents the temperature against time. The data indicate that higher capacity cells exhibit a more pronounced temperature rise with respect to lower capacity ones over a given SOC range, reflecting their greater energy content. Now, since a pronounced capacity fading is associated with cell aging, it derives that, from this point of view, the likelihood of a thermal runaway diminishes with age, along with reduced energy storage and associated thermal effects [47], due to the lower temperature reached. However, this does not necessarily diminish the likelihood of thermal runaway because, as discussed in the previous section, aged cells often exhibit increased internal resistance, leading to greater heat generation during operation and, thus, to a higher risk of thermal runaway. Therefore, despite their reduced capacity, aged cells may still be more prone to thermal runaway due to increased heat generation from higher internal resistance.

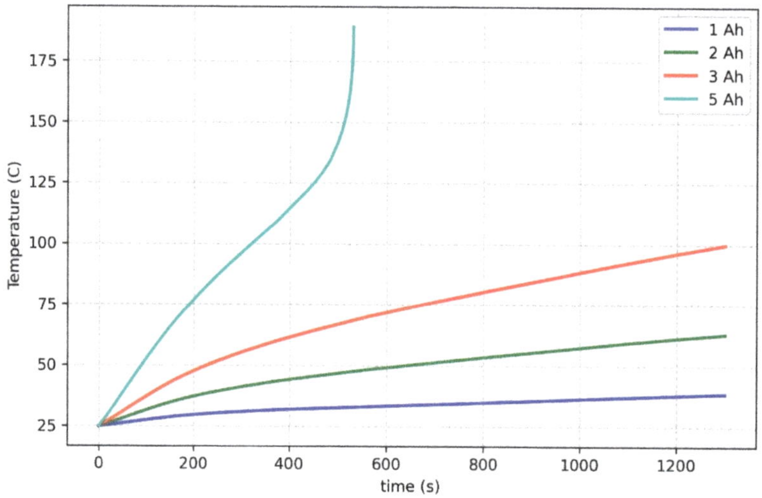

Figure 10. Temperature against time for simulations with 30 mOhm internal resistance and 3C charging rate and capacity of 1, 2, 3, and 5 Ah, in one single charge.

3.4. Combined Aging Effects

Based on the above results, the two primary aging effects, capacity fade and increased internal resistance, have opposing impacts on thermal behavior:

- An increased internal resistance exacerbates heat generation and thermal risks;
- Capacity fading mitigates thermal risk by reducing the total energy stored and released.

To fully understand these interactions and their implications for battery safety, statistical analyses are essential for quantifying their combined influence on the thermal behavior of aged cells.

3.5. ANOVA Analysis

After performing all the simulations, the mean temperature of the cell, its SOC, and the voltage of the cell against time were recorded. The necessary results obtained from the simulation for the ANOVA are shown in Table 4. This table presents data on battery thermal behavior across different test conditions, with a key focus on thermal runaway, temperature, and temperature change rate (dT/dt) at different states of charge (SOC). Run Order is the sequence in which the experiments were conducted, the thermal runaway column shows a binary indicator (0 = No thermal runaway, 1 = Thermal runaway occurred), and temperature and dT/dt at 10%, 50%, and 100% SOC reflect how the temperature and its rate of change evolve as the battery charges in a single charge phase. At 100% SOC, temperatures are often higher than at 50% and 10% SOC, which is expected as a fully charged battery generally has more stored energy to dissipate as heat. Starting from run 30, the first thermal runaway events are observed, where temperatures escalate drastically (e.g., Run 40: 169.5 °C at 100% SOC). In early runs (before thermal runaway events), dT/dt values remain low (mostly below 0.1 in different SOCs). However, once temperatures exceed certain thresholds, dT/dt rises steeply (e.g., Run 40: 2.374 at 100% SOC), indicating rapid heat generation and potential safety risks.

Table 4. Results obtained from full factorial design.

Run	TR	T at 10% SOC	dT/dt at 10% SOC	T at 50% SOC	dT/dt at 50% SOC	T at 100% SOC	dT/dt at 100% SOC	d^2T/dt^2 at 100% SOC
1	0	25.5	0.003	26.3	0.000	26.6	0.000	-3.25×10^{-07}
2	0	25.5	0.003	26.5	0.000	26.9	0.000	-3.51×10^{-07}
3	0	25.6	0.003	26.7	0.000	27.1	0.000	-3.63×10^{-07}
4	0	25.6	0.004	26.9	0.000	27.4	0.000	-3.75×10^{-07}
5	0	26.1	0.006	28.1	0.000	28.8	0.000	-7.04×10^{-07}
6	0	26.3	0.007	28.9	0.000	29.7	0.000	-7.55×10^{-07}
7	0	26.5	0.008	29.7	0.001	30.7	0.001	-3.51×10^{-07}
8	0	26.6	0.009	30.5	0.001	31.6	0.001	-4.09×10^{-07}
9	0	26.7	0.010	30.2	0.000	31.4	0.001	-3.62×10^{-07}
10	0	27.2	0.012	32.1	0.001	33.6	0.001	-5.16×10^{-07}
11	0	27.6	0.014	33.9	0.001	35.7	0.001	-1.32×10^{-06}
12	0	28.0	0.016	35.7	0.002	37.9	0.001	-1.44×10^{-06}
13	0	28.4	0.019	36.4	0.001	38.3	0.001	-1.84×10^{-06}
14	0	29.6	0.025	41.7	0.002	44.4	0.001	-2.20×10^{-06}
15	0	30.8	0.031	47.0	0.003	50.5	0.001	-2.41×10^{-06}
16	0	32.0	0.037	52.3	0.005	56.5	0.002	-1.87×10^{-06}
17	0	26.1	0.022	31.0	0.005	34.3	0.006	-7.01×10^{-06}
18	0	26.3	0.025	32.0	0.006	35.9	0.007	-8.58×10^{-06}
19	0	26.4	0.027	33.1	0.007	37.5	0.007	-1.01×10^{-05}
20	0	26.6	0.029	34.1	0.009	39.1	0.008	-1.17×10^{-05}
21	0	27.5	0.049	39.4	0.014	47.1	0.013	-2.46×10^{-05}
22	0	28.2	0.059	43.6	0.019	53.6	0.015	-3.03×10^{-05}
23	0	28.8	0.069	47.8	0.025	60.2	0.018	-3.61×10^{-05}
24	0	29.4	0.079	52.0	0.031	66.8	0.021	-4.16×10^{-05}
25	0	29.3	0.081	49.9	0.025	63.5	0.022	-4.10×10^{-05}
26	0	30.7	0.103	59.4	0.039	78.5	0.028	-5.15×10^{-05}
27	0	32.0	0.126	69.0	0.052	94.0	0.038	-3.91×10^{-05}
28	0	33.4	0.148	78.6	0.066	113.5	0.068	0.0002
29	0	33.7	0.160	77.8	0.059	111.7	0.073	0.0001
30	1	37.5	0.222	105.6	0.104	- *	- *	- *
31	1	41.4	0.284	- *	- *	- *	- *	- *
32	1	45.2	0.346	- *	- *	- *	- *	- *
33	0	26.7	0.068	36.4	0.027	46.1	0.035	-1.84×10^{-05}
34	0	27.1	0.078	38.8	0.035	50.4	0.041	-3.02×10^{-05}
35	0	27.4	0.089	41.3	0.043	54.8	0.046	-4.18×10^{-05}
36	0	27.7	0.099	43.7	0.051	59.2	0.052	-5.34×10^{-05}
37	0	29.1	0.157	52.9	0.071	76.6	0.081	−0.0001
38	0	30.4	0.198	62.8	0.103	94.8	0.109	3.20×10^{-05}
39	1	31.6	0.239	72.7	0.135	115.5	0.166	0.0006
40	1	32.9	0.279	82.6	0.168	169.5	2.374	0.4273
41	1	32.1	0.267	74.8	0.134	123.3	0.253	0.0032
42	1	34.9	0.359	97.4	0.211	- *	- *	- *
43	1	37.8	0.451	122.3	0.347	- *	- *	- *
44	1	40.6	0.543	- *	- *	- *	- *	- *
45	1	40.0	0.548	152.3	1.220	- *	- *	- *
46	1	47.9	0.805	- *	- *	- *	- *	- *
47	1	55.8	1.062	- *	- *	- *	- *	- *

Table 4. *Cont.*

Run	TR	T at 10% SOC	dT/dt at 10% SOC	T at 50% SOC	dT/dt at 50% SOC	T at 100% SOC	dT/dt at 100% SOC	d^2T/dt^2 at 100% SOC
48	1	63.7	1.318	- *	- *	- *	- *	- *
49	0	27.0	0.103	39.2	0.051	52.8	0.067	-5.32×10^{-05}
50	0	27.5	0.121	42.6	0.066	59.1	0.079	-8.28×10^{-05}
51	0	27.9	0.140	46.0	0.081	65.5	0.091	−0.0001
52	0	28.3	0.158	49.4	0.096	71.9	0.103	−0.0001
53	0	29.9	0.243	60.5	0.134	94.9	0.168	0.0001
54	1	31.6	0.316	74.3	0.196	124.9	0.294	0.003
55	1	33.3	0.390	88.1	0.258	- *	- *	- *
56	1	35.0	0.463	102.0	0.328	- *	- *	- *
57	1	33.7	0.420	89.3	0.255	- *	- *	- *
58	1	37.5	0.586	122.4	0.503	- *	- *	- *
59	1	41.3	0.751	- *	- *	- *	- *	- *
60	1	45.1	0.916	- *	- *	- *	- *	- *
61	1	43.7	0.888	- *	- *	- *	- *	- *
62	1	54.3	1.349	- *	- *	- *	- *	- *
63	1	64.9	1.811	- *	- *	- *	- *	- *
64	1	75.5	2.272	- *	- *	- *	- *	- *

* Indicates that the simulation could not proceed beyond this point due to the onset of thermal runaway.

In cells that did not experience thermal runaway (TR = 0), the temperature at 10% SOC generally remained below approximately 34 °C. In contrast, cells that underwent thermal runaway (TR = 1) typically exhibited temperatures above 35 °C at 10% SOC, with some values reaching over 40 °C. This suggests that a temperature exceeding around 35 °C at 10% SOC could serve as an early warning signal. For non-runaway cases, the rate of temperature rise at 10% SOC was usually below 0.16 °C/s. However, in cells that experienced thermal runaway, this rate exceeded approximately 0.22 °C/s. Thus, a dT/dt greater than about 0.2 °C/s at low SOC may indicate an increased risk. While non-thermal runaway cells typically maintained temperatures below around 80 °C at 50% SOC, cells prone to thermal runaway showed significantly higher temperatures, often exceeding 80 °C and, in some cases, reaching 100 °C or more. This marked increase reinforces the idea that a 50% SOC temperature above 80 °C might be a critical threshold. In cells that do not experience thermal runaway (TR = 0), the temperature at 100% SOC remains very low (typically below 40 °C). By contrast, in cases where thermal runaway occurs (TR = 1), the temperature at 100% SOC is substantially higher, often exceeding 100 °C and, in some cases, reaching values well above 150 °C. This dramatic increase indicates that once the runaway process is triggered, the battery temperature can escalate rapidly. For non-runaway cells, the first derivative of temperature (dT/dt) at 100% SOC is close to zero, reflecting a stable thermal state. However, in runaway cases, this rate becomes markedly elevated (e.g., over 2 °C/s in some instances), which suggests that the temperature is rising very rapidly (another clear sign of thermal instability). Similarly, the second derivative (d^2T/dt^2) in non-runaway scenarios remains extremely small (on the order of 10^{-5} °C/s^2), indicating that the rate of temperature change is not accelerating significantly. In contrast, when thermal runaway occurs, the second derivative increases substantially (for example, values around 0.4273 °C/s^2), demonstrating a rapidly accelerating temperature rise. In some instances, a battery may reach 100% SOC without immediately triggering a thermal runaway. However, if charging continues beyond this point, the risk of thermal runaway becomes highly probable, particularly when the temperature and its acceleration (the second derivative, d^2T/dt^2) are significantly elevated. In contrast, a non-thermal runaway case such as Run 28 shows much lower values at 100% SOC (113.5 °C with dT/dt of

0.068 °C/s and d^2T/dt^2 of 0.0002 °C/s^2), suggesting a stable thermal condition. Thus, even if a cell does not immediately experience runaway at 100% SOC, continued charging under conditions where both temperature and its acceleration are high can quickly push the battery into an unstable, runaway state.

It is important to note that these values are based on the specific dataset and experimental conditions represented in Table 4, and further analysis across a broader range of conditions would be needed to generalize these thresholds.

Tables 5 and 6 present the analytical analysis and the regression model. The R^2 values provide insight into how well the model captures the variance in temperature and dT/dt at different states of charge (SOC). The highest R^2 (0.9621) for temperature at 10% SOC suggests that the model explains 96.2% of the variation in temperature at this charge level, demonstrating strong predictive accuracy. In contrast, the lowest R^2 (0.8205) for dT/dt at 50% SOC indicates that while the model still performs reasonably well, there is a greater degree of unexplained variability in the rate of temperature change at mid-range SOCs. Overall, the model effectively predicts temperature trends, though its accuracy in capturing dT/dt is comparatively lower.

Table 5. R^2 values of the regression models.

SOC	Temp (R^2)	dT/dt (R^2)
10%	96.21%	92.35%
50%	92.92%	82.05%
100%	93.51%	87.68%

Table 6. ANOVA and linear regression coefficients.

	SOC (%)	C_T [1]	P_T [2]	C_d [3]	P_d [4]
Bias	10	24.4391	-	0.0761	-
	50	13.6308	-	−0.0655	-
	100	9.9273	-	−0.0408	-
Capacity	10	−1.8756	3.1474×10^{-17}	−0.0735	3.898×10^{-13}
	50	−0.6054	5.1716×10^{-26}	−0.0233	2.9529×10^{-24}
	100	−6.9376	1.49×10^{-22}	0.0389	0.04482
Current	10	- *	- *	−0.0496	1.1232×10^{-03}
	50	−1.9897	6.0723×10^{-04}	−0.0390	1.4285×10^{-07}
	100	−4.9313	2.098×10^{-10}	- *	- *
Resistance	10	−0.1520	9.0499×10^{-03}	- *	- *
	50	0.3777	1.0109×10^{-04}	0.0041	4.6553×10^{-08}
	100	0.8107	1.5789×10^{-04}	- *	- *
Capacity × Current	10	0.7209	1.7444×10^{-05}	0.0527	1.0182×10^{-02}
	50	4.4672	6.9643×10^{-12}	0.0373	5.3381×10^{-17}
	100	9.5669	9.0235×10^{-08}	- *	- *
Capacity × Resistance	10	0.1334	9.0783×10^{-03}	- *	- *
	50	0.0559	1.1416×10^{-05}	−0.0008	1.2349×10^{-15}
	100	0.05747	3.2543×10^{-04}	- *	- *
Current × Resistance	10	- *	- *	- *	- *
	50	- *	- *	- *	- *
	100	- *	- *	- *	- *

[1] Coefficient related to Temperature response; [2] *p*-value related to Temperature response; [3] Coefficient related to Temperature derivative over time response; [4] *p*-value related to Temperature derivative over time response; * The *p*-value exceeds 0.05, and the parameter was not included in the final analysis.

The *p*-values for capacity are extremely low ($p < 0.05$, often much lower across different conditions), confirming that capacity has a strong and statistically significant impact on both temperature and dT/dt. Current also plays a significant role, with *p*-values < 0.05 in most cases, demonstrating its strong influence on thermal behavior. Internal resistance significantly influences temperature ($p < 0.01$ in most cases), and its coefficients are consistently positive, meaning that higher resistance leads to higher temperatures—a trend that aligns well with Joule heating principles. However, its effect on dT/dt is less consistent, suggesting that while internal resistance directly contributes to steady-state temperature rise, its impact on transient thermal behavior may be more complex and dependent on additional factors.

Notably, the coefficients for capacity and current are negative. However, the interaction between them is highly significant ($p < 0.05$ in all cases) and has a large coefficient, meaning their combined influence is critical in determining both temperature and dT/dt. This suggests that the relationship between capacity and current is not simply additive—rather, their combined influence plays a key role in determining overall thermal behavior. This highlights the importance of jointly managing capacity and current when optimizing thermal performance and ensuring battery safety rather than treating them as independent factors.

The interaction between capacity and internal resistance is also significant, though its effect size is smaller, indicating that while capacity influences resistance-driven heating, its effect is secondary compared to direct resistance contributions. Interestingly, the interaction between current and internal resistance shows no significant impact, suggesting that resistance-related heating is largely independent of current within the tested operating conditions. This finding implies that internal resistance effects may be more static and intrinsic to the cell's electrochemical characteristics rather than dynamically varying with current load.

Overall, these findings highlight the complex interplay between capacity, current, and internal resistance in shaping battery thermal behavior. The insights gained from this analysis can aid in the development of more effective thermal management strategies, improve battery performance predictions, and enhance safety measures in high-energy applications.

3.6. Internal Resistance Threshold

Table 7 presents the computed values corresponding to Equation (21), which are essential for determining the internal resistance threshold. Additionally, Figure 11 illustrates the confusion matrix for the classification task performed using the logistic regression model on the test set. The model demonstrates strong predictive performance, achieving an accuracy of 95%, indicating its reliability in distinguishing between thermal runaway and non-thermal runaway cases.

Table 7. Parameters for logistic regression model.

Parameters	Values
b_0	−0.3963
b_1	−0.0491
b_2	0.1300
b_3	0.2386

The confusion matrix represents the classification performance of the model, where rows indicate actual values and columns indicate predicted values. The top-left value (8) represents true positive (TP), meaning the model correctly predicted 8 cases where thermal runaway occurred. The top-right value (0) represents false positives (FP), indicating no incorrect predictions of thermal runaway when it did not actually happen. The bottom-left value (1) represents false negatives (FN), meaning one case of not thermal runaway classified as thermal runaway. Lastly, the bottom-right value (11) represents true negatives (TN), where the model correctly identified 11 instances of not thermal runaway. From the figure, it can be concluded that the model performs well in terms of avoiding risks, i.e., it never ensures a lack of a thermal runaway when it actually occurs. By leveraging the proposed logistic regression model, the internal resistance threshold can be determined by fixing the charging current, capacity of the cell, and probability of thermal runaway occurrence. This approach allows for precise identification of the critical resistance value beyond which the likelihood of thermal runaway significantly increases. The results obtained from this analysis can be instrumental in enhancing battery safety, enabling proactive measures such as real-time monitoring, thermal management, and early fault detection.

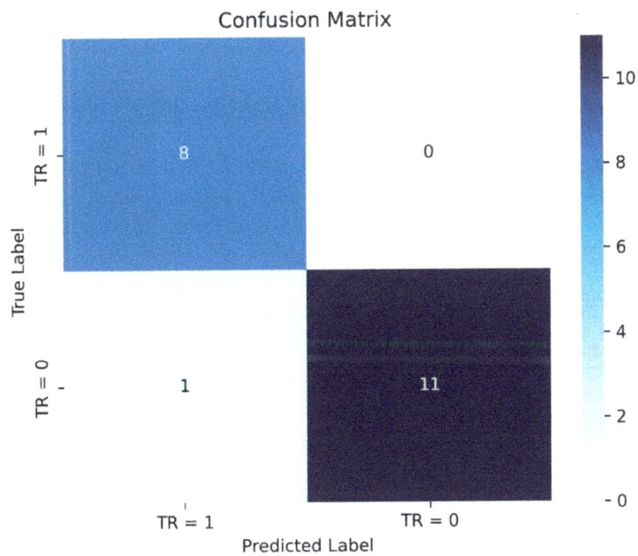

Figure 11. Confusion matrix for classification task.

4. Conclusions

One of the most critical safety hazards connected with lithium-ion batteries is thermal runaway. While mechanical abuse is often unpredictable, other types of failures are frequently caused by system faults. Battery aging is a key contributor to such failures, as it reduces thermal stability and decreases a cell's ability to tolerate faults.

To investigate how aging impacts the thermal sensitivity of lithium-ion cells, this study focused on three key parameters: capacity, internal resistance, and current. A thermal model was developed to incorporate internal resistance as a proxy for aging and thermal runaway side reactions.

The results demonstrate that the interaction between capacity, internal resistance, and charging rate significantly influences thermal behavior and the likelihood of thermal runaway. At low charging rates (1C), cells with higher internal resistance exhibit increased temperature rise, particularly in high-capacity cells. In contrast, low-capacity cells show minimal temperature deviations due to their lower energy density. As the charging rate increases, heat generation intensifies, with several high-capacity simulations exceeding the critical 90 °C threshold, a known precursor to thermal runaway. Notably, cells with both high capacity and high internal resistance experience the most severe temperature spikes, indicating a compounded effect of Joule heating.

A statistical analysis confirmed that capacity, current, and internal resistance are major determinants of battery thermal behavior. Capacity has a significant impact on temperature rise and dT/dt, while current plays a crucial role in heat generation. Internal resistance is strongly correlated with temperature increase due to Joule heating, though its effect on dT/dt is less predictable. A key finding is that the interaction between capacity and current is highly significant, meaning their combined influence is nonlinear. In contrast, the interaction between current and internal resistance was found to be negligible, suggesting that resistance-related heating remains relatively independent of applied current within the tested range. These insights reinforce the importance of an integrated thermal management strategy, particularly in aged batteries where resistance and capacity degrade at different rates.

A machine learning approach was also adopted to classify thermal runaway events based on battery conditions. This approach holds promise for real-time battery monitoring and early failure detection and offers potential applications in BMS algorithms for enhancing battery safety and longevity.

However, battery aging is a complex phenomenon that, besides the internal resistance, also affects other parameters like active material composition, thermal conductivity, and heat generation. Therefore, additional work is required to integrate these aging mechanisms into a more comprehensive model for a more accurate representation of aged battery behavior. Doing so will improve predictive capabilities, enhance thermal management strategies, and ultimately ensure safer battery operation in real-world applications.

Author Contributions: Conceptualization, R.B.; Data curation, M.T.; Funding acquisition, R.B.; Investigation, M.T.; Methodology, M.T. and R.B.; Software, M.T.; Supervision, R.B.; Validation, M.T.; Writing—original draft, M.T.; Writing—review and editing, R.B. All authors have read and agreed to the published version of the manuscript.

Funding: The results of this work have been obtained with the financial support of the National Recovery and Resilience Plan (PNRR), Mission 4 Component 2 Investment 1.3, funded by the European Union—NextGenerationEU, Spoke 6 "Energy Storage" of the Extended Partnership (PE2) NEST—Network 4 Energy Sustainable Transition.

Data Availability Statement: The raw data supporting the conclusions of this article will be made available by the authors on request.

Conflicts of Interest: The authors declare no conflicts of interest.

Abbreviations

The following abbreviations are used in this manuscript:

ANOVA	analysis of variance
ARC	Accelerating Rate Calorimetry
DOD	depth of discharge
DOE	design of experiments
EIS	Electrochemical Impedance Spectroscopy
EV	electric vehicle
HTC	heat transfer coefficient
HTR	Heat–Temperature–Reaction
P2D	pseudo-2D
SEI	solid electrolyte interface
SP	single-particle
SOC	state of charge

TR thermal runaway

References

1. Wang, J.; Azam, W. Natural resource scarcity, fossil fuel energy consumption, and total greenhouse gas emissions in top emitting countries. *Geosci. Front.* **2024**, *15*, 101757. [CrossRef]
2. Wen, J.; Yu, Y.; Chen, C. A review on lithium-ion batteries safety issues: Existing problems and possible solutions. *Mater. Express* **2012**, *2*, 197–212. [CrossRef]
3. Chen, S.C.; Wan, C.C.; Wang, Y.Y. Thermal analysis of lithium-ion batteries. *J. Power Sources* **2005**, *140*, 111–124. [CrossRef]
4. Feng, X.; Ouyang, M.; Liu, X.; Lu, L.; Xia, Y.; He, X. Thermal runaway mechanism of lithium ion battery for electric vehicles: A review. *Energy Storage Mater.* **2018**, *10*, 246–267. [CrossRef]
5. Menale, C.; D'Annibale, F.; Mazzarotta, B.; Bubbico, R. Thermal management of lithium-ion batteries: An experimental investigation. *Energy* **2019**, *182*, 57–71. [CrossRef]
6. Hendricks, C.; Williard, N.; Mathew, S.; Pecht, M. A failure modes, mechanisms, and effects analysis (FMMEA) of lithium-ion batteries. *J. Power Sources* **2015**, *297*, 113–120. [CrossRef]
7. Bubbico, R.; Greco, V.; Menale, C. Hazardous scenarios identification for Li-ion secondary batteries. *Saf. Sci.* **2018**, *108*, 72–88. [CrossRef]
8. Wang, Q.; Ping, P.; Zhao, X.; Chub, G.; Sun, J.; Chen, C. Thermal runaway caused fire and explosion of lithium ion battery. *J. Power Sources* **2012**, *208*, 210–224. [CrossRef]
9. Kim, G.H.; Pesaran, A.; Spotnitz, R. A three-dimensional thermal abuse model for lithium-ion cells. *J. Power Sources* **2007**, *170*, 476–489. [CrossRef]
10. Doyle, M.; Fuller, T.F.; Newman, J. Modeling of galvanostatic charge and discharge of the lithium/polymer/insertion cell. *J. Electrochem. Soc.* **1993**, *140*, 1526. [CrossRef]
11. Doyle, M.; Newman, J.; Gozdz, A.S.; Schmutz, C.N.; Tarascon, J.M. Comparison of modeling predictions with experimental data from plastic lithium ion cells. *J. Electrochem. Soc.* **1996**, *143*, 1890. [CrossRef]
12. Guo, M.; Sikha, G.; White, R.E. Single-particle model for a lithium-ion cell: Thermal behavior. *J. Electrochem. Soc.* **2010**, *158*, A122. [CrossRef]
13. Wu, L.; Liu, K.; Liu, J.; Pang, H. Evaluating the heat generation characteristics of cylindrical lithium-ion battery considering the discharge rates and N/P ratio. *J. Energy Storage* **2023**, *64*, 107182. [CrossRef]
14. Hatchard, T.D.; MacNeil, D.D.; Basu, A.; Dahn, J.R. Thermal model of cylindrical and prismatic lithium-ion cells. *J. Electrochem. Soc.* **2001**, *148*, A755. [CrossRef]
15. Talele, V.; Moralı, U.; Patil, M.S.; Panchal, S.; Fraser, R.; Fowler, M.; Thorat, P.; Gokhale, Y.P. Computational modelling and statistical evaluation of thermal runaway safety regime response on lithium-ion battery with different cathodic chemistry and varying ambient condition. *Int. Commun. Heat Mass Transf.* **2023**, *146*, 106907. [CrossRef]
16. Azuaje–Berbecí, B.J.; Ertan, H.B. A model for the prediction of thermal runaway in lithium–ion batteries. *J. Energy Storage* **2024**, *90*, 111831. [CrossRef]
17. Ozdemir, T.; Ekici, O.; Koksal, M. Numerical and experimental investigation of the electrical and thermal behaviors of the Li-ion batteries under normal and abuse operating conditions. *J. Energy Storage* **2024**, *77*, 109880. [CrossRef]
18. Wu, S.; Wang, C.; Luan, W.; Zhang, Y.; Chen, Y.; Chen, H. Thermal runaway behaviors of Li-ion batteries after low temperature aging: Experimental study and predictive modeling. *J. Energy Storage* **2023**, *66*, 107451. [CrossRef]
19. García, A.; Monsalve-Serrano, J.; de Vargas Lewiski, F.; Golke, D. Characterization of pristine and aged NMC lithium-ion battery thermal runaway using ARC experiments coupled with optical techniques. *Appl. Therm. Eng.* **2024**, *257*, 124244. [CrossRef]
20. Wang, Y.; Xiang, H.; Soo, Y.Y.; Fan, X. Aging mechanisms, prognostics and management for lithium-ion batteries: Recent advances. *Renew. Sustain. Energy Rev.* **2025**, *207*, 114915. [CrossRef]
21. Galatro, D.; Al-Zareer, M.; Da Silva, C.; Romero, D.; Amon, C. Thermal behavior of lithium-ion batteries: Aging, heat generation, thermal management and failure. *Front. Heat Mass Transf. (FHMT)* **2020**, *27*, 14. [CrossRef]
22. Birkl, C.R.; Roberts, M.R.; McTurk, E.; Bruce, P.G.; Howey, D.A. Degradation diagnostics for lithium ion cells. *J. Power Sources* **2017**, *341*, 373–386. [CrossRef]
23. Fleischhammer, M.; Waldmann, T.; Bisle, G.; Hogg, B.I.; Wohlfahrt-Mehrens, M. Interaction of cyclic ageing at high-rate and low temperatures and safety in lithium-ion batteries. *J. Power Sources* **2015**, *274*, 432–439. [CrossRef]
24. Röder, P.; Stiaszny, B.; Ziegler, J.C.; Baba, N.; Lagaly, P.; Wiemhöfer, H.D. The impact of calendar aging on the thermal stability of a LiMn2O4–Li (Ni1/3Mn1/3Co1/3) O2/graphite lithium-ion cell. *J. Power Sources* **2014**, *268*, 315–325. [CrossRef]
25. Guo, Y.; Cai, J.; Liao, Y.; Hu, J.; Zhou, X. Insight into fast charging/discharging aging mechanism and degradation-safety analytics of 18650 lithium-ion batteries. *J. Energy Storage* **2023**, *72*, 108331. [CrossRef]
26. Zhang, G.; Wei, X.; Chen, S.; Wei, G.; Zhu, J.; Wang, X.; Han, G.; Dai, H. Research on the impact of high-temperature aging on the thermal safety of lithium-ion batteries. *J. Energy Chem.* **2023**, *87*, 378–389. [CrossRef]

27. Zhang, G.; Shen, W.; Wei, X. Lithium-ion battery thermal safety evolution during high-temperature nonlinear aging. *Fuel* **2024**, *362*, 130845. [CrossRef]
28. Abada, S.; Petit, M.; Lecocq, A.; Marlair, G.; Sauvant-Moynot, V.; Huet, F. Combined experimental and modeling approaches of the thermal runaway of fresh and aged lithium-ion batteries. *J. Power Sources* **2018**, *399*, 264–273. [CrossRef]
29. Grandjacques, M.; Kuntz, P.; Azaïs, P.; Genies, S.; Raccurt, O. Thermal runaway modelling of Li-ion cells at various states of ageing with a semi-empirical model based on a kinetic equation. *Batteries* **2021**, *7*, 68. [CrossRef]
30. Zhao, L.; Zheng, M.; Zhang, J.; Liu, H.; Li, W.; Chen, M. Numerical modeling of thermal runaway for low temperature cycling lithium-ion batteries. *J. Energy Storage* **2023**, *63*, 107053. [CrossRef]
31. Pastor, J.V.; García, A.; Monsalve-Serrano, J.; Golke, D. Analysis of the aging effects on the thermal runaway characteristics of Lithium-Ion cells through stepwise reactions. *Appl. Therm. Eng.* **2023**, *230*, 120685. [CrossRef]
32. Kabra, V.; Karmakar, A.; Vishnugopi, B.S.; Mukherjee, P.P. Quantifying the effect of degradation modes on Li-ion battery thermal instability and safety. *Energy Storage Mater.* **2025**, *74*, 103878. [CrossRef]
33. NASA Dataset. Available online: https://phm-datasets.s3.amazonaws.com/NASA/5.+Battery+Data+Set.zip (accessed on 8 October 2024).
34. Ekström, H.; Fridholm, B.; Lindbergh, G. Comparison of lumped diffusion models for voltage prediction of a lithium-ion battery cell during dynamic loads. *J. Power Sources* **2018**, *402*, 296–300. [CrossRef]
35. Cai, L.; White, R.E. Mathematical modeling of a lithium ion battery with thermal effects in COMSOL Inc. Multiphysics (MP) software. *J. Power Sources* **2011**, *196*, 5985–5989. [CrossRef]
36. Kumaresan, K.; Sikha, G.; White, R.E. Thermal model for a Li-ion cell. *J. Electrochem. Soc.* **2007**, *155*, A164. [CrossRef]
37. Chen, S.C.; Wang, Y.Y.; Wan, C.C. Thermal analysis of spirally wound lithium batteries. *J. Electrochem. Soc.* **2006**, *153*, A637. [CrossRef]
38. Cianciullo, M.; Vilardi, G.; Mazzarotta, B.; Bubbico, R. Simulation of the Thermal Runaway Onset in Li-Ion Cells—Influence of Cathode Materials and Operating Conditions. *Energies* **2022**, *15*, 4169. [CrossRef]
39. Peng, P.; Jiang, F. Thermal safety of lithium-ion batteries with various cathode materials: A numerical study. *Int. J. Heat Mass Transf.* **2016**, *103*, 1008–1016. [CrossRef]
40. Tran, M.K.; DaCosta, A.; Mevawalla, A.; Panchal, S.; Fowler, M. Comparative study of equivalent circuit models performance in four common lithium-ion batteries: LFP, NMC, LMO, NCA. *Batteries* **2021**, *7*, 51. [CrossRef]
41. Mei, W.; Zhang, L.; Sun, J.; Wang, Q. Experimental and numerical methods to investigate the overcharge caused lithium plating for lithium ion battery. *Energy Storage Mater.* **2020**, *32*, 91–104. [CrossRef]
42. Dong, T.; Peng, P.; Jiang, F. Numerical modeling and analysis of the thermal behavior of NCM lithium-ion batteries subjected to very high C-rate discharge/charge operations. *Int. J. Heat Mass Transf.* **2018**, *117*, 261–272. [CrossRef]
43. Liu, Z.; Guo, X.; Meng, N.; Yu, Z.; Yang, H. Study of thermal runaway and the combustion behavior of lithium-ion batteries overcharged with high current rates. *Thermochim. Acta* **2022**, *715*, 179276. [CrossRef]
44. Hellwig, C.; Sörgel, S.; Bessler, W.G. A multi-scale electrochemical and thermal model of a $LiFePO_4$ battery. *Ecs Trans.* **2011**, *35*, 215. [CrossRef]
45. Yoo, K.; Kim, J. Thermal behavior of full-scale battery pack based on comprehensive heat-generation model. *J. Power Sources* **2019**, *433*, 226715. [CrossRef]
46. Han, Z.; Zhao, L.; Zhao, J.; Xu, G.; Liu, H.; Chen, M. An experimental study on the thermal runaway propagation of cycling aged lithium-ion battery modules. *Fire* **2024**, *7*, 119. [CrossRef]
47. Wei, N.; Zhang, F.; Zhang, W.; Li, X. Comparative study on the thermal runaway characteristics of Li $(Ni_xCo_yMn_z)$ O_2 batteries. *Heliyon* **2024**, *10*, e31203. [CrossRef]

MDPI AG
Grosspeteranlage 5
4052 Basel
Switzerland
Tel.: +41 61 683 77 34

Batteries Editorial Office
E-mail: batteries@mdpi.com
www.mdpi.com/journal/batteries

www.ingramcontent.com/pod-product-compliance
Lightning Source LLC
Chambersburg PA
CBHW061614210926
51850CB00009B/2317
* 9 7 8 3 7 2 5 8 7 0 0 8 0 *